Chemistry for
Environmental
Engineering and Science

Chemistry for Environmental Engineering and Science

Contributors

N. B. Raut, Dinesh Kumar Saini et al.

AURIS
Reference

www.aurisreference.com

Chemistry for Environmental Engineering and Science

Contributors: N. B. Raut, Dinesh Kumar Saini et al.

Published by Auris Reference Limited

www.aurisreference.com

United Kingdom

Chemistry for Environmental Engineering and Science

ISBN: 978-1-78154-970-4

British Library Cataloguing in Publication Data
A CIP record for this book is available from the British Library

Printed in the United Kingdom

Exclusively distributed by CBS Publishers & Distributors Pvt. Ltd.

Sales & Distribution Rights only for India, Pakistan, Bangladesh, Sri Lanka, Nepal and Bhutan. This book is not to be sold outside these territories.

Contents

List of Abbreviations

BM	Blodgett method
BHI	Brain–heart infusion
CCS	Carbon capture and storage
CG	Characteristic geomorphological
CC	Climate change
DBI	Dual band images
EOR	Enhanced Oil Recovery
GCTE	Global Change and Terrestrial Ecosystems
GEC	Global environmental change
GHG	Greenhouse gas
IPCC	Intergovernmental Panel on Climate Change
IPBES	Intergovernmental Science-Policy Platform on Biodiversity and Ecosystem Services
LUCC	Land Use and Land Cover Change
RETC	Release and Transfer Inventory
RR	Response regulator
SP	Surface pressure
TAA	Terminal amino acid
TE	Trace elements
TCE	Transitional coastal environments
TCS	Two-component systems
UNFCCC	United Nations Framework Convention on Climate Change

List of Contributors

N. B. Raut
Faculty of Engineering, Sohar University, Oman

Dinesh Kumar Saini
Faculty of Information Technology, Sohar University, Oman

G. B. Shinde
Department of Chemical Engineering, SVIT, Nashik, Maharashtra, India

Akindele O. Oyinloye
Department of Geology, University of Ado-Ekiti Nigeria

A. K. Somarin
Department of Geology, Brandon University, Brandon, Manitoba, Canada

Andreas Laake
WesternGeco Cairo Egypt

Belén Rubio
Universidad de Vigo, Vigo, Pontevedra Spain

Paula Álvarez-**Iglesias**
Universidad de Vigo, Vigo, Pontevedra Spain

Ana M. Bernabeu
Universidad de Vigo, Vigo, Pontevedra Spain

Iván León
Universidad del Atlántico, Barranquilla Colombia

Kais J. Mohamed
Universidad de Vigo, Vigo, Pontevedra Spain

Daniel Rey
Universidad de Vigo, Vigo, Pontevedra Spain

Federico Vilas
Universidad de Vigo, Vigo, Pontevedra Spain

George R. Ivanov
Department of Physics, Faculty of Hydraulic Engineering, University of Architecture, Civil Engineering and Geodesy & Advanced Technologies Ltd

Georgi Georgiev
Department of Biochemistry, Faculty of Biology, Sofia University, Sofia, Bulgaria

Zdravko Lalchev
Department of Biochemistry, Faculty of Biology, Sofia University, Sofia, Bulgaria

Julius I. Agboola
United Nations University, Institute of Advanced Studies, Operating Unit in Ishika-wa/Kanazawa, 2-1-1

Hirosaka, Kanazawa, Ishikawa,Japan
Department of Fisheries, Faculty of Science/ Centre for Environment and Science Education (CESE), Lagos State University, Ojo, Lagos Nigeria

Yoshihisa Yamashita and Yukie Shibata
Section of Preventive and Public Health Dentistry, Kyushu University Faculty of Dental Science Japan

Ali Hammood
Department of Materials Engineering-University of Kufa, Iraq

Zainab Radeef
Department of Materials Engineering-University of Kufa, Iraq
Oscar Jiménez
Comisión Federal de Electricidad México

Moisés Dávila
Comisión Federal de Electricidad México

Vicente Arévalo
Comisión Federal de Electricidad México

Erik Medina
Comisión Federal de Electricidad México

Reyna Castro
Comisión Federal de Electricidad México

Manuel Ferreira Rebelo
CLEGI, Lusíada University, Vila Nova de Famalicão, Portugal

Gilberto Santos

CLEGI, Lusíada University, Vila Nova de Famalicão, Portugal
College of Technology, Polytechnic Institute of Cávado and Ave, Barcelos, Portugal

Rui Silva
CLEGI, Lusíada University, Vila Nova de Famalicão, Portugal

Sunil Londhe
Shri Shiv Shahu College, Sarud, Kolhapur 416 214, India

Nitin Kamble
Department of Zoology, Shivaji University, Kolhapur 416 004, India

Preface

Environmental engineering is the branch of engineering that is concerned with protecting people from the effects of adverse environmental effects, such as pollution, as well as improving environmental quality. The text *Chemistry for Environmental Engineering and Science* focuses on the aspects of chemistry that are particularly valuable for solving environmental problems, and deals with groundwork for understanding water and wastewater analysis-a fundamental basis of environmental engineering practice. A solution to environmental issue on processing of cocos nucifera shell derived activated carbon for treatment of distillery spent wash has been presented in first chapter. Second chapter focuses on geology and geotectonic setting of the basement complex rocks in south western Nigeria. Third chapter deals with petrography, geochemistry and petrogenesis of late-stage granites. Fourth chapter explains how geological information can be extracted from satellite imagery and how this information can be merged with geological and geophysical data to build consistent geological models for the surface and subsurface. Fifth chapter reviews the main factors that control the incorporation of metals to the sediments in transitional coastal environments, with a focus on the forcing factors and their temporal evolution in the recent sedimentary. In sixth chapter, we propose a new class of materials–fluorescently labeled phospholipids, which can be used as chemical and biochemical sensors. Relevant issues and current dimensions in global environmental change (GEC) have been described in seventh chapter. In eighth chapter, we review those genes that have been reported to be involved in *S. mutans* aciduricity, including those participating in two-component systems and others, especially targeting the dgk homolog. Ninth chapter focuses on characterizations of environmental composites. The purpose of last chapter is to present the analysis of different geological provinces to address the possibility of storing anthropogenic CO_2 in deep underground geologic formations, particularly in eastern continental Mexico.

Chapter 1

ENVIRONMENTAL INFORMATICS AND SOFT COMPUTING PARADIGM: PROCESSING OF COCOS NUCIFERA SHELL DERIVED ACTIVATED CARBON FOR TREATMENT OF DISTILLERY SPENT WASH—A SOLUTION TO ENVIRONMENTAL ISSUE

N. B. Raut[1] Dinesh Kumar Saini[2] and G. B. Shinde[3]

[1]Faculty of Engineering, Sohar University, Oman

[2]Faculty of Information Technology, Sohar University, Oman

[3]Department of Chemical Engineering, SVIT, Nashik, Maharashtra, India

ABSTRACT

Soft computing techniques are very much needed to design the environmental related systems these days. Soft computing (SC) is a set of computational methods that attempt to determine satisfactory approximate solutions to find a model for real-world problems. Techniques such as artificial neural networks, fuzzy logic, and genetic algorithms can be used in solving complex environmental problems. Self-organizing feature map (SOFM) model is proposed in monitoring and collecting of the data that are real time and static datasets acquired through pollution monitoring sensors and stations in the distilleries. In the environmental monitoring systems the ultimate requirement is to establish controls for the sensor based data acquisition systems and needs interactive and dynamic reporting services. SOFM techniques are used for data analysis and processing. The processed data is used for control system which even feeds to the treatment systems. Cocos nucifera activated carbon commonly known as coconut shell activated carbon (CSC) was utilized for the treatment of distillery spent wash. Batch and column studies were done to investigate the kinetics and effect of operating parameter on the rate of adsorption. Since the quantum of spent water generated from the sugar industry allied distillery units is huge, this low cost adsorbent is found to be an attractive economic option. Equilibrium adsorption date was generated to plot Langmuir and Tempkin adsorption isotherm. The investigation reveals that though with lower adsorption capacities CSC seems to be technically feasible solution for treating sugar distillery spent. Efforts are made in this paper to build informatics for derived activated carbon for solving the problem of treatment of distillery spent wash. Capsule. Coconut shell derived activated carbon was synthesized, characterized, and successfully employed as a low cost adsorbent for treatment of distillery spent wash.

INTRODUCTION

The wastewater coming from distillery generally known as spent wash is dark brown in color, carries high organic load, and causes severe fouling of the atmosphere. Waste water discharged by distillery from sugarcane molasses poses problems of disposal to acceptable standards due to their high BOD, COD, and color. About 12–15 liters of spent wash is produced per liter of alcohol produced. After fermentation of molasses, alcohol is separated by distillation and the residual liquor is discharged as spent wash. The effluent as such discharged from the plant is hot, dark brown colored, acidic and possesses objectionable odor. The biological oxygen demand (BOD) value are extremely [1] high like the values of suspended solids, dissolved solid, chlorides, sulphates, and nitrogen which are also too high. The potassium content of the effluent is used for irrigation. Though the distillery effluents do not contain any toxic substances, these create toxic wastes resulting in massive fish kills, production of fouls, odors, and decolorisation of streams. The distillery effluent is generally disposed off in open ponds or lagoons. Due to this, seepage and ground water pollution occurs. It is a serious threat to soil and water quality because of the melanoid coloring compounds present in the effluent. Also the obnoxiuos odor of the spent wash which spreads a few kilometers can cause serious health concerns.

This massive quantity, approximately 40 billion liters of effluent, from the distilleries are associated with very high organic load with BOD and COD levels in the range of 35,000 to 60,000 and 60,000 to 1, 20,000 mg/liter respectively. In India the population equivalent of distillery waste to organic pollution is approximately seven times more than the entire population of India [2]. Environmental regulation in force in India requires that the effluent stress to have BOD less that 100 mg/liter for discharge on land and less that 30 mg/liter for discharge into inland surface waters. These stipulations necessitate further treatment of effluent [3].

A number of conventional treatment technologies have been considered for treatment of wastewater contaminated with organic substances. Among them, adsorption process is found to be the most effective method. Adsorption as a wastewater treatment process has aroused considerable interest during recent years. Commercial activated carbon is regarded as the most effective material for controlling the organic load [4]. However due to its high cost and about 10–15% loss during regeneration, unconventional adsorbents like coconut shells, fly ash, peat, lignite, bagasse pith, wood, saw dust, and so forth have attracted the attention of several investigations and adsorption characteristics have been widely investigated for the removal of refractory materials [5, 6] for varying degree of success. Activated charcoal is charcoal made from wood, coconut shell, or bone, which has been treated with oxygen to remove all the impurities sticking to its surface inside the pores. Activated charcoal, for example, can have a total surface area of up to 2000 square meters per gram [7]. There is lots of surface to hold molecules, which remain in place by Vander Waals forces (physisorption). The bigger its surface area, the more molecules it can trap on its surface [8].

It is felt that the shells can be used for producing an adsorbent for treating distillery spent wash from sugar industry using adsorption process [9].

In most of the coastal cities and towns across the globe, Cocos nucifera commonly known as coconut trees is plentiful and the green empty coconuts finds their way in the cities refuse. In some places the green nuts are dried and the fibers are removed for various purpose like rope making, floor mats or rubberized coir mattresses, and so forth; however, the shells are not much used other than for use as fuel. Coconut shells are made up of "stone cells" and are hard, porous, impregnated with lignins and tannin and a little oil.

Steam activation and chemical activation are the two commonly used processes for the manufacture of activated carbon. However coconut shell based activated carbon units are adopting the steam activation process to produce good quality activated carbon [10].

Lowering of pH value of the stream, increase in organic load, depletion of oxygen content, destruction of aquatic life, and bad smell are some of the major pollution problems due to distillery wastewater. One of the main advantages of COD [4] reduction by using coconut shell ash over the other chemical treatment methods is that it is available as a waste material in abundance and easy availability makes it a strong choice in the investigation of economic way of COD removal [11].

MATERIAL AND METHODS

Sample Collection and Pretreatment

The distillery spent wash sample was taken from Niphad Sahakari Sakhar Karkhana, Niphad, Dist. Nashik, and Maharashtra India. Once a sample is taken, the constituents of the sample should be maintained in the same condition, as when collected. When it is not possible to analyze collected samples immediately, samples should be preserved properly in such a way that the biological activity (microbial respiration), chemical activity (precipitation or pH change), and physical activity are the minimum. Methods of preservation [12] include cooling, pH control, and chemical addition as shown in Table 1. Freezing is usually not recommended. The length of time in which a constituent in waste water will remain stable is related to the character of the constituent and preservation method used.

Table 1: Required containers, preservation techniques, and holding times for samples

Maximum holding time	Preservative	Container	Parameter
14 days	Cool 4°C	P.G.	Acidity
14 days	Cool 4°C	P.G.	Alkalinity
48 hours	Cool 4°C	P.G.	BOD
28 days	Cool 4°C, H_2SO_4 to pH < 2	P.G.	COD
28 days	Cool 4°C H_2SO_4 to pH < 2	G.	Phenols
0 hours	None required	P.G.	pH
0 hours	None required	G.	DO
48 hours	Cool 4°C	P.G.	Nitrate
7 days	Cool 4°C	P.G.	Residue volatile
7 days	Cool 4°C	P.G.	TSS

Production of Activated Carbon

The coconut shells derived from various waste sources were cleanly shaved to remove all the fibers on its surface. Cleaned shells were cut into 3 mm to 8 mm pieces. These pieces were washed with distilled water to remove surface dust and dried in sunlight for two to three days.

The cleaned and dried coconut shells were then broken into small pieces to allow insertion into a muffle furnace {KW = 0.3, AMP = 13, TYPE = MFRA, PHASE = Single} and subjected to destructive distillation in the absence of air. The temperature of the furnace was maintained at 400°C for three hours to avoid reactions with atmospheric oxygen, thereby preventing formation of ash. This resulted in the formation of black carbonized matter. The carbonized matter was washed [13] with distilled water and dried at 100°C. These pieces were broken down into fine powder with the help of grinder and size separation which is done by screening using standard sieves [4]. Commercial activated carbon (CAC) was procured from M/s Dayo Scientific Sales, Nashik Road, Nashik, Maharashtra, India. The typical properties of CSC and CAC are summarized in Table 2.

Table 2: Characteristics of adsorbents

Sr. no.	Characteristics	CSC	CAC
1	Moisture, %	16.26	6.37
2	Ash content, %	0.33	4.51
3	Bulk density, g/cc	0.67	0.57
4	Matter soluble in water, %	1.32	1.31
5	Matter soluble in acid, %	2.25	1.39
6	pH	2.40	8.50
7	Decolorizing power, mg/g	39	78
8	Surface area, m^2/g	1.69	337

Batch Adsorption Tests

To study the effect of important parameters like initial pH, adsorbent dose (gm), contact time(t) initial concentration (C_o), adsorbent particle size, and agitation speed on the adsorptive removal of COD by CSC and CAC, batch experiments were conducted at 30°C. For each experimental run, 50 mL of distillery spent wash solution of known C_o, pH, and a known amount of adsorbent dose was taken in a 100 mL stoppered conical flask. This mixture was agitated in a temperature controlled shaking water bath at a constant speed of 90 rpm at 30°C. Samples were withdrawn at appropriate time intervals and analyzed for COD. The effect of pH on COD removal was studied over a pH range of 3 to 11. pH was adjusted by the addition of dilute aqueous solutions of NaOH (0.10 M) or HCl. For the optimum amount of adsorbent dose (gm), a 50 mL of distillery spent wash solution was contacted with different amounts of CSC and CAC till equilibrium was attained.

The kinetics of adsorption was determined by analyzing adsorptive [14, 15] uptake of COD from the distillery spent wash at different time intervals. For adsorption isotherms, distillery spent wash samples of different concentrations were agitated with the known amount of adsorbent till the equilibrium was achieved. The samples were then analyzed for COD removal. The effect of speed of agitation was studied by agitating the samples at different known speed of agitation and analyzing agitated samples for COD.

Column Adsorption Tests

Columns are mounted vertically one above the other. It is often good practice to operate columns in upflow as this reduces the opportunity for channeling. It is also preferred where suspended solids create a high-pressure drop. For downflow or percolation system an influent line should be installed at the top of column with effluent line at bottom. To develop reasonably good data from scaleup to full plant design, it is important to have operation of the pilot column system as near as possible to the anticipated conditions. The most critical factors, flow rate and feed impurity concentration, must be constant for the entire test run. The adsorbent bed should be at least 60 cm deep with a 4 cm internal diameter. A smaller column is not recommended as the wall effect becomes significant. The adsorbent bed can be supported by glass wool, wire cloth, and so forth. Columns and fixtures can be constructed from glass, plastic, reinforced fiberglass, or metal. Borosilicate glass is commonly used. It is essential that all columns used in the pilot system have at least the same diameter [16].

A bed volume is the volume occupied by the adsorbent including adsorbent volume and void volume. The quantity of effluent is expressed in the number of adsorbent volumes passing through the column per hour. As the flow rate and quantity of liquor are the most important controllable variables in developing design data, a feed pump suitable for accurate and continuous flow is required. Depending on the size of the pilot column system, the use of peristaltic, diaphragm, piston type, and centrifugal pumps is recommended if the effluent is viscous. When loading the column care should be taken to avoid entrapping air in the adsorbent bed. Entrapped air can cause channeling during operation preventing complete contact of the process effluent with the adsorbent. In small columns entrapped air can be avoided by pouring out the adsorbent in the boiling water just before loading of most of the excess water can be poured off, along with most of the adsorbent particles. Before starting the test the complete system should be checked by running on water for several hours. After setting the appropriate flow rate, the effluent to be treated can be fed to the columns and this will displace the water. Samples of effluent after each column should be taken at regular intervals of time. When the effluent from the parallel columns or last column in series exceeds the purity requirement, the test should be stopped [17].

RESULTS AND DISCUSSION

Distillery Spent

Characteristics of spent wash vary widely depending upon the quality of raw material, process technology, energy conservation, and effluent treatment strategies. The typical composition of spent wash from Niphad Sahakari Sakhar Karkhana, Niphad, Tal Niphad, Dist. Nashik, is given in Table 3. India, which is one of the largest producers of sugar, produces the poorest quality of molasses. Indian molasses contain the lowest amount of sugars and the highest amount of impurities and nonfermentable sugars. This impure molasses typically contains the higher amount of mineral matter, calcium oxide, and potash which is further degraded by poor storage facilities at the sugar factories, before it is lifted by the distilleries.

Table 3: Typical composition of distillery spent wash. From Niphad Sahakari Sakhar Karkhana, Niphad, Dist, Nashik.

Sr. no.	Parameter	Spent wash analysis
1	pH	7.87
2	Suspended solids	56,891 mg/lit
3	Total dissolved solids	2,41,066.6 mg/lit
4	B.O.D. at 27°C for 3 days	24,000 mg/lit
5	C.O.D.	39,840 mg/lit
6	Oil and grease	23.0 mg/lit
7	Sulphate	4,000 mg/lit
8	Potassium	1,3000 mg/lit
9	Chloride	1,00,000 mg/lit
10	Sodium	1,200 mg/lit
11	Calcium	2,600 mg/lit
12	Magnesium	2,700 mg/lit
13	Iron	61 mg/lit
14	Total nitrogen	1,350 mg/lit
15	Colour	Dark brown
16	Odor	Smell of burnt sugar
17	Temperature	90–95°C

During the short storage time at the sugar mills, dust and other extraneous materials of unknown origin contaminate it. Most of the components found in molasses are from the sugarcane and find their way to molasses because of the rigorous extraction [18] procedures to molasses because of the intensive extraction process. In Brazil, for example, which is also a large sugar producer, the processes of sugar extraction and recovery are mild and, as consequence, the composition of molasses is superior. Similarly, in other sugarcane producing countries, no attempt is made to recover maximum quantity of sugar from the cane. Lately sugar factories have begun to use certain specific chemicals such as biocide, flocculating agents, color, and viscosity reducing agents which all ultimately make their way into molasses, making it more difficult to be degraded by microorganism.

The composition of the effluent is a reflection of the composition of the molasses. A bad molasses produces a bad effluent. Thus, the poor quality of the cane molasses produced by the sugar mills is responsible for the complex character of the distillery effluent. The composition of the molasses from different sugar mills used in different distilleries in India also varies, due to different processes of sugar manufacture.

In matter of characteristics, effluent is normally a diluted replica of the molasses pressed in the production of alcohol. If molasses differ in composition, as a natural corollary it would yield an effluent, which would also differ in composition and as a consequence the extent of BOD reduction during primary treatment of the effluent would also differ [19]. This makes the situation more complicated as no single technology could produce similar results in all types of effluents and reduce the BOD to an identical level. The effluents of Indian molasses based distilleries have the highest BOD/COD and contain a high percentage of inorganic and organic dissolved matter, compared to the effluents of other countries. The highly dissolved organic matter, which cannot be utilized by yeast and which ultimately finds its way into the effluents, is primarily responsible for escalating the high biological oxygen demand (BOD). In addition, during the fermentation of molasses by yeast, the yeast cells multiply and also produce other organic compounds such as organic acids and so forth, in addition to alcohol. The yeast cells and organic acids also form a part of the effluent and contribute to the BOD. A molasses rich in salts and other minerals allows the yeast to produce a higher amount of organic components, some of which may not be easily degraded. In India both the composition of molasses as well as the effluent is complex primarily because of the sugar recovery process. In other countries, the BOD of the effluent is low and is therefore amenable to easy degradation. The Indian cane molasses effluent contains a very high BOD and its treatment needs special technology [14].

Adsorption Test

Effect of Temperature on COD Removal

Usually, the adsorption reactions are exothermic; high temperature or slow adsorption would seem to inhibit, but this is not usually found to be a factor in most systems. Adsorption rate is limited by diffusion; variable that influences diffusion has significant effect on adsorption rate. Higher temperatures may obstruct adsorption sight, but they significantly speed up the pace of diffusion, offending any negative temperature effect. The temperature dependence of adsorption process is of a very complex nature. Thermodynamic parameter like heat of adsorption and the energy of activation plays an important role in predicting the adsorption behavior and both are strongly temperature dependent. Heat of adsorption influences the equilibrium absorption capacity and also indicates that the nature of interaction taking place between the adsorbate and adsorbent is dependent of the energy of activation.

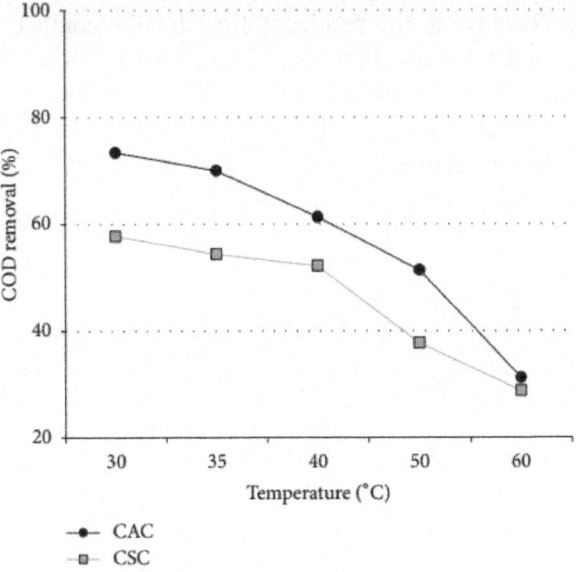

Figure 1: Effect of temperature on COD removal, (C_o), = 13120 mg/L, time = 1 hr, and adsorbent dose = 2 gms.

The decrease in adsorption with the rise in temperature may be due to weakening of adsorptive forces between the active sites of the adsorbent and adsorbate species and also between the adjacent molecules of the adsorbed phase [14, 20]. Figure 1 shows the result of variation in adsorption rate with temperature. It has been generally observed that with the increase in temperature the percent removal of organic matter (COD) due to the adsorption decreases which is also evident from our results. The rate of adsorption is optimum within 30 to 45°C.

It may be observed that the (COD) removal rate for activated carbon is higher and slightly lower for coconut shell ash. But due to the easy availability of coconut shells it is more suitable adsorbent than commercial activated carbon. It may be concluded that, at higher temperature, the decrease in the adsorption could be due to the possible exothermic nature of the adsorption process. In the present study, the optimum (COD) removal was found near at room temperature (25–35°C).

Effect of Flow Rate on Adsorption

The effect of flow rate on removal of COD is shown in Figures 2 and 3. It may be observed that in general the percentage COD removal decreases with increase in the flow rate. The rate of adsorption decreases with increase in the flow rate due to decrease in the residence time for the contact of adsorbate with adsorbent. The COD reduction was remarkable at a flow rate of 25 mL/min dropping gradually but steadily in case of CSC but it dropped sharply with CAC under the same conditions. (COD) due to the adsorption decreases which is also evident from our results.

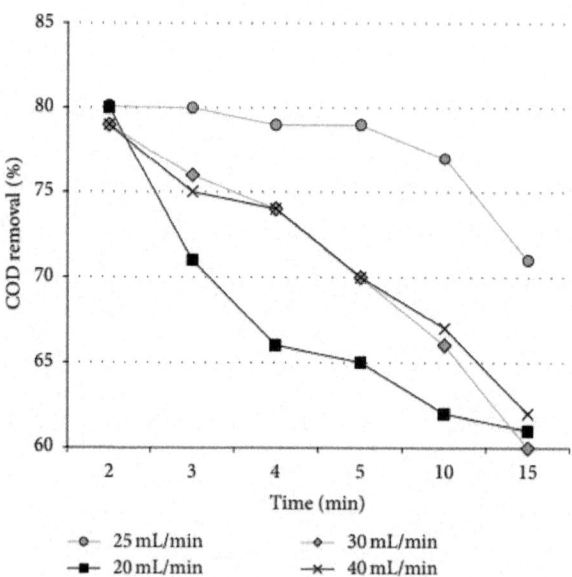

Figure 2: Effect of flow rate on COD removal (CSC), (C_o)= 13120 mg/L, bed depth = 45 Cm, and temperature = 30°C.

The rate of adsorption is optimum within 30 to 45° C. It may be observed that the (COD) removal rate for activated carbon is higher and slightly lower for coconut shell ash. But due to the easy availability of coconut shells it is more suitable adsorbent than commercial activated carbon. It may be concluded that, at higher temperature, the decrease in the adsorption could be due to the possible exothermic nature of the adsorption process. In the present study, the optimum (COD) removal was found near at room temperature (25– 35° C).

Effect of Flow Rate on Adsorption

The effect of flow rate on removal of COD is shown in Figures 2 and 3. It may be observed that in general the percentage COD removal decreases with increase in the flow rate. The rate of adsorption decreases with increase in the flow rate due to decrease in the residence time for the contact of adsorbate with adsorbent. The COD reduction was remarkable at a flow rate of 25 mL/min dropping gradually but steadily in case of CSC but it dropped sharply with CAC under the same conditions.

Effect of pH on COD Removal

During adsorption both anions and captions have been removed from waste water with adsorbent. Researchers have found that carbon exhibits preferential adsorption for ionic species in an order of preference as $Al^{+3} > Ca^{+2} > Li^+ > Na^+ > K^+$. For anions, the NO_3 ion is preferred over the Cl^- ion. It is not surprising that pH plays an important role in the adsorption characteristics of these ions as the low pH provides large quantities of the preferred H^+ ion, which may take the space of other ions on potential adsorption sites.

Even without pH variations inorganic compounds exhibit a wide range of absorbability on adsorbents. The adsorption by using coconut shell ash and commercial activated carbon was studied at different pH values from 3 to 11. Figure 4 shows the removal of COD as a function of pH. Any adsorbate adsorbent system pH of the system affects the nature of the surface change of the adsorbent, extend of ionization, and the extent and the rate of adsorption. From graph it may be seen that the COD removal increases as pH increased from 3 to 5. Thereafter, the removal decreases with further increase in pH. Similar observations have been reported for COD removal from distillery effluents using activated carbon.

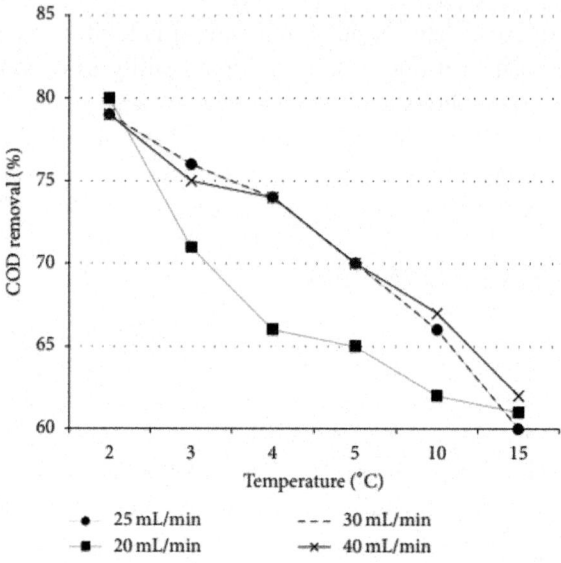

Figure 3: Effect of flow rate on COD removal (CAC), C_o = 13120 mg/L, bed depth = 45^0 Cm, and temperature = 30∘ C.

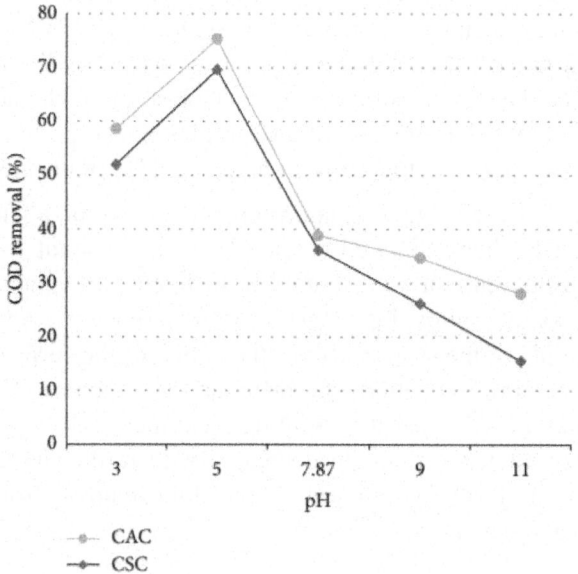

Figure 4: Effect of pH on COD removal (CAC), (C_o), = 13120 mg/L, agitation time = 1 hr., adsorbent dose = 2 gms, and temperature = 30°C.

Effect of Adsorbent Dose on COD Removal

The effect of adsorbent dose on removal of COD is shown COD in Figure 5. It may be observed that in general the percentage COD removal increases with increase in the adsorbent dose. From graph it is noted that the rate of decrease of percentage COD removal has been found to be rapid in the beginning, which slows down as the dose increased. The rate of adsorption increases with increase in dosage because of increase in surface area of the adsorbent [21].

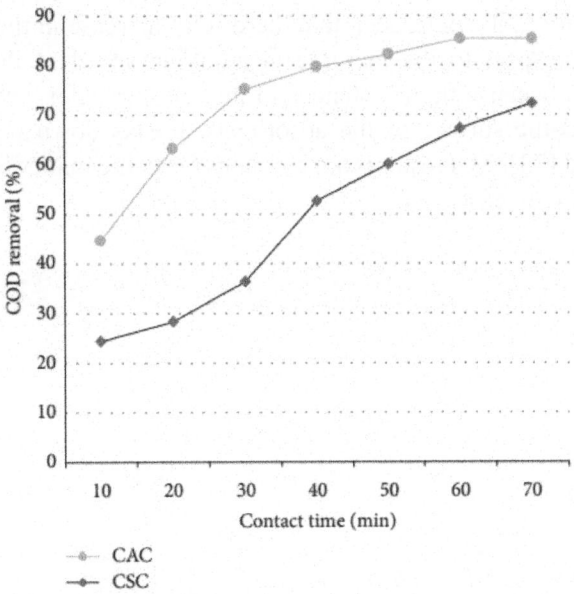

Figure 5: Effect of contact time on COD removal (CAC), (C_o), = 13120 mg/L, adsorbent dose = 2 gms, and temperature = 30°C.

EFFECT OF CONTACT TIME ON COD REMOVAL

In column adsorption the rate of adsorption goes on decreasing with the passage of time which is shown in Figure 5 because as the time passes the adsorption sites of the adsorbent are utilized for the adsorption of the organic matter and the stage is reached when all the sites are used and the adsorbent is saturated. In batch study it is concluded that the optimum time for adsorption is up to 60–70 min only. After that the removal rate was very low. Generally, the rate of adsorption increases with time and after some time it remains constant [22] due to equilibrium condition.

Effect of Adsorption on Color Removal

Colorization of distillery effluent has been a cause of major concern and even after the primary treatment and two-stage biological treatment (anaerobic followed by aeration) the color is not destroyed. Adsorption experiments were carried on distillery effluent using coconut shell ash and commercial activated carbon. Color removal of 40–50% was observed in case of coconut shell ash while in case of activated carbon as adsorbent (dose of 20 gm/L) color removal of 70–75% was observed.

From Figure 6, it can be seen that there is an increase in the rate of COD removal with respect to the increase in agitation speed of the mechanical shaker. This is because the resistances to the mass transfer, which is mainly present around the surface of the adsorbents, breaks down with increasing agitation speed [23]; as a result more amount of organic matter penetrates into the adsorbents with ease.

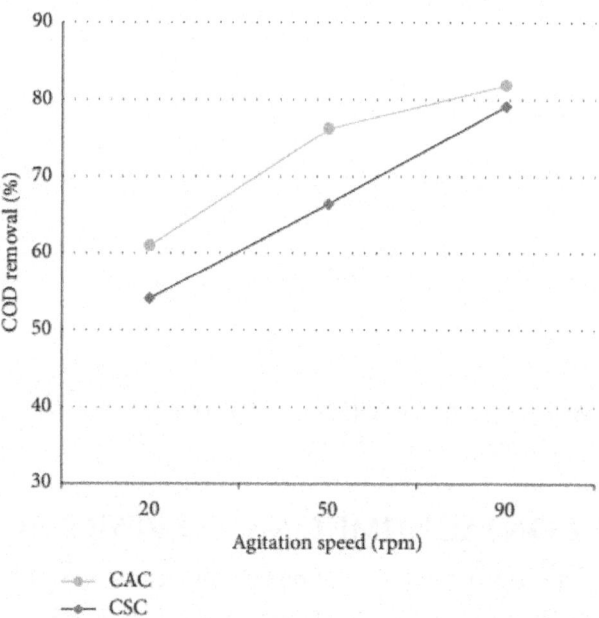

Figure 6: Effect of agitation speed on COD removal (CAC), (C_o) = 13 120 mg/L, adsorbent dose = 2 gms, agitation time = 1 hr., temperature = 30°C, and V= 50 mL.

Effect of Concentration on Adsorption

The effect of concentration on adsorption is shown in Figure 7. It can be concluded that from highly concentrated effluent the COD removal quantity is more than dilute. So this indicates that the effluent can be directly used for the treatment without dilution. At low concentration COD removed quantity is low. The COD reduction is optimum at a concentration above 8500 mg/lit for both CAC and CSC.

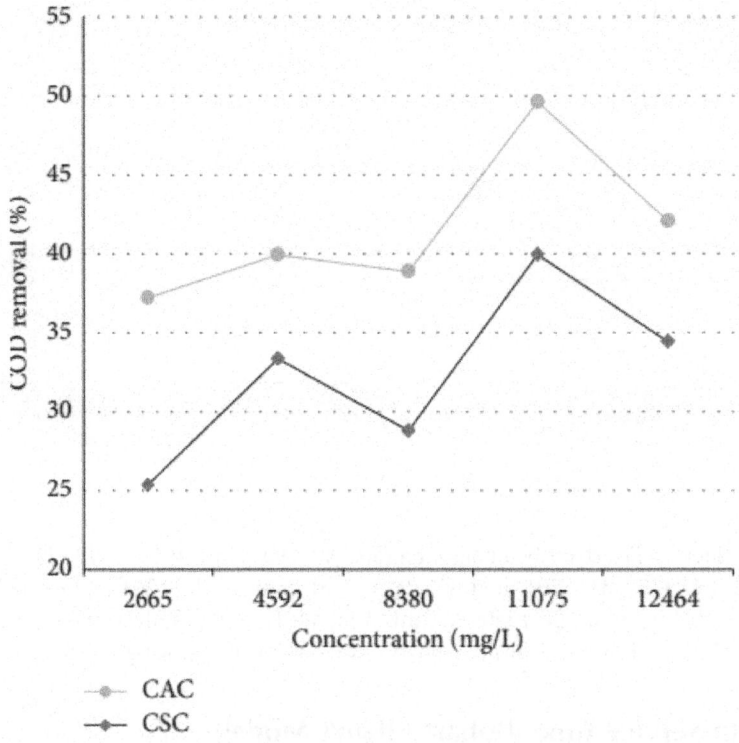

Figure 7: Effect of concentration on COD removal, adsorbent dose = 2 gms.

Effect of Adsorbent Bed Depth on COD Removal

The effect of adsorbent bed depth on adsorption, that is, COD removal in this case, was studied by packing the adsorption column with different depths of the adsorbent (20 cm, 30 cm, and 45 cm). The results are shown in Figure 8. The adsorption increased with increasing the adsorbent bed depth in agreement with the earlier work [18, 20]. The effect of adsorbent bed depth on COD removal was studied with respect to different flow rates 20 mL/min., 30 mL/min., and 45 mL/min. of the effluent.

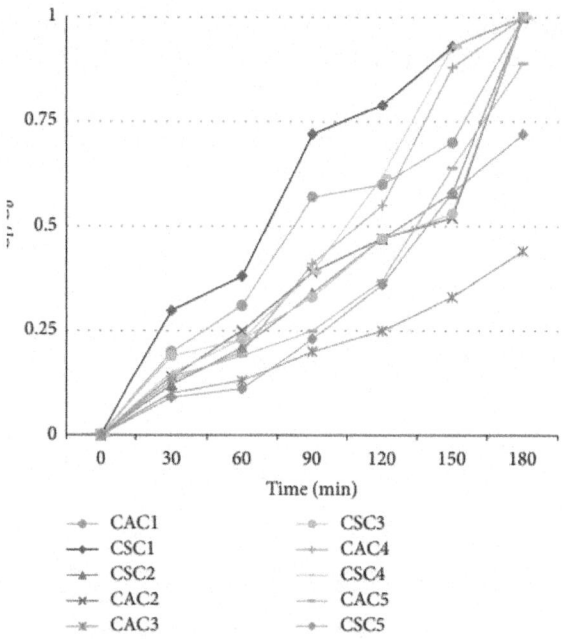

Figure 8: Effect of adsorbent bed height on rate of COD removal $^{(C_a)}=$ 13 120 mg/L, temperature = 30°C 1 = H = 20 Cm, flow rate = 20 mL/min, 2 = H = 30 Cm, flow rate = 20 mL/min, 3 = H = 45 Cm, flow rate = 20 mL/min, 4 = H = 20 Cm, flow rate = 30 mL/ min, 5 = H = 30 Cm, flow rate = 30 mL/min, 6 = H = 45 Cm, and flow rate = 30 mL/min.

Bed Depth Service Time (Bohart Adam's Model)

In continuous flow experiments, it is essential to predict the exhaustion rate of adsorbent bed on how long the bed will last before regeneration is necessary. In this method the service time of a fixed bed adsorbent, treating a solution of single adsorbate, can be expressed as a function of operational variables as shown by (1).

The straight line plots of breakthrough time versus bed height were obtained as shown in Figure 9 indicating the applicability of Bohart Adam's model:

$$\ln\left(\frac{C_0}{C_b} - 1\right) = \ln\left(\frac{KN_oH}{V}\right) - 1 - K_bC_0T_b,$$

$$T_b = \left(\frac{N_0}{C_0V}\right)H - \left(\frac{V}{K_bN_0}\right)\ln\left(\frac{C_0}{C_b} - 1\right),$$

$$(1)$$

where C_b = breakthrough concentration {mg/L}, c_0 = initial concentration {mg/L}, H = critical bed depth {m}, V = linear flow velocity {m/min.}, K_b = rate constant, t_b = breakthrough time {Min.}, and N_0 = adsorptive capacity {mg/m}.

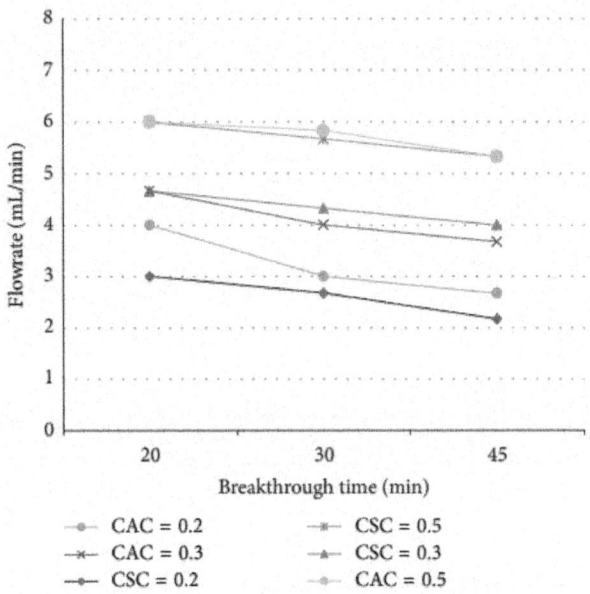

Figure 9: Bohart Adam's model.

The breakthrough time is found to increase with increase in the adsorbent bed height. Thus, column test shows the applicability of BDST model [8].

Effect of Particle Size of Adsorbent on COD Removal

The effect of using various particle sizes of CSC on COD removal was studied by using CSC adsorbent of three different particle sizes, namely, 425 microns, 610 microns, and 750 microns for the column experiments. The Co was kept constant at 20480 mg/lit. and all the experiments were carried out at natural pH of 7.87 and temperature 30°C.

From the results (Figure 10) it is clear that the rate of adsorption increases with decrease in particle size. This is because increase in the surface area with the decreasing particle size of adsorbents and since adsorption process is a surface phenomenon the rate of adsorption increases with decrease in particle size. Thus the smaller the particle size, the greater the interfacial area for adsorption of COD. But with the reduced sizes, there may be other problems like clogging during the operation particularly with continuous operations resulting in high- pressure drops.

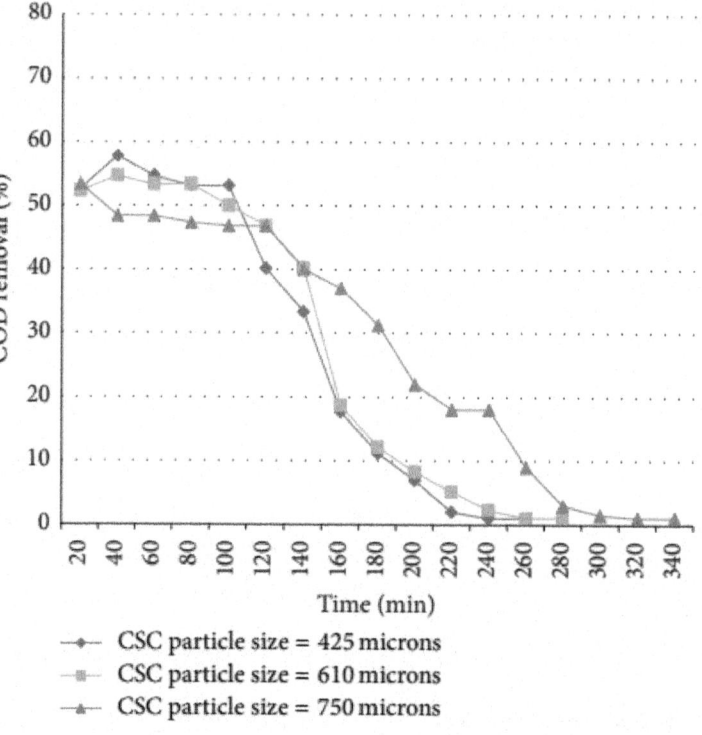

Figure 10: Effect of particle size on COD removal (CAC), c_0= 20480 mg/L, H = 12 cm, flow rate = 45 mL/min.

Effect of Series Arrangement of CSC and CAC Columns

Figure 11 shows that if the CSC column is arranged in series with CAC column for adsorption operation, the efficiency of carbon utilization is increased. Thus, more quantity of CSC adsorbent and less quantity of CAC adsorbent can be effectively used for treating distillery spent. Though the percentage COD removal using series combination is slightly less than that using the two columns separately but the low cost of CSC makes it an attractive option.

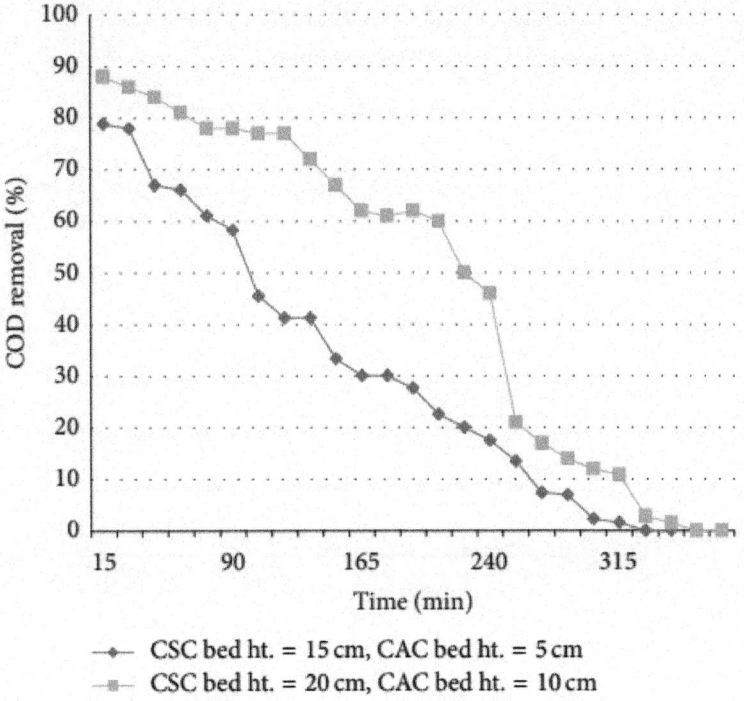

—●— CSC bed ht. = 15 cm, CAC bed ht. = 5 cm
—■— CSC bed ht. = 20 cm, CAC bed ht. = 10 cm

Figure 11: Effect of CSC column in series with CAC bed on COD removal,   =
20480 mg/L.

Adsorption Isotherms

Langmuir Adsorption Isotherm

The equilibrium data for removal of COD by using CSC and CAC in the
present investigations were analyzed using Langmuir adsorption model.

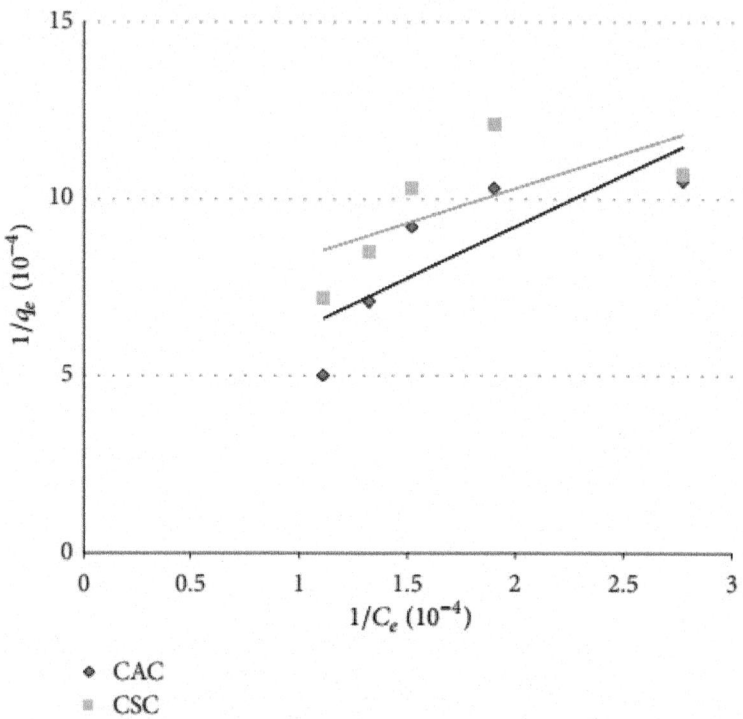

Figure 12: Langmuir adsorption isotherm Co = 13120 mg/L.

The linear plots of $1/qe$ versus $1/C_e$ are shown in Figure 12. The values of θ and b were determined as intercept = $1/\theta^0$ and slope = $1/\theta^0\, b$. The adsorption data were fitted to the linear form of Langmuir isotherm. The essential characteristics of Langmuir isotherm may be expressed in terms of dimensionless equilibrium parameter R using following equation:

$$R = \frac{1}{(1 + bC_0)}.$$

$$(2)$$

The values of R lie between "0" and "1" at initial adsorbate concentrations showing favorable adsorption of organic matter on the adsorbent.

For CSC, b_{CSC} = 4.9685 × 10^{-4} R_{CSC} = 0.133.
For CAC, b_{CSC} = 1.24 × 10^{-5} R_{CAC} = 0.860.

Tempkin Adsorption Isotherms

A plot of qe versus ln Ce enables the determination of the isotherm constants B_1 and K_t from the slope and the intercept, respectively. K_t is the equilibrium binding constant (1/mol) corresponding to the maximum binding energy and constant B_1 is related to the heat of adsorption. The Tempkin isotherm plots for CSC and CAC are shown in Figure 13. From Figure 13, slope = B_1 = 31.389 and intercept = 75. Now,

$$q_e = B_1 \ln K_t + B_1 \ln C_e. \tag{3}$$

For CSC, B_1 = 31.389 and K_t = 10.90. From Figure 13, slope = 57.93 and intercept = 90. For CAC, B_1 = 119.94 and K_t = 4.72.

Adsorption Kinetics.

The rate of adsorption of organic matter on to CSC and CAC was studied and the rate constant

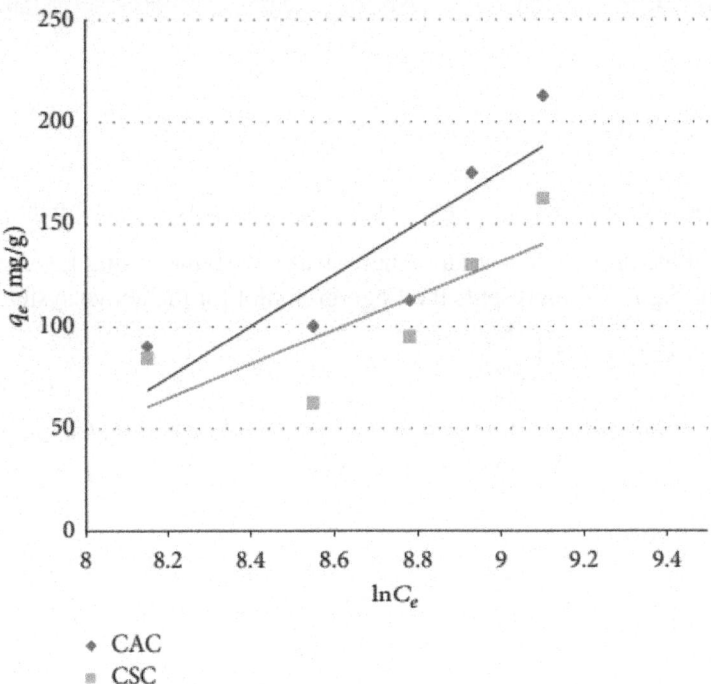

Figure 13: Tempkin adsorption isotherm Co = 13120 mg/L.

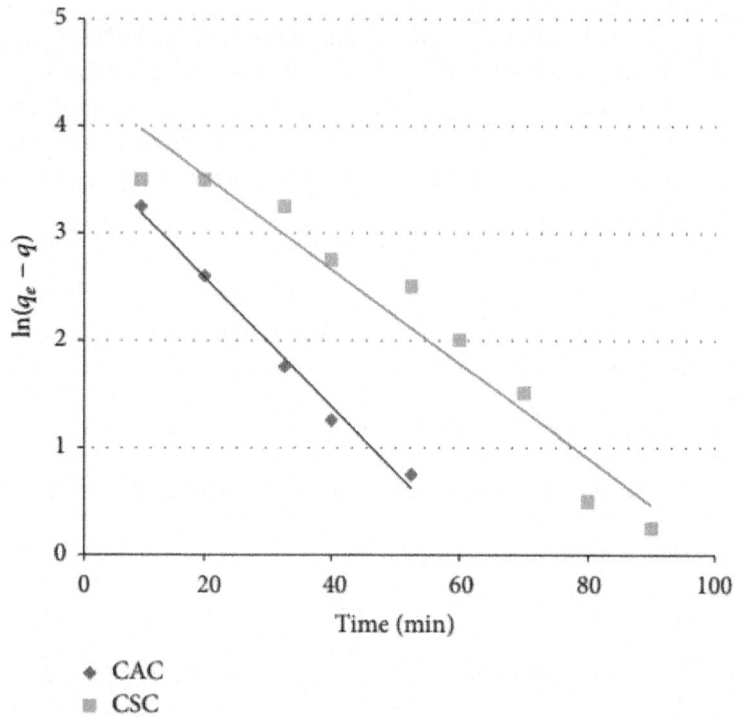

Figure 14: Lagergren plot C_o = 13120 mg/L, adsorbent dose = 2 gms, Temp = 30° C.

K_{ad} of the process for organic matter was determined using Lagergren rate equation. Figure 14 represents the Lagergren plot for the above system:

$$\ln (q_e - 1) = \ln q_e - \left(\frac{K_{ad}}{2.303}\right) t,$$

(4)

where q = amount adsorbed at time "t", K_{ad} = adsorption constant, and q_e = equilibrium uptake. At equilibrium, COD removal is Nil. So, $q_e = 0$:

$$\ln (q_e - q) = \ln q_e - \left(\frac{K_{ad}}{2.303}\right) t.$$

(5)

For CSC, from Figure 14, slope is $[K_{ad}/2.303]$ = 5.111 − 3.084/20 − 70 = −0.04054:

$$K_{ad csc} = 0.09336 \text{ min}^{-1}.$$

(6)

For CSC, from Figure 14, slope is $[K_{ad}/2.303]$ = 4.892 − 2.324/10 − 50 = −0.0642:

$$K_{ad\,cac} = 0.1478\,min^{-1}.$$

(7)

The linear plots of $\ln(q_e - q)$ versus t for COD removal shown in Figure 14, indicate the applicability of Lagergren first order equation.

Cost Analysis

One of the most challenging elements in a design will be estimating the cost to build and operate the wastewater treatment facility that is designed. Cost estimation is also an important element when selecting between alternative designs [24].

For treating 1 liter of distillery spent wash, using CAC as adsorbent, time required = 22.22 minutes, amount of CAC required = 32 gms, cost of adsorbent CAC = Rs. 300 per kg, and cost of adsorbent CAC required to treat 1 liter of distillery spent wash = Rs. 9.60.

For treating 1 liter of distillery spent wash, using CSC as adsorbent, time required = 22.22 minutes, amount of CSC required = 36 gms, and cost of adsorbent CSC = Rs. 100 per kg.

Cost of adsorbent CSC required to treat 1 liter of distillery spent wash = Rs. 3.60. On comparing the cost of treatment of distillery spent wash using both CSC and CAC, it can be observed that treatment using CSC as adsorbent is much cheaper than that using CAC as adsorbent.

However, as far as the production of sugar is concerned, 0.4–0.45 tone of molasses is produced per ton of sugar. About 90% of the total molasses produced is utilized for production of alcohol. For producing 1 liter of alcohol about 10–15 liters of spent wash is produced. Hence for producing 1 liter of alcohol, treatment cost of spent wash is about Rs. 96 by using commercial activated carbon as adsorbent and Rs.36/- by using coconut shell as adsorbent. Thus it can be observed that for treating 1 liter of distillery spent wash, there is a difference of Rs. 6 between the treatments by two adsorbents {CSC and CAC}.

CONCLUSIONS

Coconut shell activated carbon was synthesized and characterized for comparison of adsorption capacities with commercial activated carbon. The results revealed that CAC gives the best result between the two adsorbents, that is, coconut shell carbon {CSC} and commercial activated carbon {CAC} tested for COD reduction from distillery spent wash. However, CSC also showed quiet effective adsorbent capacities. Though the adsorption capacity of

CSC is lower than that of commercial grade activated carbon, the low material cost makes it an attractive option for the treatment of effluents. The treatment of distillery spent with CSC becomes highly effective, as it is a readily and locally available low cost absorbent which needs not be regenerated.

With the increase in adsorbent dosage there is an increase in percentage COD removal from the distillery spent wash. Change in natural pH of the distillery spent wash caused color removal due to the structural changes being affected in the adsorbent molecules. COD removal by both adsorbents, that is, CAC and CSC, is the maximum at a pH between 3 and 5. Effect of initial concentration on COD removal by both CSC and CAC shows that the percentage COD removal increases with increase in initial concentration (above 8580 mg/lit) for initial contact time intervals and decreases with increase in initial concentration as the saturation of adsorbent is reached. Adsorption capacity decreases with increase in temperature for both the adsorbents. Breakthrough time increases with increase in bed height and decrease in flow rate for the column adsorption. By operating CSC column in series with CAC column the efficiency of carbon utilization is increased.

The equilibrium time was nearly one hour for both CAC and CSC adsorbents. The correlation for equilibrium adsorption data were well fitted by the Langmuir and Tempkin adsorption isotherm for both the adsorbents. Bohart Adam's equation was found to be applicable for the column study. The adsorption kinetics was well in agreement with the Lagergren plots revealing first order rate.

From the results obtained it can be stated that though CAC is more efficient adsorbent than CSC, but due to high cost and 15 to 20% loss of CAC during regeneration CSC is more preferable than CAC for COD removal from sugar-distillery spent wash. However, for evaluation of the technical and economical feasibility, more research is required on pilot plant scale, so that this low cost adsorbent can be used on larger commercial scale for removal of COD from distillery spent wash.

FUTURE INSIGHTS

Expansion of this technology to large-scale applications should be encouraged. As used coconut shells find their way as urban and rural refuse; exploitation of this waste for waste water treatment will dispose this waste in a safer manner.

Besides the gold mining industries, the breweries, and soft drink industries that currently use activated carbon, other potential users such as the textile industries, soap manufacturing industries, vegetable oil nulls, and the Water and Sewerage Corporation can also be sensitied to use this commodity, thereby creating more market for the product. The pilot studies to develop a national capability for the production of activated carbon form coconut shells will, among several benefits, contribute to measures for abating the environmental degradation caused by dumping of agricultural wastes. Improved public education to ensure awareness of the technology and its benefits, both environmental and economic, is recommended.

More experimental work on removal of organic matter from other industrial effluents by using coconut shell ash as adsorbent should be encouraged in order to prove the suitability of this low cost adsorbent for waste water treatment.

Adsorption column studies should be carried out for different column configurations to get the characteristic dependence of the adsorption capacity and other parameters. Design parameters for scaled up adsorption column should thereafter be fixed for the removal of COD from distillery spent wash. Costing of the adsorption based on industrial scale treatment system should be carried out to popularize the adsorption technique. Computing, modeling, and simulation will be carried out for the large-scale plants based on the computations proposed.

CONFLICT OF INTERESTS

The authors declare that there is no conflict of interests regarding the publication of this paper.

REFERENCES

1.   B. Subba Rao, Aerobic Composting of Spent Wash, Environmental Protection Research Foundation, Sangli, India, 2008.

2.   G. Patil, V. V. Deshpande, and P. L. Kulkarni, Distillery Spent Wash: A Source for Production of Chemicals, Deccan Sugar Institute, Pune, India, 2009.

3.   S. Rao, S. V. Ranade, and J. M. Gadgil, "Wealth from waste," in Proceedings of the 3rd International Conference on Appropriate Waste Management Technologies for Developing Countries, pp. 25–26, NEERI, Nagpur, India, 1995.

4.   L. M. Arulanantham, T. V. Ramkrishna, and S. N. Bal, "Studies on fluoride removal by Coconut shell carbon," Indian Journal of Environmental Protection, vol. 12, no. 7, 1992.

5.   G. N. Pandey and G. S. Carney, Environmental Engg, TATA McGraw-Hill, New York, NY, USA, 2010.

6.   A. Rahman, Y. Y. Singh, M. F. Bari, and B. Saad, "Adsorption of paraquat by treated and untreated rice husks studied by flow injection—analysis," Research Journal of Chemistry and Environment, vol. 9, no. 1, pp. 17–22, 2005.

7.   R. Shyamala, S. Sivakamasundari, and P. Lalitha, "Comparison of the adsorption potential of biosorbents-waste tealeaves and rice husk, in the removal of chromium (VI) from waste water," Journal of Industrial Pollution Control, vol. 21, no. 1, pp. 31–36, 2005.

8.   S. D. Foust and O. M. Aly, "Adsorption process for water treatment," Science Report, 1986.

9.   M. K. B. Gratuito, T. Panyathanmaporn, R.-A. Chumnanklang, N. Sirinuntawittaya, and A. Dutta, "Production of activated carbon from coconut shell: optimization using response surface methodology,"Bioresource Technology, vol. 99, no. 11, pp. 4887–4895, 2008.

10. S. D. Khattri and M. K. Singh, "Adsorption of basic dyes from aqueous solution by natural adsorbent,"Indian Journal of Chemical Technology, vol. 6, no. 2, pp. 112–116, 1999.

11. S. Dadhich, S. K. Beebi, and G. V. Kavitha, "Adsorption of Ni (II) using agrowaste, rice husk," Journal of Environmental Science and Engineering, vol. 46, no. 3, pp. 179–185, 2004.

12. R. J. Krupadam and A. V. S. Prabhakar Rao, "Removal of lead from industrial effluents using coconut shell carbon," Environmental Pollution Control Journal, vol. 56, no. 2, pp. 52–53, 1997.

13. S. A. Raj, "Adsorption behaviour of nickel on activated carbon," Indian Journal of Environmental Protection, vol. 24, no. 7, pp. 530–533, 2004. ·

14. K. Vasanth Kumar and K. Subanandam, "Studies on decolorisation of basic dye onto carbonized agro based waste adsorbent," Environmental Pollution Control Journal, vol. 5, no. 2, p. 8, 2002.

15. K. V. Kumar, "Studies on adsorption of basic dyes on to agro-based wastes—part I: kinetic studies,"Indian Journal of Environmental Protection, vol. 22, no. 11, pp. 1236–1240, 2002.

16. K. Palanivelu and N. Elangovan, "Phosphate removal studies using Al impregnated coconut shell carbon," Indian Journal of Environmental Protection, vol. 16, no. 3, pp. 183–185, 1996.

17. S. T. M. L. D. Senevirathna, S. Tanaka, S. Fujii et al., "Adsorption of perfluorooctane sulfonate (n-PFOS) onto non ion-exchange polymers and granular activated carbon: batch and column test," Desalination, vol. 260, no. 1–3, pp. 29–33, 2010.

18. S. Sohail Ayub, S. Iqbal Ali, N. A. Khan, and H. S. Danish, "Treatment of waste water by agricultural waste—a review," Environmental Pollution Control Report, 1998.

19. S. Rengaraj, B. Arabindoo, and V. Murugesan, "Preparation and characterisation of activated carbon from agricultural wastes," Indian Journal of Chemical Technology, vol. 6, no. 1, pp. 1–4, 1999. ·

20. S. Dahiya and A. Kaur, "Studies on fluoride removal by coconut coir pith carbon," Indian Journal of Environmental Protection, vol. 19, no. 11, pp. 811–814, 1999.

21. D.S. Ramteke, S.R. Wate, and C.A. Moghe, "Comparative adsorption studies of distillery waste on ativated carbon," Indian Journal of Environmental Health, vol. 31, no. 1, pp. 17–24, 1989.

22. S. Satyanarayan, A. Juwarkar, and S. N. Kaul, "Distillery Waste Water Treatment—A case Study," NEERI, Nagpur-20, 2009.

23. S. Gupta and R. P. Singh, "Comparative adsorption study of toxic metal by waste product," Indian Journal of Environmental Protection, vol. 24, no. 11, pp. 863–866, 2004. ·

24. G. G. Stavropoulos and A. A. Zabaniotou, "Minimizing activated carbons production cost," Fuel Processing Technology, vol. 90, no. 7-8, pp. 952–957, 2009.

Chapter 2

GEOLOGY AND GEOTECTONIC SETTING OF THE BASEMENT COMPLEX ROCKS IN SOUTH WESTERN NIGERIA: IMPLICATIONS ON PROVENANCE AND EVOLUTION

Akindele O. Oyinloye

Department of Geology, University of Ado-Ekiti Nigeria

INTRODUCTION

Regional geology of Nigeria Nigeria

Nigeria lies approximately between latitudes 4oN and 15oN and Longitudes 3oE and 14oE, within the Pan African mobile belt in between the West African and Congo cratons. The Geology of Nigeria is dominated by crystalline and sedimentary rocks both occurring approximately in equal proportions (Woakes et al 1987). The crystalline rocks are made up of Precambrian basement complex and the Phanerozoic rocks which occur in the eastern region of the country and in the north central part of Nigeria. The Precambrian basement rocks in Nigeria consist of the migmatite gneissic –quartzite complex dated Archean to Early Proterozoic (2700-2000 Ma). Other units include the NE-SW trending schist belts mostly developed in the western half of the country and the granitoid plutons of the older granite suite dated Late Proterozoic to Early Phanerozoic (750-450Ma).

Geology of southwestern

Nigeria basement complex The area covered by the southwestern Nigeria basement complex lies between latitudes 7oN and 10oN and longtitudes 3oE and 6oE right in the equatorial rain forest region of Africa (Fig.1). The main lithologies include the amphibolites, migmatite gneisses, granites and pegmatites. Other important rock units are the schists, made up of biotite schist, quartzite schist talk-tremolite schist, and the muscovite schists. The crystalline rocks intruded into these schistose rocks. For the purpose of this chapter, discussion is limited to the crystalline basement rocks of southwestern Nigeria.

The Amphibolite and the hornblende gneiss

The amphibolite and the hornblende gneiss are the mafic and intermediate rocks in south western Nigeria. The amphibolites are made up of the massive melanocratic and foliated amphibolites. In Ilesha and Ife areas these amphibolites occur as low lying outcrops and most are seen in riverbeds. The massive melanocratic amphibolite is darkish green and fine grained. Commonly hornblende gneiss outcrops share common boundaries with the

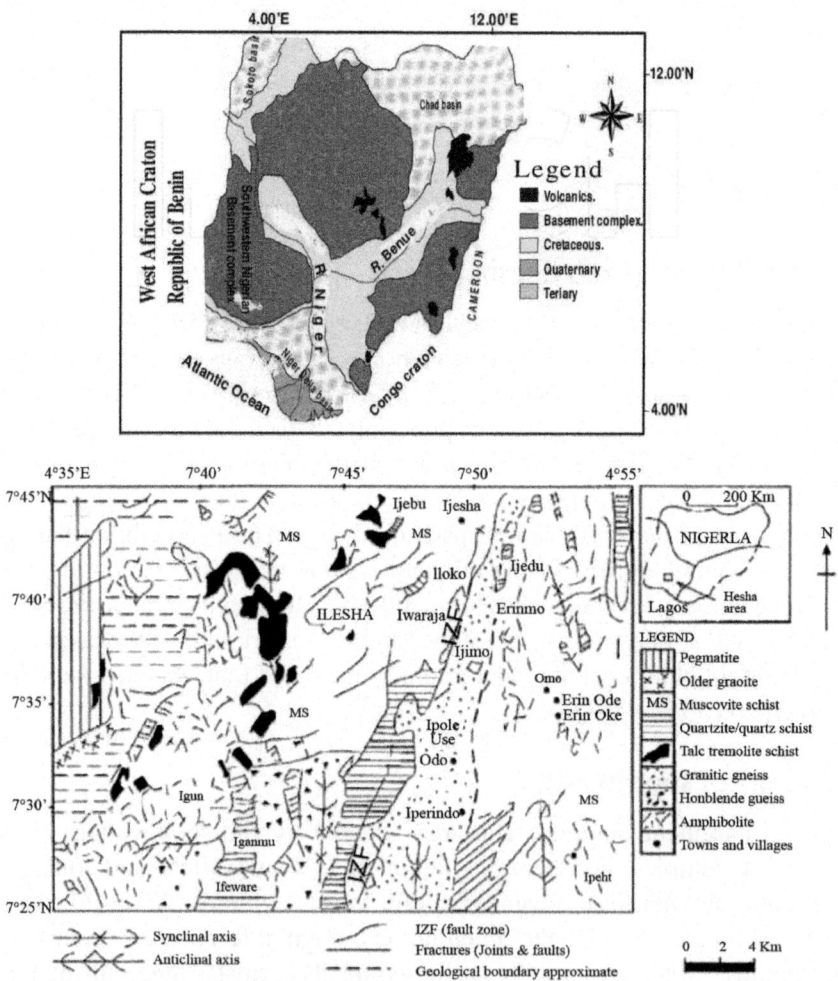

Figure 1.a): Geological map of Nigeria; b) Geological map of Ilesha schist belt Southwestern Nigeria (modified from Elueze, 1982)

melanocratic amphibolite. This rock (hornblende gneiss) crops out at Igangan, Aiyetoro and Ifewara, along Ile-Ife road as low lying hills in southwestern Nigeria. The hornblende gneiss is highly foliated, folded and faulted in places.

The Magmatite -gnessic complex

This geotectonic complex which constitutes over 75% of the surface area of the southwestern Nigerian basement complex is said to have evolved through 3 major geotectonic events: · Initiation of crust forming process during the Early Proterozoic (2000Ma) typified by the Ibadan (Southwestern Nigeria) grey gneisses considered by Woakes et al; (1987) as to have been derived directly from the mantle. · Emplacement of granites in Early Proterozoic (2000Ma) and · The Pan African events (450Ma-750Ma). Rahaman and Ocan (1978) on the basis of geological field mapping reported over ten evolutionary events within the basement complex with the emplacement of dolerite dykes as the youngest. On the basis of wide geochemical analyses and interpretation, geotectonic studies, field mapping and plumbotectonics, Oyinloye (1998 and 2011) had suggested a modified Burke et al; (1976) sequence of evolutionary events in the Southwestern Nigeria basement complex as detailed in Table1.

Table 1: A modified sequence of events in the basement rocks in Ibadan-Ile-Ife-Ilesha area (modified from Burk et al; 1976)

S/N	Sequence of Evolutionary Events in the South Western Nigeria basement complex
11	Shearing, Chloritic and Zeolite mineralization of uncertain age.
10	Emplacement of dolerite dykes and gold mineralization at about 550Ma (Oyinloye, 2006b)
9	Formation of unsheared pegmatite, unfolded granitic veins of mid-Pan African age.
8	Major remobilization and deformation in Early Pan-African
7	Minor metamorphic deformation in Kibaran
6	Emplacement of microdiorite
5	Emplacement of Ibadan-Ile-Ife-Ilesha Granite gneiss: F2 folding fabrics in Granite, Gneiss.
4	Emplacement of microdiorite dykes of uncertain age
3	Emplacement of semiconcordant aplite sheets in the banded gneiss, collision of plates subduction of ocean slab in to the mantle (Oyinloye, 2002b)
2	Deposition of ocean sediments covering the whole basement complex (Oyinloye 2002a)
1	Generation and differentiation of wet basaltic magma and formation of proto continent, (Oyinloye, 2004a).

Metamorphism in the southwestern

Nigeria basement rocks On the basis of petrology a medium pressure Barrovian and Low-medium pressure types of metamorphism had been suggested for the Precambrian basement rocks in south western Nigeria (Rahaman 1988). These metamorphic types are based on the occurrence of index minerals like chlorite, biotite and sillimenite in the basement rocks of southwestern Nigeria. Rahaman (1988) therefore concluded that metamorphism in all Nigerian Precambrian complex rocks especially that of Ife-Ilesha (Southwestern Nigeria) ranges from green schist to lower amphibolite metamorphic facies. However, Oyinloye (1992) on the basis of petrology, field mapping and structural analyses reported that the prominent gneissic foliations observed on some of the gneisses suggest that metamorphism actually reached an upper amphibolite facies in the rocks of the basement complex in Southwestern Nigeria. Egbuniwe (1982) suggested 3 phases of metamorphism (M1, M2 and M3) associated with 3 phases of deformation (D1, D2, D3) within the crystalline rocks of the basement complex in northern Nigeria. According to this author M1 represents a period of progressive metamorphism to lower amphibolite facies. M2 is described as retrogressive and reached only green schist grade as did M3. In the southwest Boesse and Ocan (1988) recognized 3 phases of metamorphism but only 2 phases of deformation. M1 is considered to be a syntectonic progressive phase of metamorphism to amphibolite facies with isoclinal folding, mineralogical banding and development of staurolite, sillimanite and garnet. M2 is described as syntectonic and associated with shear deformation and M3 being static retrogressing the earlier formed garnet and biotite to chlorite. Oyinloye (1992) however suggested that M1 is syntectonic and perhaps synchronous with the formation of the large scale major fault zone indicated by formation of mylonite outcrops at Iwaraja, Southwestern Nigeria. M2 is also syntectonic and contemporaneous with D2 as indicated by the development of micro faulting folding (Plate 1), fracturing, shearing, formation of phyllonite and mylonite with distorted garnet crystals surrounded by syllimanite crystals and mylonitised granite gneisses.

GEOCHRONOLOGY OF THE BASEMENT ROCKS OF SOUTHWESTERN NIGERIA

It has been established that the Precambrian basement complex of Nigeria including Southwestern Nigeria is polycyclic in nature, (Ajibade and Fitches 1988). The southwestern Nigeria basement complex had undergone 4 major orogenesis in:- i. Liberian (Archaean) 2500Ma-2750± 25Ma ii. The Eburnean orogeny (Early Proterozoic), 2000Ma-2500Ma iii. The Kibaran orogeny (Mid Proterozoic), 1100Ma - 2000Ma iv. The Pan African Orogeny, 450Ma-750Ma.

Of all the above, the Eburnean and the Pan-African are major events which modified the Precambrian Geology of Nigeria including the Southwestern Nigerian basement complex. The Eburnean event is marked by the emplacement of the Ibadan granite gneiss in Southwestern Nigeria which has been dated 2500±200Ma (Rahaman 1988) and a pink granite gneiss at Ile-Ife Southwestern Nigeria dated 1875Ma using U-Pb on Zircon. Thus Archaean to Pan African ages had been suggested for the basement rocks of the Southwestern Nigeria. Oyinloye (2006b), based on Pb-Pb model dating suggested 2750±25Ma (Archaean age) for the gneisses in Ilesha area Southwestern Nigeria. Few studies have been carried out on the basement complex due to its assumed monotonous petrology and mineralogy and the erroneous belief that it contains no mineralization. This current chapter will therefore contribute immensely to the debate on geology of the basement complex of Southwestern Nigeria.

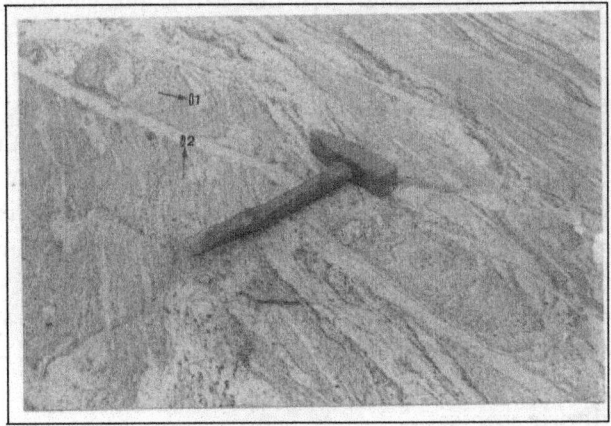

Plate 1: Deformation structures in the blotite granite gneiss, Iperindo area, Ilesha schist belt southwestern Nigeria, D2 (qtz vein) cuts across foliation planes (D1), The Hammer (2cm wide) marks a minor fault displacement

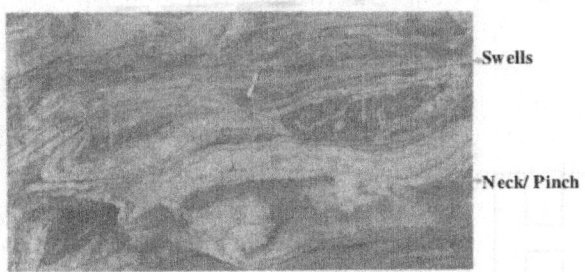

Plate 2: Showing Pinch and Swell Stuctures

GEOLOGICAL SETTING OF SOUTHWESTERN NIGERIA BASEMENT COMPLEX

The basement rocks which occur in southwestern Nigeria are all duplicated in the Ilesha area of southwestern Nigeria and samples of rocks here were analyzed and used as a case study of the basement rocks in Southwestern Nigeria to avoid repetition. These rocks are amphibolite, the hornblende gneiss and the granite gneisses. These rocks are described in that order.

The massive melanocratic amphibolite

Amphibolite occurs widely in southwestern Nigeria in Ile-Ife area, Ibodi, Itagunmodi in Ilesha area. Most outcrops of the massive melanocratic amphibolites are exposed in streams and river channels in these areas. The overburden soil here is strikingly red due to the presence of hematite and magmatite liberated during the weathering of the amphibolites to form the overburden soil. Two major textural varieties of amphibolites occur in this region. These are the leucocratic amphibolites and (not discussed in this study) the massive melanocractic amphibolites. The massive melanocratic amphibolite is darkish green and fine grained without any obvious folds or foliations. In places thin colourless quartz veins occur on the outcrops. This amphibolite variety is composed of hornblende, actinolite and tremolite. In thin section the mineral composition includes (apart from the above) magnetite, sphene calcite and minor monazite and zircon. The skeletal olivine contains small, opaque inclusions which are probably magnetite

The hornblende gneiss

The hornblende gneiss shares a common boundary with the massive melanocratic amphibolite in Ilesha area, southwestern Nigeria. This rock crops out as low lying hills in Ife-Ilesha area Southwestern Nigeria. It is composed predominantly of porphy- roblastic plagioclase and hornblende phenocrysts almost in equal proportion. This rock is highly foliated folded and faulted in places and varies from medium to coarse in texture. These outcrops trend in a NE-SW direction and dip to the east at an average angle of between 50-70°. The apparent character varies from intermediate to acid. Microbands of foliation rich in plagioclase and some K-feldspars alternate with bands rich in amphiboles. In thin section this foliated hornblende gneiss consists largely of hornblende and plagioclase porphyroblats in a ground mass of ilmenite fine grained recrystallized quartz and pyroxene fragments. Brown coloured epidote (with dark cracks) apartite, sphene, zircon and monazite constitute major accessory minerals in this rock. Foliations defined by parallel arrangement of feldspars alternating with amphiboles are conspicuous in thin sections. Fine grained quartz and orthoclase feldspars are observed in the felsic microband, garnet, monazite, calcite and microcline containing well formed zircon crystals (as inclusions) occur in this rock as observed in thin sections.

The biotite granite gneiss complex

This rock group occurs widely in every part of the southwestern Nigerian basement complex. Again description is restricted to Ilesha area to avoid repetition. Biotite granite gneiss complex occurs in the southern part of Ilesha schist belt. Outcrops of this rock group consist of high and low lying hills with myriads of flat boulders on top in places and roundish boulders on tops of hills elsewhere. This rock complex is foliated and folded with prominent synclinal and anticlinal axes. In places microfolds and microfaulting are observed (Plate1). Wide and narrow quartz veins are commonly seen on the rock and some of these are deformed to form folds and micro faults as described above. Foliations are defined by mafic (biotite rich) and felsic (quartz and feldspars) mineral bands. Drilled core samples from the biotite granite gneiss revealed that microfolds,pinkish garnet rich mylonite, greenish friable schistose phyllonite,occur in this rock. In thin section the mylonite contains fine grains of biotite and sillimanite surrounding large crystals of garnet which show some evidence of distortion. The mylonite contains little quartz and the biotite flakes form thin foliation bands which are closely packed around garnet crystals. In some of the cores examined recovery failures are recorded indicating fracturing as described above. This biotite granite gneiss contains deformation fabrics which may be regarded as D2 and probably contemporaneous with the M2 phase of metamorphism following D1 and M1 (Plate 1). These later events may be due to movements along the major Ifewara-Zungeru fault system. The biotite granite gneiss are surrounded by muscovite-quartzite schists and in places the later are in-foled into the gneisses where they occur as reminants. At outcrop scale, the biotie granite gneiss is composed of biotite, Kfeldsper, quartz and garnet. In thin section the biotite flakes are pencil-like as a result of metamorphic deformation and are aligned in parallel to sub parallel manner. The K-feldspar is mostly microcline and is porphyroblastic in texture. Well formed zircon crystals occur in association with some of the microcline grains. Apartite, monazite, magnetite, ilmenite and sphene are other accessory minerals. In places distorted and fractured garnet grains due to metamorphic deformation are observed. Continuous well defined foliation bands of micas and felsic minerals are also common features of this gneiss. These gneissic fabrics probably indicate that metamorphism here was perhaps higher than the green schist-lower amphibolite facie regarded as the meramorphic grade for rocks in the basement complex in southwestern Nigeria. The presence of mylonite, mylonitised granite and gneissic banding are probable indications of a localized dynamic metamorphism possibly reaching an upper amphibolite facie.

The pink granite gneisses

This variety of gneiss occurs widely in the southwestern Nigeria basement complex at IleIfe, Ibadan, Iseyin, Eruwa and Iwaraja and in Ilesha area. The granite gneiss is pinkish with large pherocrysts of K-feldspar and porphyroblasts of hornblende. The texture of the pink granite gneiss varies from medium grained to very coarse almost becoming pegmatitic in places. Augen structures are commonly observed on the pink granite gneiss. This pink granite gneiss is fractured in places and elsewhere folded. Augen structures with clear elongate lozenges (boudins) and neck or pinch structures (pinch and swell) as a result of stressing are commonly seen on the pink granite gneiss in this region (Plate 2). In thin section, foliation is defined by elongate hornblende and drawn out K-feldspar porphyries. Other minerals include quartz, plagioclase, some biotite flakes, garnet, apartite and zircon. Monazite forms an important accessory mineral in this rock. Commonly, phenocrysts of orthoclase occur within a matrix of recrystallised quartz and microcline. At Iwaraja, a major fault marked by a mylonite outcrop is observed within the pink granite gneiss terrain. This mylonite marks the southern extension of the Ifewara – Iwaraja-Zungeru major fault which runs in a NE-SW direction across the country. Deformation fabrics in the southwestern Nigeria basement complex are commonly aligned parallel to the direction of the Ifewara-Zungeru fault zone implying that this fault has a profound and wide influence on fabric and metamorphism in this region. According to Boesse and Ocan (1988) this major fault (marked by the mylonite outcrops) marks a break between the granite gneissic complex and the metasediments in this region.

The grey granite gneiss

The grey granite gneiss occurs prominently at Ibadan, Oyan, and in Ilesha areas of southwestern Nigeria basement complex. Usually outcrops consist of high and low hills and at Erinmo in Ilesha area occur very close to the pink granite gneiss and only separated by a narrow strip of muscovite quartzite schist. The overall colour is greyish. The texture of this variety of gneiss is fine to medium grained with well developed foliation defined by preferred orientation of biotite. This rock is mostly composed of quartz, biotite, plagioclase, K-feldspar and hornblende. In thin section recrystallised fine grained quartz covers the surface of microcline phenocrysts as overgrowths. This is a common phenomenon in all the granite gneisses investigated in this study. Mosaic textures formed by fragments of plagioclase, biotite and recrystallised quartz are also observed. Intergrowths of orthoclase and microcline forming a perthitic texture occur in places. Quartz crystals consist of fractured and recrystallised fine varieties. Well formed rod-like and fragmented zircon crystals, apatite, monazite plus minor garnet form important accessory minerals.

GEOCHEMISTRY

The geochemical data described in this chapter are presented in the following order. 1. Massive Melanocratic Amphibolite 2. The Hornblende Gneiss 3. The Biotite Gneiss 4. The Pink Granite Gneiss Note:- The average geochemical data discussed here are not included in this write up because of space. These are available from the author on request.

The massive melanocratic amphibolite

Major elements In this study it is observed that element concentrations in the massive melanocratic amphibolite vary little even between samples collected from outcrops almost 1km apart. The mean SiO_2 concentration in this rock is 49% (17samples) alumina 15%, total iron 11%, MgO 10% and CaO 12%. The high iron concentration in the melanocratic amphibolite reflects the abundance of titanomagnetite and the high CaO content is an indication of the preponderance of Ca-rich pyroxene. TiO_2 content (average 1%) reflects some sphene in addition to titanogmagnetite . The total alkaline concentration is very low reflecting the subalkaline nature of this rock. Na_2O is consistently higher than K_2O in this rock perhaps reflecting the dominance of albite in the massive melanocratic amphibolite. MgO/ Fe_2O_3+MgO ratios vary between 0.45 to 0.48 with a mean of 0.46. This is considerably lower than that of a pure primitive upper mantle which has a range of 0.68-0.75 and a mean of 0.70 (Wilson, 1991).

Trace elements

In the massive melanocratic amphibolite Rb is characteristically low, 11ppm on average indicating low K-feldspar concentration as observed in the thin section studies. Sr with an average of 169ppm is relatively high in the massive melanocratic amphibolite due to substitution of Sr for Ca in the pyroxene and amphiboles as is Zr (58ppm) due to minor zircon. Zr can also substitute for Ti in accessory phase in sphene and rutile. Y concentrations are appreciable (mean 19ppm) since this element is readily accommodated in amphiboles which are the dominant minerals of the amphibolite. The low Th in this rock reflects fractionation into more felsic magmatic franctions.

The average concentrations of compatible elements (Ni, Cr, and Co) in the massive melanocratic amphibolite are 102ppm, 81ppm and 54pm respectively. These values are too low for an amphibolite originating from a pure primitive upper mantle. The low concentrations of compatible elements suggest that the precursor rock of the amphibolite is from a depleted or metasomatised mantle and this has a significant implication on provenance and geotectonic setting in which the rock was formed.

Rare earth elements

Rare earth elements are significantly recorded in the massive melanocratic amphibolite. The average total REE in the massive melanocratic amphibolite is 71ppm. The dominance of light rare earth elements (59ppm) over the heavy ones average 12ppm reflects the relative abundance of monazite in the amphibotite and further suggests that this mafic rock is not from a pure primitive mantle.

The horblende gniess

Field and petrological studies revealed that this rock consists of intermediate to acid varieties. The average SiO_2 (63%) in the hornblende gneiss is much higher than that of the massive melanocratic amphibolite. TiO_2 and Fe_2O_3 averages are lower than that of the amphibolite reflecting the less mafic character of the hornblende gneiss. The relatively higher total alkalis (mean 7%) and K_2O/Na_2O ratios (0.8) are indicative of more abundant feldspars in the hornblende gneiss than the amphibolite which is consistent with field and thin section observations. The MgO/Fe_2O_3+MgO mean ratio is 0.32 and this is lower than in the amphibolite and thus further from pure upper mantle value. In this rock the average concentration of Rb (68ppm) is more than 6 times its concentration in the amphibolite paralleling the increase K-feldpar content. Sr and Ba are also strongly enriched (1266ppm and 1493ppm respectively). This is perhaps due to substitution of Ba for K in the K-feldspar and Sr for Ca in plagioclase. The higher Y (mean 37ppm) concentration in the hornblende gneiss compared with the amphibolite may be due to the presence of more hornblende (dominant mineral in the hornblende gneiss) which often concentrates this element. Low Th concent of the hornblende gneiss average (6ppm) might be due to minimal crustal contribution. In the hornblende gneiss, the mean concentraction of Ni, Cr, are less still (26ppm and 39ppm respectively) reflecting the less mafic character of this rock.

Rare earth elements (REE)

In the hornblende gneiss, there is a high concentration of REE most especially the light ones. The average total REE in the hornblende gneiss is 3232ppm. The light REE in this rock has a mean of 2174ppm. The mean concentration of the heavy REE in the hornblende gneiss (58ppm) is relatively higher than in the amphibolite reflecting more abundant REE concentrating minerals e.g. sphene and monazite.

THE BIOTITE GRANITE GNEISS

Major elements

The biotite granite gneiss is one of the series of granitic rocks with SiO_2 higher than 70% in southwestern Nigeria basement complex. TiO_2 average concentration in this rock (0.42%) is slightly higher than in the hornblende gneiss. Al_2O_3 mean (15%) is slightly higher in this rock than in the hornblende gneiss. Unlike the amphibolite and the hornblende gneiss, the Na_2O concentration is consistently less than K_2O in the biotite granite gneiss. K_2O/Na_2O ratios are consistently higher than 1 (one) in the biotite granite gneiss although the average total Na_2O and K_2O (8%) in the biotite gneiss is only 1% higher than that of the hornblende gneiss.

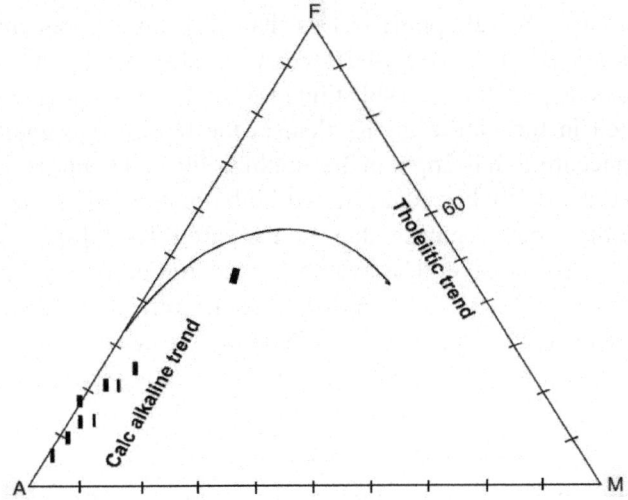

Figure 2: A=Al2O3; F=FeO (total iron); M=MgO (AFM) diagram for the biotite granite gneiss, from Ilesha area

The concentration of SiO2, Al2O3, Na2O and K2O in the biotite granite gneiss indicates an abundance of felsic silicates e.g. feldspars and quartz. The consistently higher concentration of K2O than Na2O (thus K2O/Na2O ratios greater than 1) in the biotite granite gneiss reflects the abundance of K bearing rock forming silicates (i.e microcline and biotite). This trend is also characteristic of Archaean granitic rocks (Martin (1986). In the biotite granite gneiss, the concentrations of Fe2O3, MnO, MgO, CaO and P2O5 are much less than the average values in the amphibolite and hornblende gneiss. This reflects its less mafic character. Although major elements of gneisses are sensitive to metamorphic alteration, AFM diagrams can be used to study the enrichment of these rocks in alkalis and Fe in a general way. When the AFM diagram is plotted for the hornblende gneiss and the biotite granite gneiss, both rocks plot in the calc alkali fractionation trend (Fig.2) reflecting enrichment in Al2O3, Na2O and K2O due to development of biotite and K-feldspars in the biotite granite gneiss and plagioclase in the hornblende gneiss as observed in the petrological studies.

Trace elements

The average concentration of Rb in the biotite granite gneiss (182ppm) is more than double its average concentration in the hornblende gneiss. The high concentration of Rb in the biotite granite gneiss is due to substitution of Rb for K in the microcline and biotite which are abundant in this rock. The average concentration of Sr (299ppm) is less than 25% of its concentration in the hornblende gneiss. This reflects the low concentration of plagioclase, hornblende and pyroxene in which Sr can substitute for Ca in the biotite gneiss. Zr is more concentrated in the biotite gneiss than in the hornblende gneiss. Th whose average concentration is 2ppm in the amphibiolite and 6ppm in the hornblende gneiss is relatively highly concentrated in the biotite gneiss (average 33ppm) in the biotite granite gneiss due to increased K-feldspar content which concentrates this element. The average concentration of Y (32ppm) is lower than its concentration in the hornblende gneiss reflecting less hornblende in the biotite gneiss. The concentration of compatible heavy elements (Cr, Ni and Co) are relatively minor in the biotite gneiss reflecting its acid nature.

RARE EARTH ELEMENTS (REE)

The average total REE in the biotite granite gneiss is 328ppm of which the light ones account for 320ppm, and heavy rare earth elements only 18ppm. As in the amphibolite and hornblende gneiss, higher amount of light rare earth elements are concentrated in the biotite granite gneiss than the heavy rare earth elements. However, the total REE in the biotite granite gneiss is lower than that of the hornblende gneiss due to lower abundance of sphene, plagioclase and hornblende in the biotite granite gneiss than in the hornblende gneiss.

The pink granite gneiss

The average concentrations of SiO_2 (76%) and Na_2O (2.40%) recorded in the pink granite gneiss in this region reflect higher concentration of K-feldspar in the pink granite gneiss than in the biotite granite gneiss. Al_2O_3, K_2O, CaO, P_2O_5 average concentrations in this rock are lower than in the biotite granite gneiss. The averages of K_2O/Na_2O ratios which are greater than 1 and the total K_2O+Na_2O (7%) on the average in the pink granite gneiss show the same trends as in the biotite granite gneiss. AFM plot for this rock also show a calc-alkali franctionation trend.

Trace elements

Ba and Sr concentrations in the pink granite gneiss are lower than in the biotite granite gneiss indicating lower plagioclase in the former than in the later. Higher concentration of Rb in the pink granite gneiss than in the biotite granite gneiss reflects more abundant Kfeldspar in the pink granite gneiss. Zr has lower average in the pink granite gneiss than in the biotite granite gneiss which corresponds to lower zircon occurrence in the pink granite gneiss. Th mean concentration in the pink granite gneiss is higher than in the biotite granite gneiss which may be due to higher sedimentary contribution to the precursor of this rock. In the pink granite gneiss an increase of Y concentration is recorded compared with its average concentration in the biotite granite gneiss. This may be due to higher hornblende component of the pink granite gneiss than in the biotite granite gneiss. The average Ni in the pink granite gneiss has diminished compared with the hornblende gneiss and biotite granite gneisses and Cr concentration in the pink granite gneiss is below the detection limit (3ppm) of the XRF used for these analyses due to implied less concentration of mafic minerals in this rock.

Rare earth elements (REE)

The total absolute REE on the average is 235ppm out of which light Rare Earth Elements account for 215 ppm and heavy rare earth elements 20ppm. The average ratio of LREE/HREE: 11, recorded in the pink granite gneiss is lower than those of the biotite granite gneiss. However, the average HREE in the pink granite gneiss is higher than in the biotite granite gneiss. The lower concentration of REE in the pink granite gneiss than in the biotite granite gneiss reflects the less abundant REE concentrating minerals in the pink granite gneiss.

GEOCHRONOLOGY

There had been some reported determination of the age of the basement complex rocks generally in Nigeria including the southwestern Nigeria basement rocks. On the basis of isotopic studies, Archaean and Proterozoic ages had been suggested as the ages of emplacement of the basement rocks in Nigeria by Dada et al; (1998) and Annor (1995). According to Ajibade et al, (1987) the southwestern Nigeria basement complex are of two age generations, one represented by migmatite gneiss complex probably of Archaean to Early Proterozoic age while the other is believed to be of Late Proterozoic age. Age determination of the southwestern Nigeria basement complex rocks has not been completed as much work needs to be done to actually date these rocks satisfactory. However Oyinloye (2006b, this author) carried out a Pb-Pb, 2-stage model age based on Stacey and Kramers (1975) on the granite gneisses in Ilesha area of southwestern Nigeria and part of the result is reproduced here.

Lead (Pb-Pb) model dates

The whole rock and feldspar samples analysed for lead isotopes in this study were from the biotite granie gneiss in Ilesha area of southwestern Nigeria. On plotting, these samples revealed limited scatter points on the Pb-Pb isochron (Fig.3) but with a well defined trend. Pb-Pb data for the six K-feldspar seperates (plotted in addition) are from the same biotite granite gneiss and are comparable to the equivalent whole rock. These results fit well to the indicated best fit line which corresponded to a two-stage isochron 2750±25Ma with an initial ratio of 12.809 and MSWD of 16 (Fig.3). On the Stacey and Kramers (1975) growth curve, the biotite granite gneiss whole rock and feldspar Pb experimental points plotted to the left of the geochron Q-P (Fig.5) crossing the growth curve at point N giving an initial ratio of 12.809 which was due to geochemical differentiation. The experimental Pb-Pb isochron yields a model age of 2750±25Ma (Fig.3). This implies that Pb was withdrawn from the unradiogenic reservoir and incorporated into the feldspars and the protolith of the biotite granite gneiss at about 2750±25Ma . This Pb-Pb age which is Archaean is therefore the age of emplacement of the precursor rock which gave rise to the biotite granite gneiss in Ilesha area of southwestern Nigeria. On ploting the Pb-Pb data on Zartman and Doe (1981) evolutionary curve, (Fig.4) five out of the six whole rock samples and five out of the feldspar samples plot between the two curves OR and UP, (Fig.4). While only one sample of each (feldspar and whole rock) of the feldspar and whole rock samples plot outside the curves. Samples which plot within the two curves UP and OR (Orogen) in Figure 4 indicate that their precursor rocks were derived from a tectonic environment where crustal/sedimentary and mantle materials were partially metted to generate the initial magma from which the protolith of this biotite granite gneiss was formed, (Cf. Zartman and Doe 1981). Furthermore, Pb-Pb isotope data show that the whole rock samples from the biotite granite gneiss are extremely homogenerous with only very slight deviations from the mean values. The feldspar separates show more isotopic homogeneity.(Oyinloye 2006b). This type of extreme isotopic homogeneity in rocks and feldspars is characteristic of rocks derived from a subduction related environment like a back arc or an island arc, where mantle and upper crustal materials are thoroughly mixed to generate a magma (Billstrom 1990). Burke and Dewey (1972) had earlier described the Ilesha area in southwestern Nigeria as one that evolved in an island arc marginal basin but Oyinloye (2006b) showed that it evolved in a back arc tectonic setting.

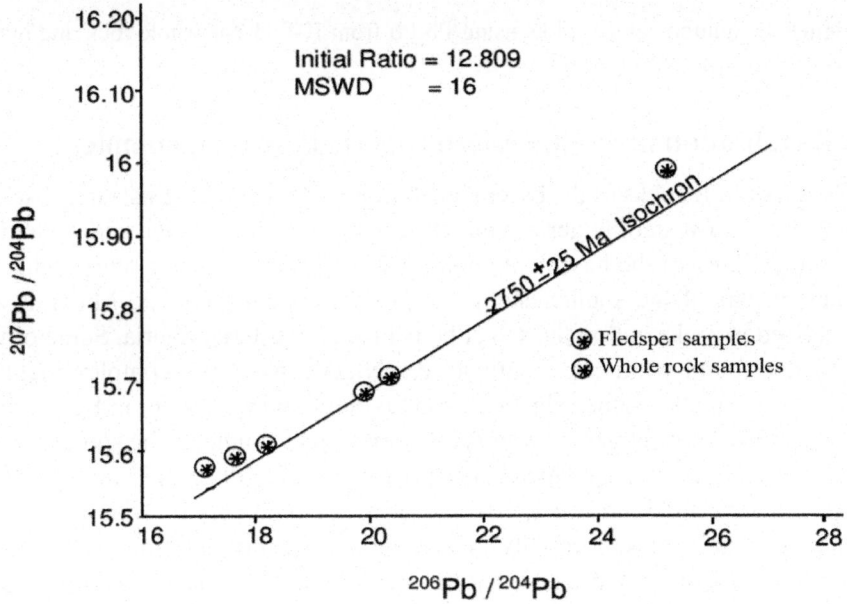

Figure 3: Lead (Pb-Pb) whole rock and K-feldspar isochron diagram for biotite granite gneiss and K-feldspar from Ilesha schist belt , southwestern Nigeria

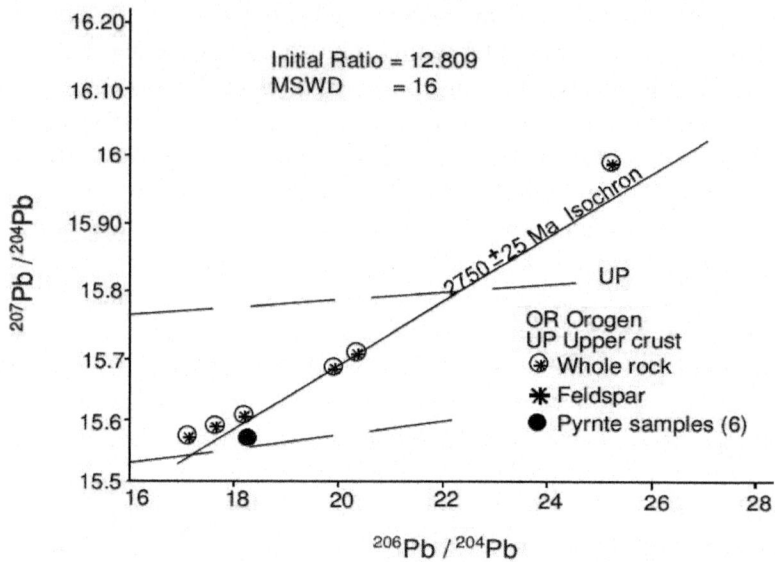

Figure 4: Plumbotectonic plots using Pb-Pb from K-feldspar whole rock and pyrite samples (based on Zartman and Doe, 1981)

Mineralisation in Southwestern Nigeria basement complex

The crystalline rocks of the basement complex intruded the schists in the schist belts in southwestern Nigeria. The schist belts are critical to the understanding of the geology of the basement complex in Nigeria. Infact the schist belts are integral part of the southwestern Nigerian basement complex. Minerals are localized in rocks within the schist belts in southwestern Nigeria. Some of the minerals found within the southwestern Nigeria basement complex include, gold in Oyan and Ilesha schist belts, in Okolom and Gurungaji in Egbe schist belt southwestern Nigeria. Gold is the major metal found in the southwestern Nigeria basement rocks. However unproven reserves of cassiterite, columbite and tantalite are found in Ijero-Ekiti area of southwestern Nigeria. Others include gem stones – amythyst, tourmaline and quartz. Gold is the only metallic mineral of substance that has been studied in this area especially by this author. A summary of the geology and geochemistry of the gold deposit at Ilesha in southwestern Nigeria is reported here. There are two (2) types of gold mineralization in Ilesha area. The first type is an alluvial form which occurs within the amphibolite terrain in Ilesha area southwestern Nigeria. This has not been well studied. The second one is a primary gold deposit found as auriferous quartz veins localized in a shear zone about 4km from the major Ifewara-Zungeru fault zone described within the pink grnite gneiss in this study. The biotite granite gneiss described in this report is the host rock of the Ilesha primary gold deposit, known as the Iperindo primary gold deposit. This gold deposit occurs as a system of auriferous quartz veins infillings, structurally localized at a folded boundary between biotite granite gneiss host rock and the adjacent metasedimentary complex. The granite gneiss host rock was altered by an invading hydrothermal fluid. The alteration selvages which form the hanging and footwall rocks are dominantly phyllic in nature with a minor chlorite overprinting in the foot wall rocks. These alteration selvages are relatively narrow but prominent and intensive around all the mineralized quartz veins at Iperindo. Geochemical analyses of the country and selvage rocks of this lode gold deposit show that the altered rocks are enriched in Cu, Zn, Pb and rare earth elements generally and heavy rare earth elements in particular relative to the country wall rocks reflecting presence of chalcopyrite, sphalerite, galena, and rare earth elements concentrating minerals and development of secondary alteration products such as sericite in the alteration selvages. General studies carried out using, stable carbon 13, oxygen 18 isotopes and plumbotectonics show that mineralization of gold at Iperindo near Ilesha was meteoric in origin. Ore fluid inclusion studies of selected samples from the auriferous quartz veins from Iperindo gold deposit indicate that the ore fluid was rich in carbondioxide. Microthermometric measurements show that there are two types of fluid inclusions in this gold deposit and these two homogenized at high temperatures but underwent phase separation at low temperatures. These two fluids are: 2-phases carbondioxide, (gas and liquid) and 3-phases, carbondioxide gas, carbondioxide liquid and water. Fluid inclusion studies also show that mineralization of gold took place at a temperature in excess of 286oC. The data obtained from Pb isotope studies from pyrite which is found in Iperindo gold deposit show an extreme homogenous relationship. On Zartman and Doe (1981) evolutionary plot all the pyrite Pb plot within the two curves just like the whole rock and feldspar separates (Fig.4) indicating genetic relationship (note: only one of the pyrite Pb samples appears in Figure 4 because they all clustered at a point). The model ages calculated for each pyrite sample varies from 559Ma-573Ma with a mean of 550Ma (Oyinloye 2006b).

Also, the pyrite lead data showed that Pb isotopes in pyrites from the Iperindo lode gold are extremely homogeneous and very similar in value to those obtained from the whole rock and feldspar separates. Therefore going by the earlier interpretation of Pb homogeneity in feldspar, and hole rocks samples the Pb homogeneity observed in pyrite samples from this gold deposit might indicate derivation from a mixed crustal and mantle sources (Volcanoproto-continent precursor rocks of amphibolites and amphibolite schists) for the Pb in pyrite which forms a prominent gangue in Iperindo gold deposit. The component of the ordinary Pb was probably withdrawn from its reservoir before 2750±25Ma as a result of magma generation and protocontinent rock formation. There was an hydrothermal invation of the volcanics leading to leaching of Au from these rocks, removal of Pb from the reservoir and incorporation into pyrite at about 550Ma, the age of gold mineralization in Ilesha area of southwestern Nigeria

PROVENANCE AND EVOLUTION OF THE SOUTHWESTERN NIGERIA BASEMENT ROCKS

A controversial aspect of the geology of the Nigeria basement complex is its geotectonic origin. Only very few workers had applied geotectonics to interprete the origin of the basement rocks in southwestern Nigeria. In my research studies, I was able to use sophisticated equipment like scanning electron microscope Cambridge 250model in the United Kingdom to determine the spot chemical composition and empirical formulae of nearly all rock forming minerals in the rocks of the basement complex of southwestern Nigeria as represented by the amphibolite and granite gneisses in Ilesha area. A mineral known as monazite was discovered in this process. This mineral is present as a notable accessory mineral in all the crystalline rocks of the basement complex here in Ilesha area even in the amphibolite which is supposed to be purely igneous. Hither to except in the Younger Granites in the north central Nigeria and in sedimentary rocks in Lokoja and Auchi areas, monazite has not been described by any worker in the rocks of the basement rock of southwestern Nigeria in general and in Ilesha area in particular. It is in my research that monazite is being described for the first time in the rocks of the southwestern Nigeria and in Ilesha area in particular. Monazite is a phosphate of the rare earth elements, especially the light ones e.g. (La, Ce, Nd) PO_4. Monazite is known to be a crustal or sedimentary mineral. Its presence in a supposedly igneous rocks of mantle origin therefore raises a petrogenetic question. The petrogenetic implication of the presence of monazite in the crystalline rocks of the southwestern Nigeria is that the initial magma from which the precursor rocks were formed contain some input from a crustal or sedimentary source. As described in this text. The $MgO/(Fe_2O_3+MgO)$ ratios recorded for the amphibolite are lower than that of the basalts derived from a pure primitive mantle and this ratio decays further from the hornblende gneiss to the granite gneisses. In order to further unravel the provenance of these rocks in southwestern Nigeria, normative corundum of the hornblende gneiss and the granite gneisses were plotted against the $Mol.\%Al_2O_3/(Na_2O+K_2O+CaO)$. Also, the histogram of the $Mol.\%Al_2O_3/(Na_2O+K_2O+CaO)$ distribution for the gneisses were plotted (Figs.6 A and B). Most of the gneisses samples plot in S-type field while few samples plot in the I-type field (Fig. 6A). In Figure 6B, the gneisses sample show a bimodal histogram with a mode at I-type field and another at the S-Type field. Also the hornblende gneiss samples occur in both I-Type and S-Type fields. These plots imply that the magma which gave rise to the precursor rocks of these gneisses originated from a mixed source, containing igneous and sedimentary materials. Generally, all the plots show that the granitoids are very homogenous, related petrogenetically and are derived from a mixed source. The index trace elements were used to plot many discriminating diagrams (only one is shown here because of space). On plotting Ti versus Zr (Fig.8 based on Pearce et al; 1984) the massive melanocratic amphibolites data plot in the arc lavas field indicating a volcanic arc (similar to a back arc tectonic setting). Further more, it is observed that the average concentration of the compatible elements, (Ni, Cr, Co) in these rocks are extremely lower than that of the normal rocks derived from a primitive upper mantle source implying that the magmatic source had been metasomatised. Chondrite normalized REE were plotted for all the crystalline rocks described in this study including the massive melanocratic amphibolite. The massive melanocratic amphibolite shows a slight negative Eu/Eu* anomaly and high LaN/YbN ratios. The gneisses show similar REE patterns and a higher negative Eu/Eu* anomally. These trends show a progressive differentiation from the basalts to the gneisses (Fig.7). The implication of these is that the precursor of these rocks originated from a basalt that differentiated progressively to the granite precursor rocks of the gneisses as shown in Figure 7. The extreme Pb isotopic homogeneity as observed in the biotite granite gneiss samples and its feldspar separates indicates derivation from a subduction related environment like a back arc or an island arc where mantle and upper crust materials are thoroughly mixed to generate a magma (Billstron 1990). The southwestern Nigeria basement complex as typified by the Ilesha schist belt had earlier been described as one that evolved in a subduction related environment of island arc and marginal basin, Burke and Dewey (1972). But Oyinloye (2002a), Oyinloye and Steed (1996), and Oyinloye and Odeyemi (2001) on the bases of petrological, geochemical, plumbotectonics and structural analyses showed that the environment of the emplacement of the protocontinent which were the precursors of the rocks of the southwestern Nigeria basement rocks was a back arc basin. The linear trends displayed by the whole rock Pb data on the growth curve (Fig.5) reflects a mixing process between varying amounts of upper crust and mantle materials as in an island arc or a back arc environment, Billstrom (1989).

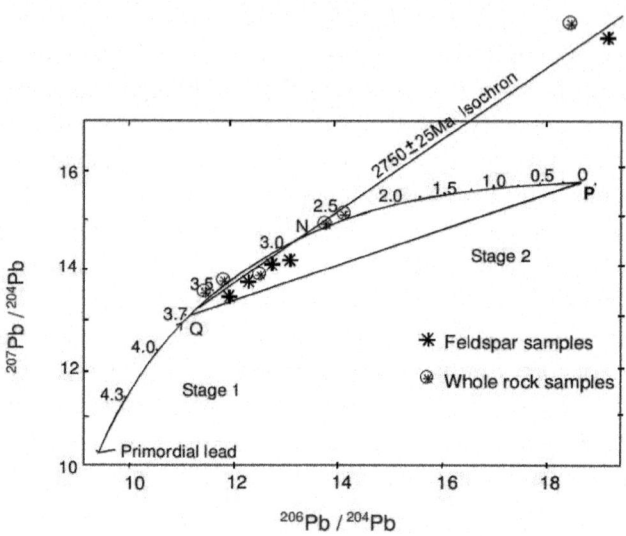

Figure 5: Whole rock and K-feldspar Pb experimental data points on the two stage growth curve of Stacey and Kramer (1975)

The various discrimination diagrams based on the trace elements considered immobile during metamorphic alteration show that the rocks of the basement complex of southwestern Nigeria may possibly be derived from a low-KTholeritic magma, (Oyinloye and Odeyemi 2001). These plots also indicate a possible volcanic arc for these rocks. (Fig.8) both back arc and island arc are grouped in volcanic arch in Fig.8. A volcanic arc characteristic of the massive melanocratic amphibolite suggests that subduction tectonics was important in the formation of its parent magma. The flat shape of the REE curve (Fig.7A) with slight Eu anomaly is typical of a back arc basic rock (Wilson, 1989). Also the spider diagrams for the massive-melanocratic amphibolites (not shown) are similar to those of a spreading tectonic settings (e.g. Mid Ocean Redge basalt, or a back arc setting). But none of the samples in the discriminating diagrams plotted in the Mid Ocean Ridge Besalts Field (Fig.8). Furthermore the development of a negative Eu anomaly shown by these rocks especially the massive melanocratic amphibolite (Fig.7A) is alien to a midocean ridge basalt or an island arc. A back arc tectonic setting will adequately account for the characteristics displayed by these rocks in the discrimination and REE fractionation trends as described above.

The basement complex rocks of the southwestern Nigeria are believed to have developed in a back arc basin (Oyinloye and Odeyemi 2001). Rahaman et al; (1988), had suggested that an ocean was closing and opening at the West African margin. Holt et al (1978) based on geotectonic studies explained that the southwestern Nigeria basement complex resulted from the opening and closing of an ensialic basin with consequent extensive subduction during the Pan African events. Burke and Dewey (1972) from structural point of view believed that components of the schist belts containing the crystalline complex rocks in southwestern Nigeria had been formed in a back arc basin caused by the collision between the continental margin of the Tuareg shield (Hoggar belt in northwest Africa).

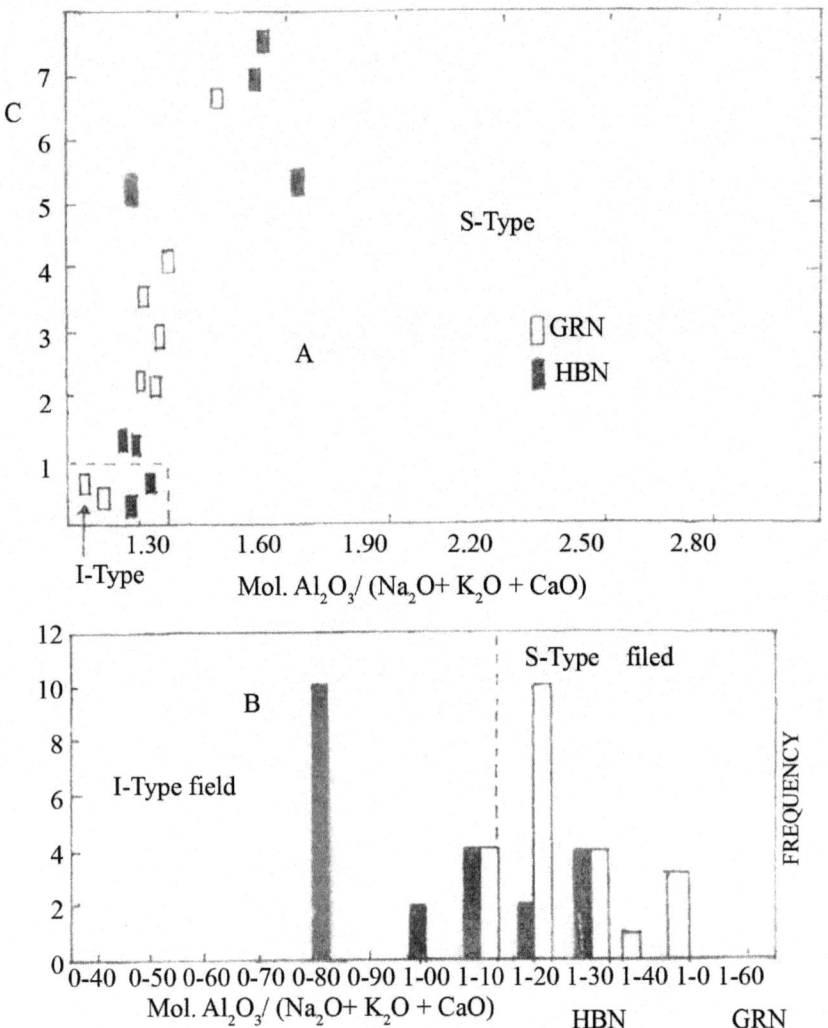

Figure 6 a): Normative Corundum versus mol. Al2O3/(Na2O+K2O+CaO) for classification of IType and S-type igneous rocks, b) Histogram of mol. Al2O3/ (Na2O+K2O+CaO) distribution for the Hornblende gneiss (HBN) and Granite gneiss (GRN). (Method based ov Vivaldo Waldo and David Rickard (1990)).

But in this study and in the previous ones the petrological, geochemical and plumbotectonic studies revealed that these rocks originated from a mixed magma containing both mantle and sedimentary materials.

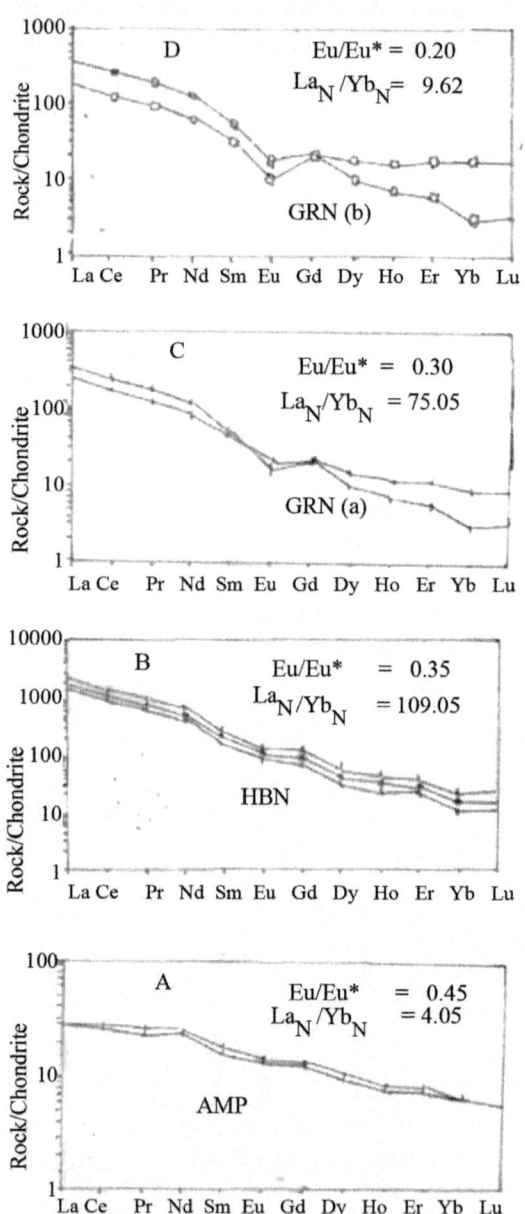

Figure 7: Chondrite normalized REE patterns for AMP, HBN, GRN from the Ilesha schist belt Southwestern Nigeria, a) AMP chondrite normalized REE patterns showing

essentially flat patterns, slight EU depletion and low LaN/LbN implying little or no differntation, b) HBN chondrite normalized REE patterns showing high LREE, low HREE, stepped patterns and moderate Eu depletion and very high LaN/LbN implying little or no differentation, c) GRN (a) chondrite normalized REE patterns showing high LREE, low HREE, pronaunced Eu depletion and high LaN/LbN showing very high differentiation of the source, d) GRN (b) GRN (a) chondrite normalized REE patterns showing high LREE, low HREE, high Eu depletion and moderate LaN/LbN showing little differentiation (last magmatic phase) These rocks shows an increase from AMP to HBN and decrease from HBN to GRN. These trends probably suggest a possible differentiation trends implicating differentiation of a basalitic magma.

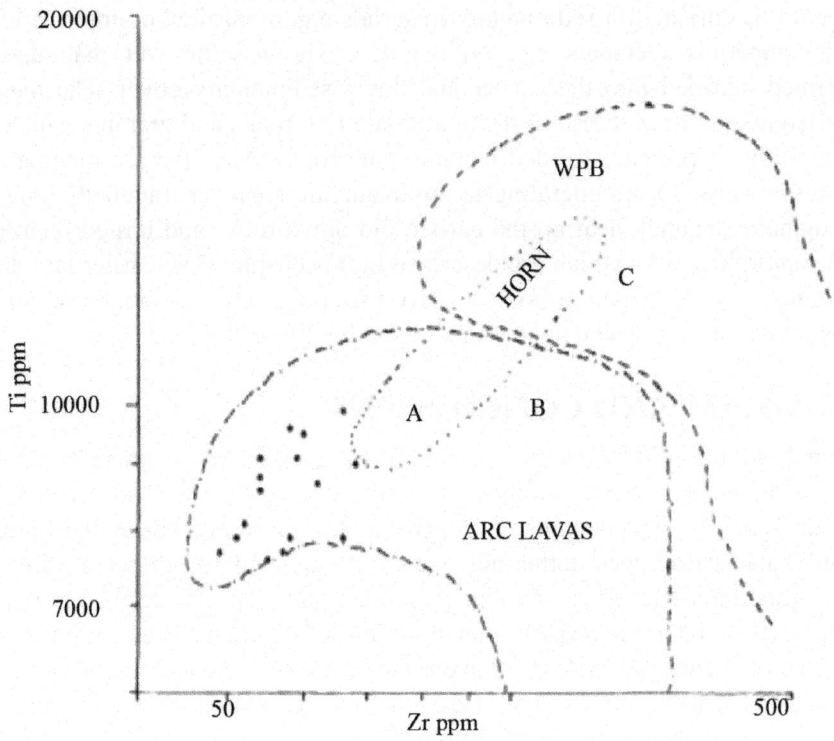

Figure 8: Plot of Ti against Zr for the massive ampibolites in Ilesha schist belt Southwestern Nigeria. A) Mid-Ocean Ridge Basalt (MORB) field, B) Arc Lava field, C) Within Plate Basalt (WPB), (after Pearce et al. (1984)).

Reviewing all the known tectonic environments especially island arcs and back arcs (which had been suggested as the geotectonic setting in which the rocks of the southerwestern Nigeria basement complex originated), the petrology, geochemistry and plumbotectonic studies of the rocks understudy implicate a back arc tectonic setting in which an ocean slab was subducted into the mantle. This subduction was due to a collision between an ocean slab and a continental shelf. In such an environment, the ocean slab would be subducted into the mantle with sedimentary materials and water which makes a wet mixed magma formation possible.

During collision between the continental shelf and the ocean plate, materials are scraped from the descending ocean slab and spread all over the area in southwestern Nigeria. Meanwhile, the descending ocean slab would carry sedimentary materials including water into the mantle. This is responsible for metasomatism of the mantle materials. There would be an exchange of materials in which the mantle portion of the wet magma formed would be enriched in sedimentary materials e.g. monazite and inproveriched in compatible elements e.g. Ni, Cr, Co. The basaltic wet magma thus formed intruded into the earlier laid down sedimentary cover. The magma differentiated to give rise to the amphibole rich rocks and granites which are protoliths of the later formed metamorphic rocks. After the arc magmatism, transpsressive forces operating in the magmatic chamber travelled along the magmatic channel, heat up the earlier laid down rocks and turned them into metamorphic rocks which are described in this chapter. The earlier laid down sediments were metamorphosed to give rise to the schists and metasediments found within the basement complex of southwestern Nigeria.

SUMMARY AND CONCLUSIONS

Amphibolites, hornblende gneiss and granite gneisses are the main crystalline rocks in the basement complex of southwestern Nigeria. These rocks had undergone a polycyclic metamorphism which is mostly pervasive in Eburnean and Pan-African tectonothermal events. As a result of these, a series of deformation fabrics and evolutionary episodes had been recorded in these rocks. However, M1 and M2 phase of metamorphic deformation corresponding to two D1 and D2 phases of deformation are mostly discernible as recorded on these crystalline rocks. A two stage lead model age determination for the gneisses revealed that the protoliths of the basement rocks in southwestern Nigeria were emplaced in the Archaean (2750 ±25Ma). In this region gold mineralization was effected by the invation of a meteoric ore fluid at a temperature above 286oC. Au (gold) was probably leached from the metavolcanics in the belt and deposited as a system of auriferous quartz veins in a shear zone at about 550Ma in this region. Geochemical (major, trace elements and REE), geological and petrological studies revealed that all these crystalline rocks are genetically related (comagmatic) and had evolved by progressive differentiation of a parent basaltic magma to give rise to the protoliths of the amphibolites probably represent the parent basaltic magma. Chemical studies also revealed that the magma of the protoliths of these rocks were from a metasomatised mantle. Plumbotectonics, petrological, geological and geochemical analyses and interpretations carried out in this study implicate a back arc tectonic setting as the environment of emplacement of these rocks. In this type of tectonic setting an ocean slab was subducted into the mantle after colliding with a continental shelf. The subduction of an ocean slab into the mantle would enhance the formation of a mixed wet basaltic magma, consisting of both mantle and ocean sediments thoroughly mixed to form a basaltic magma. This magma extruded and intruded the earlier laid down sediments in the region, differentiated and gave rise to the protoliths of these crystalline rocks in southwestern Nigeria. Post-magmatic transpressive forces operating in this region were responsible for the metamorphism of the protoliths of the amphibolites, hornblende gneiss and the granite gneiss (Oyinloye 2007) . Further deformation of these rocks led to faulting, fracturing, shearing, folding gneissic banding and foliation fabrics observed on some of the rocks especially, the leucocratic amphibolites, hornblende gneiss and granite gneisses in the basement complex of southwestern Nigeria.

REFERENCES

1. Ajibade, A.C and Fitches W.R. (1988): The Nigerian Precambrian and the Pan –African Orogeny,Precambrian Geology of Nigeria, pp. 45-53.

2. Ajibade, A.C.,Woaks, M., and Rahaman, M.A. (1987): Proterozoic crustal development in Pan-African regime of Nigeria: In A. Croner (ed) Proterozoic Lithospheric Evolution Geodynamics Vol. 17, pp 259-231.

3. Annor , A.E. and Freeth, S.J., (1985): Thermotectonic evolution of the basement complex around Okene, Nigeria with special refrence to deformation mechanism, Precambrian Research, 28, pp. 269-281.

4. Billstrom, K.A. (1989): A model for the Lead isotope evolution of Early Proterozoic Svecofennian sulphide ores in Sweden and Finland. Isotopic Geology 79, pp 307-316.

5. Billstrom, K.A. (1990): A lead isotope study of two sulphide deposits and adjacent igneous rocks in south-central Sweden. Mineralium Deposita 25: pp 152-159.

6. Boesse, T.N. and Ocan, O.O. (1988): Geology and evolution of the Ife-Ilesha Schist belt southwestern Nigeria. Symposium on Benin-Nigeria geo-traverse of Proterozoic geology and tectonics of high grade terains pp. 87-107.

7. Burke, K.C. Dewey, J.F (1972): Orogeny in Africa . In African Geology

A.J. Dessauragie, T.F.J. Whiteman (eds), pp 583-608, University of Ibadan Press, Nigeria.

8.   Burke, K.C., Freeth S.J. and Grant, N.K. (1976): The structure and sequence of geological. Events in the basement complex of Ibadan area Western Nigeria Precamb. Res.3, pp 537-545

9.   Dada, S.S, Briqueu, K.L., Birck. J.L. (1998): Primodial crustal growth in northern Nigeria Preliminary Rb-Sr and Sm-Nd constraints from Kaduna migmatite gneiss complex J. Min. Geol. 34, pp1-6.

10.   Egbuniwe, I.G. (1982): Geotectonic evolution of Maru Belt, northwestern Nigeria, unpublished Ph.D thesis of the University of Wales, U.K.

11.   Holt, R.W, Egbuniwe , I.G., Fitches, W.R. and Wright J.B. (1978): The relationship between low grade metasedimentary calc-alkaline volvanics and Pan-African Orogeny, Geol. Rundsh, 67 (2), pp 631-646.

12.   Martins, H. (1986): Progressive alteration associated with auriferous massive sulphide bodies at The Dumagami Mine, Abitibi Greenstone Belt, Quebec. Econ. Geol. Vol. 85, pp 746-764.

13.   Oyinloye, A.O. (1992): Genesis of the Iperindo gold deposit, Ilesha schist belt, Southwestern Nigeria. Unpublished thesis of the University of Wales, Cardiff, U.K. pp. 1-267.

14.   Oyinloye, A.O. and Odeyemi, S.B. (2001): The geochemistry, tectonic setting and origin of theMassive melanocratic amphibolites in Ilesha schist belt Southwestern Nigeria, Global Journal, Pure and Appl. Sci. (7) (1), pp.55-66.

15.   Oyinloye,A.O. (1998): Geology, Geochemistry and origin of the banded granite gneisses in the basement complex of the Ilesha Area Southwestern Nigerian. J. Africa Earth Science, London 264, pp633-641.

16.   Oyinloye, A.O. and Steed, G.M. (1996): Geology and Geochemistry of the Iperindo Primary gold deposits Ilesha schist belt Southwestern Nigeria. Inferences from stable carbon isotope studies. Africa J. Sc. Tech. 8 (1) pp 16-19.

17.   Oyinloye,A.O. (2002a): Geochemical Studies of granite gneisses: the implication on source determination. Jour. Chem. Soc. Nigeria (26) (1) 131-134

18.   Oyinloye,A.O. (2002b): Geochemical characteristics of some granite gneisses in Ilesha area southwestern Nigeria: Implication on evolution of Ilesha schist belt, southwestern Nigeria. Trends in Geochemistry India vol.2, 59-71

19. Oyinloye, A.O. (2004a): Petrochemistry, pb isotope systematic and geotectonic setting of granite gneisses in Ilesha schist belt southwestern Nigeria Global Jour. Geol. Sci. 2(1) 1-13.

20. Oyinloye, A.O. (2006b): Metallogenesis of the lode gold deposits in Ilesha Area of Southwestern Nigeria: Inferences from lead isotope systematic, Pak. J. Sci. Ind. Res. 49 (11) pp 1-11.

21. Oyinloye, A.O. (2007): Geology and Geochemistry of some Crystalline Basement Rocks in Ilesha area sourthwestern Nigeria: Implications on Provenance and Evolution Pak. Jour. Sci. Ind. Res. Vol. 50, No.4, 223-231.

22. Oyinloye, A.O. (2011): Beyond Petroleum Resources: Solid Minerals to the rescue: 31st Inaugural Lecture of the University of Ado-Ekiti, Nigeria Press, 1-36.

23. Pearce J.A., Harris N.W. and Tindle A.G. (1984): Trace elements discrimination diagrams for tectonic interpretation of granite rocks. Journal Petrology. Vol. 25 Par 4 956-983.

24. Rahaman, M.A and Ocan, O.O. (1978): On relationship in the Precambrian migmatitic gneisses of Nigeria J. Min. and Geol. Vol. 15, No.1 (abs).

25. Rahaman, M.A (1988):Recent advances in the study of the basement complex of Nigeria.

26. Symposium on the Geology of Nigeria, Obafemi Awolowo University, Nigeria.

27. Stacey, J.S., Kramers, J.D. (1975): Approximation of terrestrial lead isotope evolution by a two-stage model. Earth Planet. Leit. 26, pp 206-221.

28. Vivalo, W. and Rickard D. (1990): Genesis of an early Proterozoic zinc deposit in high grade Metamorphic terrare, saxberget, central Sweden Sco. Geo Vol.85, 714-736.

29. Wilson, M. (1991): Igneous Petrogenesis Global Tectonic Approach, Harpar Collins Academy, London Second impression pp. 227-241. Woakes, M. Ajibade C.A., Rahaman, M.A., (1987): Some metallogenic features of the Nigerian Basement, Jour. of Africa Science Vol. 5 pp. 655-664. Zartman, R.E. Doe, B.R. (1981): Plumbotectonics Tectonophysics, 75, 135-162.

Chapter 3

PETROGRAPHY, GEOCHEMISTRY AND PETROGENESIS OF LATE-STAGE GRANITES: AN EXAMPLE FROM THE GLEN EDEN AREA, NEW SOUTH WALES, AND AUSTRALIA

A. K. Somarin

Department of Geology, Brandon University, Brandon, Manitoba, Canada

INTRODUCTION

The Glen Eden area is located within the New England Orogen (also known as New England Fold Belt). This orogen is one of the major structural elements within the extensive Tasman Orogenic Province which comprises the eastern part of the Australian continent (Hensel, 1982). The present length of this orogen is about 1500 km from Townsville to Newcastle. It is separated from the Thomson and Lachlan fold belts to the west by the Permian and Triassic strata of the Bowen-Gunnedah-Sydney Basin. The Mesozoic ClarenceMoreton and Great Artesian basins separate the northern and southern parts of this orogen. The New England Orogen was the site of the extensive episodic calc-alkaline magmatism related to west-dipping subduction from middle Paleozoic to Early Cretaceous time. The oldest rocks might have formed at least partly in a volcanic island arc, but from the Late Devonian, the orogen developed as a convergent Pacific-type continental margin. During Late Devonian-Carboniferous time, parallel belts representing continental margin, volcanic arc, forearc basin and subduction complex assemblages can be recognized (Murray, 1988). More than one hundred plutons were emplaced from the Late Carboniferous to the Triassic in the southern NEO. These intrusions have been attributed to two major periods of plutonism, the first during Late Carboniferous time and the second during the Late Permian and Triassic. The resulting plutons comprise the New England Batholith. Although volcanogenic massive sulfides and volcanic-hosted epithermal gold-silver ore deposits occur in older rock sequences (Murray, 1988), almost all of the other ore deposits of this region, including the Glen Eden Mo-W-Sn deposit, have a genetic or paragenetic relationship with plutons of the New England

Batholith which is one of the largest Paleozoic-Mesozoic batholiths in eastern Australia. It underlies an area of almost 20000 km2 and is composed of more than one hundred N-S-trending plutons which include all of the granitoids in the southern part of the NEO. These granitoids intruded into the tectono-stratigraphic terranes (Flood and Aitchison, 1993a, b) and deformed trench-complex metasedimentary rocks (Shaw and Flood, 1981). The composition of this batholith is 80% monzogranite, 18% granodiorite, 1% diorite and tonalite, 1% quartz-bearing monzonite and a group of leucoadamellites. They pointed out that the differences between these six groups reflect differences in their source rock types. The Glen Eden Granite (GEG) occurs as dykes at depths of more than 80 m and is not exposed at the surface (Fig. 1). Mineralogical studies and field evidence indicate that the observed dykes have intruded after initiation of the hydrothermal activity. Based on petrographic studies, three types of GEG can be recognized: microgranite porphyry, micrographic granite, and aplite. Petrographic features of these granites are discussed below

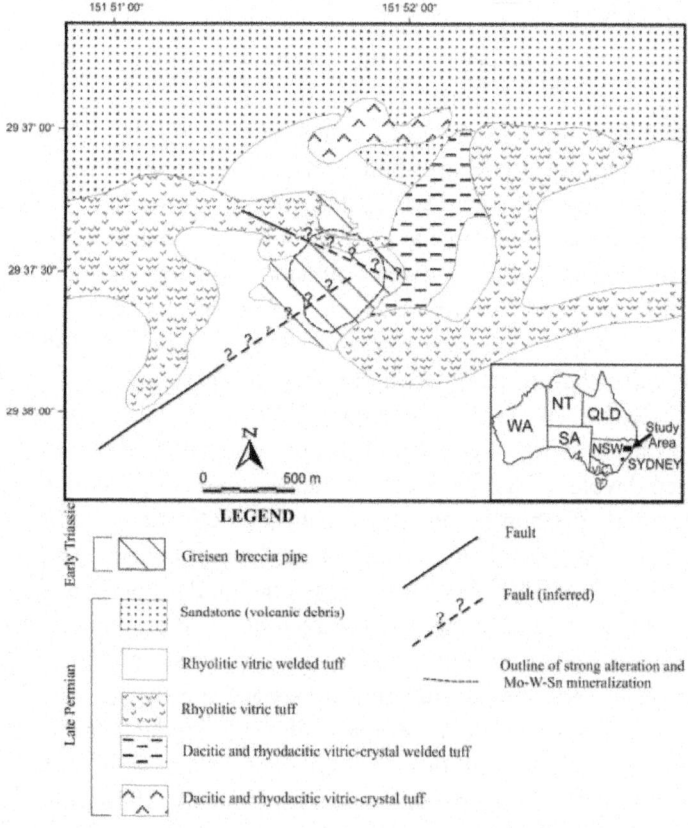

Figure. 1: Geological map of the Glen Eden area (after Somarin and Ashley, 2004).

PETROGRAPHY OF THE GLEN EDEN GRANITE

Microgranite Porphyry

Microgranite porphyry of GEG is composed of quartz, K-feldspar and plagioclase as major minerals and biotite, zircon, xenotime, monazite and fluorite as accessory phases. Its texture is granular with quartz, K-feldspar, plagioclase and biotite as phenocrysts up to 8 mm in size. The groundmass is composed of quartz, K-feldspar and plagioclase, typically 50 to 300 μm, average 200 μm, in size. Most of the phenocrysts have irregular margins due to resorption and replacement by the groundmass. The cracks and embayments in these phenocrysts have been filled by the groundmass. Quartz occurs as anhedral to euhedral grains, commonly rounded in shape and forming mosaics within feldspathic matrix. Some quartz grains form well-developed euhedral crystals, possibly due to secondary overgrowth. The presence of quartz as inclusions within biotite, plagioclase and fluorite and replacement of quartz phenocrysts by groundmass suggest that quartz crystallized relatively early. K-feldspar is mostly orthoclase ($Or_{86}Ab_{14}$ to $Or_{98}Ab_2$) and mainly occurs as cloudy or perthitic anhedral crystals up to 5 mm in size. Rare microcline occurs as anhedral to subhedral grains 200-300 μm across. Perthitic hydrothermal K-feldspar in veins is common. Plagioclase is mostly albitic ($Ab_{85}Or_{13}An_2$ to $Ab_{99}Or_1$) and occurs as subhedral to euhedral crystals and varies in size from 80-200 μm in groundmass up to 1.5-2.2 mm as phenocrysts. There is no zoning. In altered samples, the presence of K-feldspar as replacement rims around plagioclase implies sub-solidus alteration of plagioclase.

Biotite is dark brown to brown in color, strongly pleochroic and is mainly siderophyllite in composition. This mineral occurs as euhedral flakes, 50 μm up to a few millimeters in size. Commonly, biotite flakes have inclusions of magmatic quartz, zircon, xenotime, monazite and, in some samples, fluorite, rutile and secondary goethite accompany these flakes. These features are similar to those of Climax-type intrusives (e.g. White et al., 1981). Biotite is inferred to have been the most unstable mineral during hydrothermal alteration and commonly is replaced by sericite and goethite. Mostly, due to this replacement, only relicts of biotite can be seen and its color changes from brown to cream. Based on textural criteria, the position of biotite in the crystallization sequence cannot be determined unequivocally. However, the interstitial nature of biotite in GEG and occurrence of other minerals as inclusions within it are indicative of its late crystallization which is consistent with a high activity of F during crystallization (see below; Munoz and Ludington, 1974; Tischendorf, 1977;

Collins et al., 1982). Locally, muscovite occurs as flakes in samples adjacent to hydrothermal veins; they are interpreted to be of hydrothermal origin. Fluorite occurs as anhedral, interstitial grains with a purple tint in plane-polarized light and 70 μm up to 1 mm across. Locally, it occurs as inclusions within biotite flakes where biotite is unaltered. This indicates that fluorite in granite porphyry of GEG has a magmatic origin and reflects high activity of fluorine in the GEG magma. Micrographic intergrowth of quartz and K-feldspar in granite porphyry is common. Commonly the contact between a granitic dyke and surrounding rhyolitic volcanic rocks is marked by quartz veins. It seems that these contacts had the role of conduits for later hydrothermal fluids from the dykes or a deeper source.

Micrographic Granite

Mineralogy and appearance of micrographic granite is similar to that of microgranite porphyry, however, the former can be distinguished by lower biotite contents and finer grain size. Its K-feldspar (Or87Ab13 to Or95Ab5) and plagioclase (Ab90Or2An8 to Ab98An2) composition is similar to those in the microgranite porphyry. The intensity of micrographic growth varies. In some samples, there are discrete crystals of quartz and K-feldspar in addition to micrographic intergrowths, whereas in other samples almost all of the rock is composed of micrographic intergrowth of quartz and K-feldspar, and biotite and plagioclase are less abundant.

Aplite

Aplite at Glen Eden occurs as dykes up to 10 cm wide at a depth of ~85 m. It has granular texture and is composed of quartz, plagioclase and K-feldspar with grain size ranging from 50-400 μm, average 150 μm. No biotite or other accessory phases occur in aplite samples. Plagioclase is albitic ($Ab_{97}Or_2An_1$ to Ab100) and K-feldspar (orthoclase, $Or_{86}Ab_{14}$ to $Or_{94}Ab_6$) grains are cloudy. These dykes have experienced potassic alteration and contain quartz-Kfeldspar veins. The contact of aplite dykes with volcanic wall rock is sharp. Along these contacts, rhyolite groundmass has recrystallized, suggesting interaction of hot aplitic magma with cooler wall rock. Aplitic materials, in addition to aplite dykes, occur also in crenulate quartz layers and parting veins.

Crenulate Quartz Layers (Comb Layering)

Comb layering was defined by Moore and Lockwood (1973) as 'relatively unusual type of layering in granitoid rocks in which constituent crystals (plagioclase and hornblende in their study) are oriented nearly perpendicular to the planes of layering'. The types with ductile deformation are called 'crenulate quartz layers' (White et al., 1981; Kirkham and Sinclair, 1988). Comb layers are also referred to as ribbon rock, ribbon banded structures, rhythmically banded textures, brain rock, ptygmatic veins, wormy veins, vein dykes, unidirectional solidification textures and Willow Lake-type layering. Because of its deformed character, comb layering is called crenulate quartz layers, herein. The crenulate quartz layers mainly occur within 5-10 m from the GEG dykes at depths >300 m. They are composed of quartz layers ranging in thickness from 2 mm to 3 cm. Quartz crystals in these layers are anhedral and they do not show perpendicular growth against the layer walls, possibly due to deformation and recrystallization. The quartz layers typically alternate with layers of aplitic material 1 mm to 2 cm thick. Some aplitic layers are discontinuous and terminate sharply within quartz layers. This suggests that the relative content of melt in the comb layer-forming system was low. Ptygmatic folding does not occur. Some quartz crystals in quartz layers are bent and elongate due to deformation. Similar deformation has been reported from Climax, Colorado (e.g. White et al., 1981), Hall, Nevada (Shaver, 1984a) and Anticlimax, British Colombia (Kirkham and Sinclair, 1988). Aplitic layers are composed of fine-grained quartz and feldspars, including orthoclase ($Or_{91}Ab_9$ to $Or_{97}Ab_3$) and albite ($Ab_{98}Or_1An_1$ to $Ab_{99}An_1$). Locally, quartz phenocrysts, up to 2 mm across, occur in aplitic layers. Based on microscopic and macroscopic studies, these conclusions can be made.

- The broken and bent quartz and aplite layers imply formation in a dynamic environment. Although subsequent deformation could produce partly similar features in GEG and wall rock, such features are not seen in these rocks. Also, if subsequent deformation was the main cause of bending, all layers should show this bending, whereas some of them are undeformed.

- Ductile deformation of these layers indicates that they were not completely solidified at the time of deformation. Also, deformation of some layers while the others are undeformed, suggests successive precipitation and deformation.

- The absence of a sharp boundary between quartz and aplite layers, and replacement of aplitic material by quartz suggest disequilibrium conditions during formation of quartz layering.

- The magma or fluid from which aplitic material precipitated was saturated with the components of sodic plagioclase and K-feldspar. The presence of some quartz phenocrysts in aplitic layers indicates that the magma crystallized in at least two stages, in which formation of groundmass followed crystallization of phenocrysts. The finegrain size of aplitic material shows that the temperature difference between magma and the surrounding environment was large and magma crystallized rapidly.

- Delicate aplitic layers and close spatial relationship between crenulate quartz layers and parting veins indicate that the parent magma had very low viscosity. A similar conclusion was reached by Kirkham and Sinclair (1988).

- The low volume of aplitic materials and their mineralogical composition, which is similar to GEG, may imply that they represent a small portion of highly fractionated melt, possibly carried by escaping hydrothermal fluids. The association of aplitic material of crenulate quartz layers with quartz pods, parting veins, breccia zone and resorbed crystals suggests overlapping of magmatic processes by hydrothermal activity. Association of the crenulate quartz layers with Mo mineralization and silicification has been reported by other investigators and these layers have been considered as a prospecting guide (e.g., Povilaitis, 1978).

- The presence of primary two-phase fluid inclusions within quartz layers and quartz phenocrysts in the aplitic layers indicates the presence of hydrothermal fluid at the time of formation of these layers. Also, the similarity of formation temperature and salinity of these layers to those of other hydrothermal assemblages (Somarin and Ashley, 2004) indicates that at least the quartz layers and quartz phenocrysts in the aplitic layers have precipitated from fluid, not melt.

- Common occurrence of crenulate quartz layers in the apical parts (close to contact) of felsic intrusions related to porphyry deposits (White et al., 1981; Carten et al., 1988; Kirkham and Sinclair, 1988) may indicate that the main body of GEG is in the vicinity of these layers.

Generally there are two ideas regarding the genesis of crenulate quartz layers.

- They have crystallized from the melt (White et al., 1981)
- They have precipitated from the aqueous phase (Moore and Lockwood, 1973; Stewart, 1983; Shaver, 1984 a, b).

White et al. (1981) proposed that P_{H_2O} and P_{HF} increased during crystallization of the magma due to lack of hydrous minerals. The increased P_{H_2O} and P_{HF} would expand the quartz field in the ternary Q-Ab-Or system and lower the thermal minimum. They suggested that the combined effect of increasing P_{H_2O} and P_{HF} caused the precipitation of quartz without feldspar. Release of volatiles due to fracturing of wall rocks shrinks the quartz field and allows the crystallization of feldspar with quartz. This cycle occurs repeatedly to produce crenulate quartz layers. Based on this model, crenulate quartz layers have a magmatic source. It appears that even under high P_{H_2O} and P_{HF}, precipitation of pure quartz cannot be expected and some feldspar will crystallize as well. However, no feldspar occurs in quartz layers. Furthermore, 7 to 19 wt% F is needed in the system to destabilize feldspars (Glyuk and Aufiligov, 1973). This amount of F should cause movement of the eutectic point toward the Ab apex in the Q-Ab-Or system, which is not evident in the Glen Eden Granite. Therefore, it is unlikely that the formation of crenulate quartz layers of the Glen Eden Mo-W-Sn deposit can be explained by this model. Based on observations mentioned above, it seems that for aplitic and quartz layers, two different sources should be considered. Aplitic layers indicate evidence of crystallization from a very low-viscosity melt, whereas quartz layers have crystallized from an aqueous fluid. It is more likely that aplitic material represents the relicts or parts of the highly fractionated low-viscosity melt in a dynamic moving, mainly upward, fluid which has separated from the melt. Kirkham and Sinclair (1988) suggested that the rapid drop in fluid pressure due to brecciation and fracturing of surrounding rocks quenches the adjacent silicate melt along the roof and walls of the magma chamber. This results in the formation of aplitic or porphyritic aplitic layers between the comb quartz layers. Occurrence of this process, successively, explains the rhythmic repetition of layers. The successive brecciation at Glen Eden was able to release pressure alternately and cause upward quenching of the melt. High fluid pressure and continued movement of magma, probably, resulted in the ductile deformation of the layers (Kirkham and Sinclair, 1988). The absence of thick comb quartz layers and pegmatitic lenses may indicate trapping of a large volume of volatiles (testified by pervasive hydrothermal brecciation) and relatively rapid build-up of fluid pressure. This could prevent the growth of layer crystals before fluid escape. However, occasionally coarse-grained K-feldspar and ore minerals can be seen in the breccia pipe, indicating less rapid build-up of fluid pressure, permitting the growth of these minerals.

GENETIC IMPLICATIONS OF MICROGRAPHIC TEXTURE

In the Glen Eden Granite, micrographic texture occurs as the main texture of the micrographic granite and as a texture of some phenocrysts in microgranite porphyry. Generally there are two ideas about the genesis of graphic texture.

- Infiltration and replacement of one mineral (host) by another mineral (guest) (e.g. Augustithis, 1973).

- Eutectic crystallization of intergrowth-forming minerals (e.g. Fenn, 1979; Kirkham and Sinclair, 1988). Graphic textures most commonly develop in water-rich magmas, generally in the presence of a separate aqueous phase (Nabelek and Russ-Nabelek, 1990), even though studies by Fenn (1979) have shown that a separate aqueous phase is not always required. In experiments using crushed glass from bulk samples of Spruce Pine pegmatite, Burnham (1967) found that in the presence of H_2O alone, the melts crystallized to an assemblage of alkali feldspar, quartz and muscovite. However, with a solution containing 6.2 wt% total dissolved alkali feldspar, muscovite did not appear and the melt crystallized to a graphically intergrown assemblage of alkali feldspar and quartz. Based on these studies, White et al. (1981) concluded that graphic textures represent zones of accumulation of a separate, Cl-rich aqueous phase. However, the presence of F in magma, which increases the amount of water in the separate phase by decreasing its solubility in the melt, may also help the formation of graphic texture. Also, pressure-quenched crystallization is able to produce micrographic texture (Kirkham and Sinclair, 1988).

Petrographic studies show that, genetically, there are two kinds of graphic texture at Glen Eden.

- Graphic texture in the GEG. The following observations imply that this texture is the result of eutectic crystallization rather than replacement.

  a. Absence, in fresh rocks, of replacement of other minerals, such as plagioclase, by either quartz or K-feldspar.

  b. The occurrence of micrographic granite in which the entire rock is composed of micrographic intergrowth of quartz and K-feldspar.

  c. Absence of evidence of infiltration of quartz-forming solutions and replacement of K-feldspar. Although there are some low-temperature fluid inclusions in quartz phenocrysts of GEG, there is no clear evidence of replacement of other minerals by quartz.

  d. Absence of reaction margins in host K-feldspar or other minerals.

  e. The presence of graphic grains in which K-feldspar patches occur as inclusions within quartz. In the replacement model, in which quartz has been introduced by a solution, euhedral quartz grains should have formed by progressive replacement, rather than a groundmass for K-feldspar patches (Augustithis, 1973). Furthermore there is no evidence of infiltration of K-feldspar-forming solutions into quartz grains and replacement of quartz by K-feldspar.

- Graphic texture in potassic alteration zone. Microscopic studies show that infiltration of quartz-forming solutions into fractures, intergranular spaces and cleavages of Kfeldspar resulted in the replacement of K-feldspar by quartz. This replacement looks like a graphic intergrowth and clearly is the result of post-magmatic hydrothermal activity.

It seems that the presence of crenulate quartz layers, micrographic texture and hydrothermal breccia at Glen Eden indicates saturation of magma from water and the presence of a fluid-rich environment. The presence of free vapor and aqueous phases during graphic crystallization of quartz and K-feldspar is proved by the presence of fine (2-5 µm) primary two-phase fluid inclusions within quartz of the graphic texture.

EMPLACEMENT OF THE GLEN EDEN GRANITE

The presence of topaz, fluorine-rich biotite and widespread occurrence of fluorite in all alteration assemblages indicate that the Glen Eden Granite magma was uncommonly fluorine-rich. Since fluorine has significant effects on the physico-chemical properties of granitic magma, these effects are discussed below.

Effects of Fluorine on the Magma

High F content of GEG and presence of magmatic fluorite provide links between this granite and other F-rich rocks, such as topaz granite, ongonites and topaz rhyolites (Kovalenko et al., 1971; Pichavant and Manning, 1984; Taylor, 1992; Kontak, 1994). The effects of fluorine in magma have been studied by many investigators. These effects can be summarized as follows.

- Fluorine decreases the solubility of water in the melt (Dingwell, 1985, 1988), so water exsolution may occur earlier during crystallization of F-rich melts (Strong, 1988). The presence of breccia pipes testifies that magma had become saturated in water and volatiles.

- Both fluorine and water lower the crystallization temperature of granitic magmas (Bailey, 1977). Manning (1981) has documented the persistence of melt at 550°C in a granite with 4% F. This effect of fluorine would allow melt to fractionate more. The occurrence of mineral deposits similar to that at Glen Eden with highly fractionated granitic rocks may suggest that this factor (more fractionation) is important for the evolution of ore-bearing vapor phase from melt, since incompatible elements, including metals, would concentrate in residual melt.

- Fluorine changes the order of crystallization by promoting quartz, topaz and feldspars above biotite (Bailey, 1977; Hannah and Stein, 1990). This could be the result of increasing the thermal stability of hydrous phases by fluorine (Hannah and Stein, 1990).

- Fluorine lowers the viscosity of melt (Dingwell et al., 1985; Hannah and Stein, 1990). At 1000°C, addition of 1 wt% F to a melt of albitic composition results in an order of magnitude decrease in melt viscosity

(Dingwell, 1988). Because of the smaller temperature dependence of viscosity in F-bearing melts versus F-free melt, the effects of F on melt viscosity is greatest at 600° to 800°C (Dingwell, 1988). The lower viscosity could cause higher migration of melt and replacement into shallow levels. Approach of the melt to shallow levels in the crust and hence decreasing pressure and the escape of water and volatiles may lead to increasing viscosity. High-level emplacement of the Glen Eden Granite, along with the presence of crenulate quartz layers and parting veins indicate low viscosity of the melt.

- By decreasing the solidus temperature of the magma, the assimilation ability of the magma may be increased (Keith and Shanks, 1988).

- Due to the decreased solidification temperature of the melt, F-bearing magmas may show extreme differentiation. The solidification temperature could be as low as 550- 600°C in the presence of various volatiles (Strong, 1988). The solidus of an acid melt will decrease by 60°C in the presence of a vapor phase containing 5% HF (Schroecke, 1973). The association of Sn, Mo and W ore deposits with highly fractionated granites implies that extreme differentiation is essential for the concentration of these elements in evolved aqueous phase. This explains why intrusions at high levels have more potential to associate with rare-element mineralization in comparison to those intruded at low levels, since the high-level intrusions have been differentiated more than deep-level ones (Tischendorf, 1977). Also, water saturation develops through extreme differentiation. Intrusions without high concentration of magmatic water are typically barren (Strong, 1988). So it seems that the presence of volatiles, which affect the physical and chemical properties of the magma, is crucial for the formation of rare-metal ore deposits. A strongly depolymerized F-rich melt is more capable of hosting incompatible elements than a polymerized volatile-poor melt (Webster and Holloway, 1990).

- Fluorine increases cation diffusion in silicate melts (Dingwell, 1985) which is important for the transportation of the constituents necessary for ore deposition.

- The various effects of F could cause changes in commencement of the late-magmatic metasomatic processes (Tischendorf, 1977).

- Fluorine could change the solid/melt partition coefficients of elements because the stability of each element's site within the melt is altered (Hannah and Stein, 1990).

- Fluorine increases the solubility of silicate melt in the fluid phase (Hannah and Stein, 1990).

- Fluorine increases Ab content of the near-minimum melts (Manning and Pichavant, 1988).

EMPLACEMENT OF GEG

Field evidence, including presence of the breccia pipe, crenulate quartz layers and parting veins, which commonly occur in the roof of the intrusion, indicate that the Glen Eden Granite, like other leucogranites of the New England Batholith, is a high-level intrusive body. The high-level emplacement of GEG indicates that the magma was water-poor, since the main control on depth of crystallization of a rising body of granitic magma is its H_2O content (Burnham, 1979; Wyllie, 1979). Burnham (1979) pointed out that for felsic magma to attain a volcanic or sub-volcanic environment, the initial water content cannot be greater than about 3 wt%. Magma with higher initial water content would become completely crystallized after boiling of its volatiles at a depth of several kilometres (Sheppard, 1977). In addition to water, fluorine also affects the emplacement of granitic magmas. Fluorine depolymerizes the structure of the melt and decreases its viscosity which would allow higher migration and shallow-level emplacement of the magma. Also, fluorine decreases the solubility of water in the melt. This water can escape from melt as a result of pressure drop, but fluorine does not, because it enters the OH sites of biotite and possibly exists in the melt as alkali–LILE–fluoride complexes (Collins et al., 1982) or alkali-aluminium-fluoride complexes (Velde and Kushiro, 1978). Therefore, the viscosity of magma will decrease progressively while water is released, and this magma can reach epizonal environments (Plimer, 1987). The formation of massive greisen (Somarin and Ashley, 2004) before intrusion of the Glen Eden granitic dykes might be due to this released water.

The path of movement, initially, is mainly dependent on the direction of weak zones, such as faults. A velocity of 1-2 cm/year, as proposed by Bankwitz (1978), may be enough to cause upward and outward movement of melt without complete crystallization. The prolonged period of tectonic activity in the New England area during Permo-Triassic compression and extension (Collins et al., 1993) could produce suitable structures, such as faults, for the rise of plutons. Also, fracturing of roof rocks by heat flow from the melt, which increases the amount of elastic energy, helps the movement of the melt (Bankwitz, 1978). As mentioned above, high content of F would retard crystallization of melt and allow it to move away from the magma chamber. The intense veining of parts of GEG, while the other parts show less or no veining, may reflect that the outer vein-bearing parts became colder than inner parts due to encountering cold wall rock. The intrusive body utilizes structures which are later utilized by metal-bearing hydrothermal fluids (Plimer and Kleeman, 1985). The presence of quartz veins at boundaries between granitic dykes and wall rock at Glen Eden supports this idea and indicates that these boundaries were relatively weak zones, along which hydrothermal fluids could easily move. On the whole, the high-level emplacement of the GEG and its highly differentiated character reflect high content of fluorine in the magma. Phosphorus, like F, also decreases the liquidus and solidus temperatures of the melt by modifying the silica network with the formation of phosphate-oxygen-metal complexes (London, 1987; Hannah and Stein, 1990). However, the low concentration of P in the GEG and absence of apatite in the hydrothermal assemblages indicate low P content of magma.

GEOCHEMISTRY

Major Element Geochemistry

The Glen Eden Granite is highly felsic, as indicated by SiO_2 contents between 76 and 78 percent (Table 1). Aplite samples show potassic alteration. The chemical compositions of microgranite porphyry and micrographic granite are similar, however micrographic granite has lower K_2O and higher F, Nd and, in some samples, Ce (Table 2). The characteristics of GEG are similar to granites associated with Climax-type molybdenum ore deposits (White et al., 1981). The GEG, like Climax-type rocks, shows enrichment in silica and depletion in Ca, Al, total Fe, and Mg with respect to both the average calc-alkaline and alkaline granites of Nockolds (1954). Average K_2O/Na_2O in microgranite porphyry is close to that of average alkaline granite (Nockolds, 1954) and average normal granite (Le Maitre, 1976), but this ratio in micrographic granite is less than that of alkaline granite and is more than the average of normal granite. Normative Ab/An ratio is high and reflects the low Ca content of the GEG. The samples of microgranite porphyry and micrographic granite show little chemical variation and no clear trends on Harker-type diagrams using SiO_2 or MgO as an index of possible fractionation (Fig. 2). Although the least-altered samples were chosen for analysis,

Table 1: Major element analyses and CIPW norms of the Glen Eden Granite, compared with average calc-alkaline and alkaline granites of Nockolds (1954), and average granite of Le Maitre (1976) and average I-, S-, and A-type granites of Whalen et al. (1987) and biotite porphyry of Climax (White et al., 1981).

	Granite porphyry										Micrographic granite			
	R75283	R75284	R75285	R75286	R75287	R75288	R75289	R75290	R75291	Average	R75292	R75293	R75294	Average
SiO_2	76.27	76.24	76.97	76.33	76.77	77.60	76.70	76.12	77.23	**76.69**	76.66	77.78	77.72	**77.39**
TiO_2	0.06	0.07	0.08	0.06	0.05	0.09	0.05	0.06	0.05	**0.06**	0.10	0.06	0.04	**0.07**
Al_2O_3	12.82	12.34	12.15	13.04	13.15	12.63	12.66	12.43	12.39	**12.62**	12.53	12.29	12.34	**12.39**
Fe_2O_3	0.11	0.18	0.30	0.13	0.08	0.10	0.17	0.18	0.16	**0.16**	0.25	0.08	0.06	**0.13**
FeO	0.43	0.82	0.70	0.38	0.21	0.22	0.60	0.63	0.55	**0.50**	0.71	0.48	0.63	**0.61**
MnO	0.02	0.03	0.03	0.02	0.01	0.01	0.03	0.03	0.02	**0.02**	0.03	0.02	0.03	**0.03**
MgO	0.12	0.04	0.06	0.08	0.07	0.08	0.05	0.05	0.04	**0.06**	0.06	0.07	0.10	**0.08**
CaO	0.27	0.33	0.40	0.26	0.20	0.23	0.31	0.35	0.39	**0.30**	0.43	0.38	0.36	**0.39**
Na_2O	2.78	3.17	3.67	3.05	3.22	3.04	3.58	3.46	3.71	**3.30**	3.86	3.45	2.94	**3.42**
K_2O	5.25	5.29	4.56	5.94	5.43	4.83	4.54	4.96	4.70	**5.05**	4.32	4.17	4.20	**4.23**
P_2O_5	0.01	0.01	0.01	0.01	0.01	0.01	0.01	0.01	0.01	**0.01**	0.01	0.00	0.02	**0.01**
S	0.03	0.01	0.01	0.02	0.02	0.02	0.05	0.01	0.01	**0.02**	0.01	0.02	0.02	**0.02**
LOI	1.01	1.40	0.54	0.70	0.54	0.84	0.92	0.78	0.53	**0.81**	0.54	0.87	1.33	**0.91**
Total	99.15	99.92	99.47	100.00	99.74	99.68	99.62	99.06	99.78	**99.58**	99.50	99.65	99.77	**99.66**
K_2O/Na_2O	1.89	1.67	1.24	1.95	1.69	1.59	1.27	1.43	1.27	**1.53**	1.12	1.21	1.43	**1.24**
Q	39.02	36.25	36.83	34.97	36.66	40.81	37.47	35.82	36.45	**37.14**	36.24	40.52	43.20	**39.99**
C	2.10	0.82	0.47	1.14	1.63	2.01	1.33	0.77	0.52	**1.20**	0.73	1.42	2.34	**1.50**
Or	31.03	31.27	26.95	35.11	32.09	28.55	26.80	29.32	27.75	**29.87**	25.53	24.65	24.82	**25.00**
Ab	23.52	26.78	31.05	25.81	27.25	25.68	30.25	29.24	31.39	**27.89**	32.66	29.19	24.88	**28.91**
An	1.27	1.60	1.92	1.22	0.93	1.08	1.45	1.65	1.88	**1.44**	2.10	1.86	1.66	**1.87**
Di	0.00	0.00	0.00	0.00	0.00	0.00	0.00	0.00	0.00	**0.00**	0.00	0.00	0.00	**0.00**
Hy	0.86	1.38	1.06	0.70	0.38	0.35	0.95	1.07	0.89	**0.85**	1.11	0.89	1.30	**1.10**
Mt	0.16	0.26	0.43	0.19	0.12	0.14	0.25	0.23	0.26	**0.23**	0.36	0.12	0.09	**0.19**
Ilm	0.10	0.12	0.14	0.10	0.09	0.16	0.09	0.10	0.09	**0.11**	0.19	0.11	0.08	**0.13**
Ap	0.02	0.01	0.02	0.02	0.02	0.02	0.02	0.02	0.01	**0.02**	0.01	0.00	0.05	**0.02**
Py	0.06	0.02	0.02	0.03	0.04	0.04	0.09	0.02	0.02	**0.04**	0.02	0.04	0.04	**0.03**
Total	98.15	98.51	98.92	99.30	99.21	98.84	98.71	98.24	99.26	**98.79**	98.96	98.79	98.45	**98.73**
Ab/An	18.52	16.74	16.17	21.16	29.30	23.78	20.86	17.72	16.70	**19.37**	15.55	15.69	14.99	**15.46**
100Mg/Mg+Fe	40	9	16	34	49	61	16	13	14	**23**	17	24	24	**22**
A.S.	1.19	1.07	1.04	1.09	1.14	1.19	1.12	1.06	1.04	**1.10**	1.06	1.13	1.23	**1.14**
DI	94	94	95	96	96	95	95	94	96	**95**	94	94	93	**94**

	1	2	3	4	5	6	7
SiO_2	75.70	72.08	73.86	72.04	73.39	73.39	73.81
TiO_2	0.56	0.37	0.20	0.30	0.26	0.28	0.26
Al_2O_3	12.70	13.86	13.75	14.42	13.43	13.45	12.40
Fe_2O_3	0.47	0.86	0.78	1.22	0.60	0.36	1.24
FeO	0.57	1.67	1.13	1.68	1.32	1.73	1.58
MnO	NA	0.06	0.05	NA	0.05	0.04	0.06
MgO	0.37	0.52	0.26	0.71	0.55	0.58	0.20
CaO	1.07	1.33	0.72	1.82	1.71	1.28	0.75
Na_2O	3.10	3.08	3.51	3.69	3.33	2.81	4.07
K_2O	5.60	5.46	5.13	4.12	4.13	4.56	4.65
P_2O_5	ND	ND	ND	ND	0.07	0.14	0.04
S	ND	ND	ND	ND	ND	ND	ND
LOI	ND	ND	ND	ND	ND	ND	ND
Total	100.14	99.29	99.39	100.00	98.84	98.62	99.06
K_2O/Na_2O	1.81	1.77	1.46	1.12	1.24	1.62	1.14
Q	33.63	28.80	31.34	29.13	33.20	35.24	30.19
C	0.00	0.46	1.11	0.58	0.54	1.90	0.00
Or	33.10	32.27	30.32	24.35	24.41	26.95	27.48
Ab	26.23	26.06	29.69	31.22	28.18	23.78	34.44
An	4.20	6.60	3.57	9.03	8.03	5.44	1.83
Di	0.87	0.00	0.00	0.00	0.00	0.00	1.40
Hy	0.52	3.15	1.84	3.35	2.96	3.94	1.34
Mt	0.21	1.25	1.13	1.77	0.87	0.52	1.80
Ilm	1.06	0.70	0.38	0.57	0.49	0.53	0.49
Ap	0.00	0.00	0.00	0.00	0.17	0.33	0.09
Py	0.00	0.00	0.00	0.00	0.00	0.00	0.00
Total	100.14	99.29	99.39	100.00	98.84	98.62	99.06
Ab/An	6.25	3.95	8.32	3.46	3.51	4.37	18.82
100Mg/Mg+Fe	100	48	42	60	53	43	30
A.S.	0.97	1.03	1.09	1.04	1.03	1.13	0.95
DI	93	87	91	85	86	86	92

NA = Not analyzed   DI = Differentiation index
ND = No data   LOI = Loss on ignition
A.S. = Degree of aluminum saturation (molecular proportion of $Al_2O_3/CaO+Na_2O+K_2O$)

1) Biotite porphyry-Climax (White et al., 1981)	5) Average of felsic I-type granites (Whalen et al., 1987)
2) Average of calk-alkaline granite (Nockolds, 1954)	6) Average of felsic S-type granites (Whalen et al., 1987)
3) Average of alkali granite (Nockolds, 1954)	7) Average of A-type granites (Whalen et al., 1987)
4) Average of granite recalculated to 100% (Le Maitre, 1976)	

Table 2: Trace element abundances in the Glen Eden Granite and average I-, S-, and A-type granites of Whalen et al. (1987). F values of less than 2500 ppm are from sodium peroxide fusion/SIE method whereas values shown by

	Granite Porphyry										Micrographic Granite						
	R752 83	R752 84	R752 85	R752 86	R752 87	R752 88	R752 89	R752 90	R752 91	Aver age	R752 92	R752 93	R752 94	Aver age	1	2	3
Nb	27	40	27	26	38	26	42	41	36	34	28	23	13	21	12	13	37
Zr	90	103	96	89	113	115	126	110	96	104	107	86	81	91	144	136	528
Y	54	83	76	58	58	66	114	106	109	80	79	83	87	83	34	33	75
Sr	11	3	4	17	3	4	3	6	5	6	5	5	17	9	143	81	48
Rb	488	510	365	552	434	408	309	458	427	439	379	324	321	341	194	277	169
Th	50	53	48	44	50	52	52	50	48	50	50	46	44	47	22	18	23
Pb	37	38	33	43	43	38	38	36	39	38	35	36	22	31	23	28	24
As	2	4	5	3	4	8	3	2	8	4	5	5	5	5	ND	ND	ND
U	9	14	15	12	11	12	12	12	17	13	17	11	7	12	5	6	5
Ga	24	25	21	23	25	25	24	27	26	24	21	25	20	22	16	17	25
Zn	5	17	47	6	3	7	26	8	8	14	25	13	19	19	35	44	120
Cu	5	8	3	<2	<2	<2	<2	<2	<2	3	<2	2	3	2	4	4	2
Ni	20	18	6	3	3	25	10	19	11	13	4	2	7	4	2	4	<1
Ce	38	29	51	39	40	41	43	34	37	39	58	50	62	57	68	53	137
Nd	21	18	22	22	26	21	24	24	23	22	28	34	38	33	ND	ND	ND
Ba	38	19	10	105	20	16	15	16	10	28	31	19	42	31	510	388	352
V	<2	2	4	5	<2	<2	3	5	<2	3	<2	<2	4	3	22	23	6
La	16	12	18	19	15	13	13	18	10	15	25	16	22	21	ND	ND	ND
Sc	<2	4	6	6	<2	6	<2	7	3	4	8	4	5	6	8	8	4
Sn	<3	4	3	<3	<3	<3	<3	<3	<3	<3	3	5	3	4	ND	ND	ND
Mo	1	2	10	5	2	2	1	2	1	3	2	1	2	2	ND	ND	ND
W	10	65	18	286	17	16	21	32	24	54	12	14	8	11	ND	ND	ND
F	NA	1580	1530	NA	1180	480	NA	1430	NA	1240	3200	3400	<2500		ND	ND	ND
K/Rb	89	86	104	89	104	98	122	90	91	96	95	107	109	103	177	137	229
Rb/ Sr	44	170	91	32	145	102	103	76	85	71	76	65	19	38	1.36	3.42	3.52
Rb/ Ba	13	27	37	5	22	26	21	29	43	16	12	17	8	11	0.38	0.71	0.48
Ga/ Al	3.54	3.83	3.27	3.34	3.59	3.74	3.59	4.11	3.97	3.66	3.17	3.85	3.07	3.36	2.25	2.39	3.75

NA = Not analyzed   ND = No data

1) Average Felsic I-type (Whalen et al. 1987)   2) Average Felsic S-type (Whalen et al. 1987)   3) Average A-type (Whalen et al. 1987)

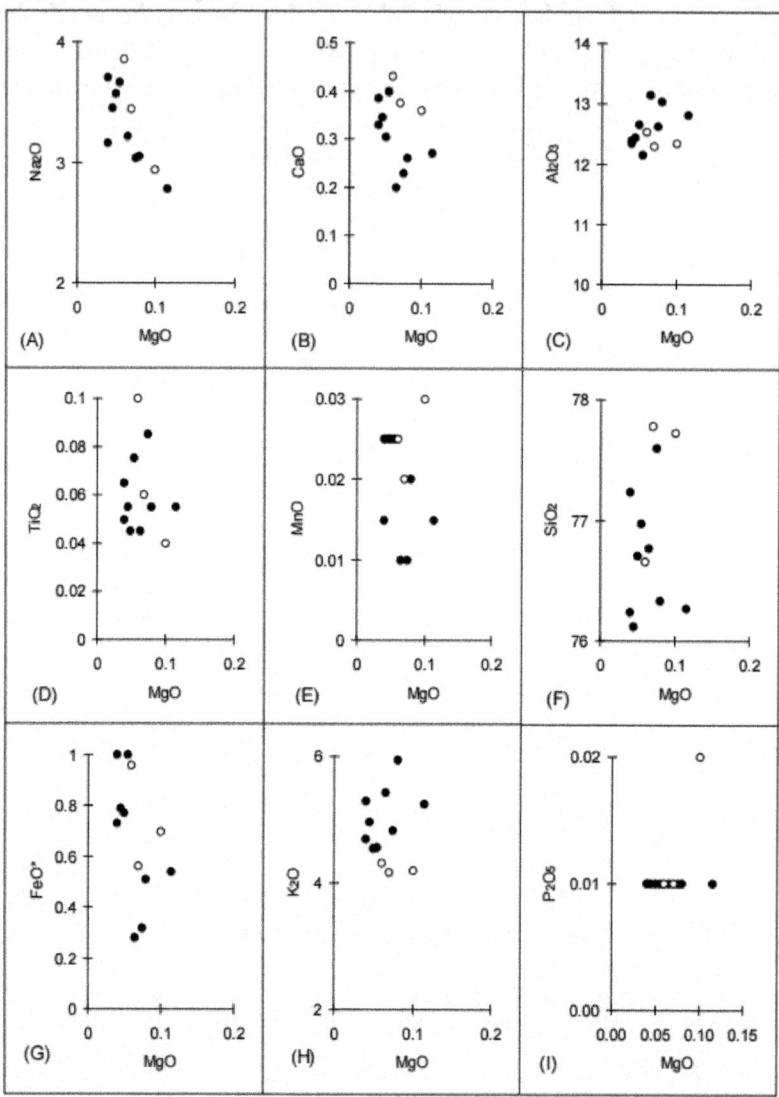

Figure. 2: Chemical variation diagrams for major elements of the GEG. None of these elements defines a well-developed trend. All oxides are in percent. Open circle = micrographic granite, closed circle = granite porphyry.

some scattering in Harker diagrams may be due to slight hydrothermal alteration. However, one of the features of Climax-type rocks is that almost none of them has completely escaped interaction with hydrothermal fluids (White et al., 1981). The GEG contains very low concentrations of P_2O_5 and CaO. Although low concentrations of CaO, Sr and Ba (Table 2) may be due to post-magmatic hydrothermal alteration, it seems that they reflect strong fractionation of the GEG magma. The geochemical changes accompanying progressive fractionation include enrichment of melt in alkali elements (Fig. 3A).

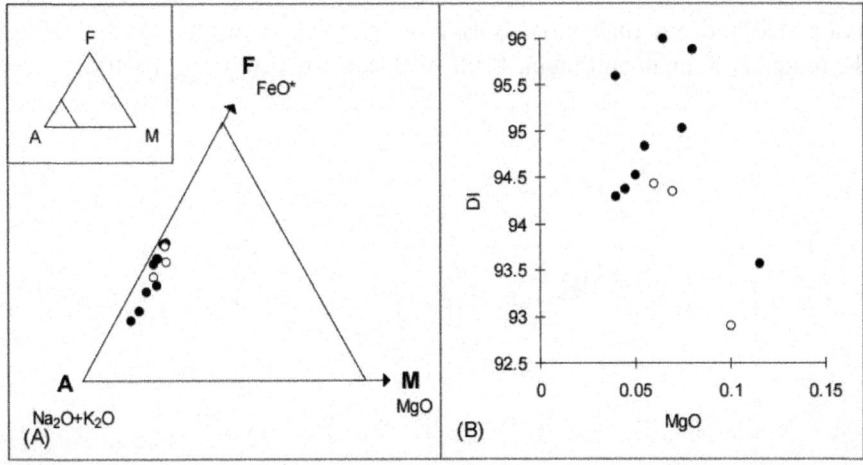

Figure. 3: A - FeO* - Na_2O+K_2O - MgO diagram for GEG samples showing evolution of the GEG toward the alkali apex. B - Plot of DI versus MgO for GEG samples showing no clear trend. All oxides are in percent. Open circle = micrographic granite, closed circle = granite porphyry.

Differentiation indices (DI= normative quartz + albite + orthoclase) for the GEG range from 93 to 97 (Table 1) which is like that of Climax granite (91-94, White et al., 1981). Inasmuch as the differentiation index represents the degree of magmatic evolution, and the normative constituents considered represent minerals with low entropies of melting (Carmichael et al., 1974; White et al., 1981), the high differentiation indices of the GEG suggest crystallization from highly differentiated, low-temperature melts. Like major elements, DI does not show any definite trend when plotted against MgO (Fig. 3B). Although the GEG, like Climax-type granites (White et al., 1981), is alkali rich, its molecular proportion of Al_2O_3 is a little more than its molecular proportion of $CaO+Na_2O+K_2O$, and so it is corundum normative (Table 1). Thus GEG is Al-saturated rather than peralkaline, like other Climax-type rocks (White et al., 1981). The GEG has peraluminous nature, however the samples show a trend toward the peralkaline field (Fig. 4A). This along with trends in Fig. 4B-C suggests that with increasing fractionation (i.e. decreasing MgO) peralkalinity increases and peraluminousity decreases. This trend (enrichment in alkali elements with fractionation), also can be seen in Fig. 3A. A low initial H_2O content for the GEG magma can be inferred from the high-level emplacement of this granite. Furthermore, chemical composition of the GEG shows low CaO, FeO and MgO contents of the magma. Under these conditions, high fluorine contents would not crystallize as fluorite nor substitute in the structure of ferromagnesian minerals, such as biotite. This would indicate that extreme enrichment in F (>4%) and Cl (>5000 ppm) could occur in the magma and in associated hydrothermal fluids during the late stages of the crystallization of the magma (Hannah and Stein, 1990; Webster and Holloway, 1990).

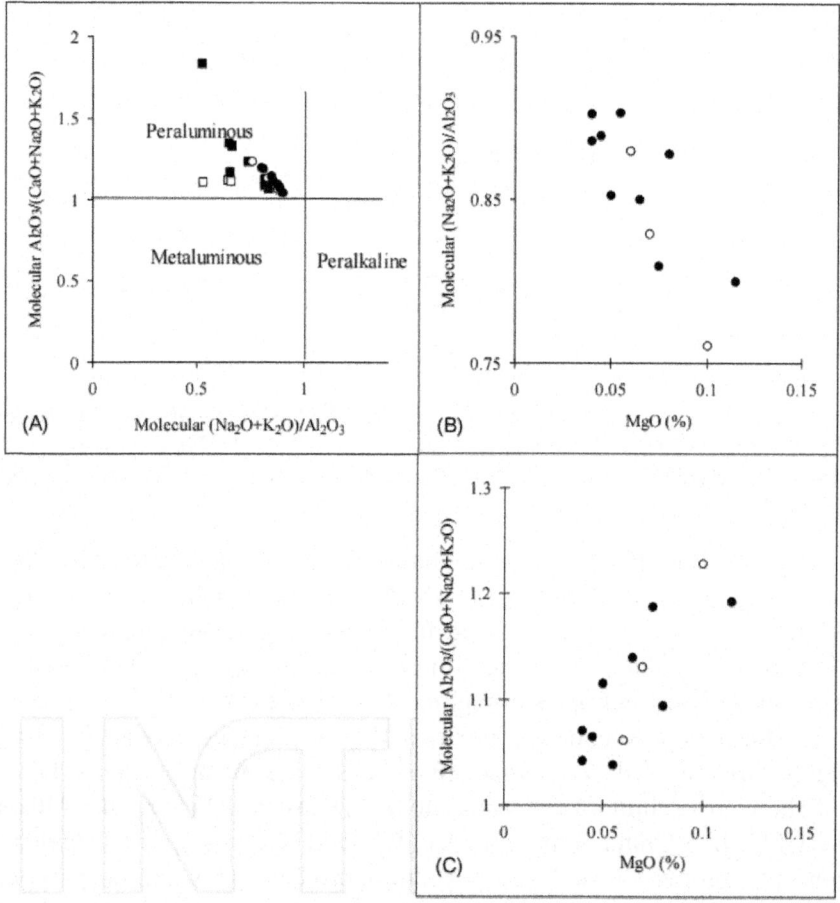

Figure. 4: A- Plot of peraluminousity index (molecular $Al_2O_3/(CaO+Na_2O+K_2O)$) versus peralkalinity index (molecular $(Na_2O+K_2O)/Al_2O_3$) for the GEG and volcanics samples showing peraluminous nature of these rocks. B- Plot of peralkalinity index versus MgO showing increasing peralkalinity with fractionation. C- Plot of peraluminousity index versus MgO showing decreasing peraluminousity with fractionation. Closed circle = granite porphyry, open circle = micrographic granite, closed square = rhyolite, open square = dacite and rhyodacite (date for volcanic rocks from Somarin, 1999).

Trace Element Geochemistry

Trace element abundances in GEG are presented in Table 2. Some trace elements such as Nb, Y, Ga, Zr, U, Nd, La and Ce show poorly developed trends in Harker-type diagrams (Fig. 5).

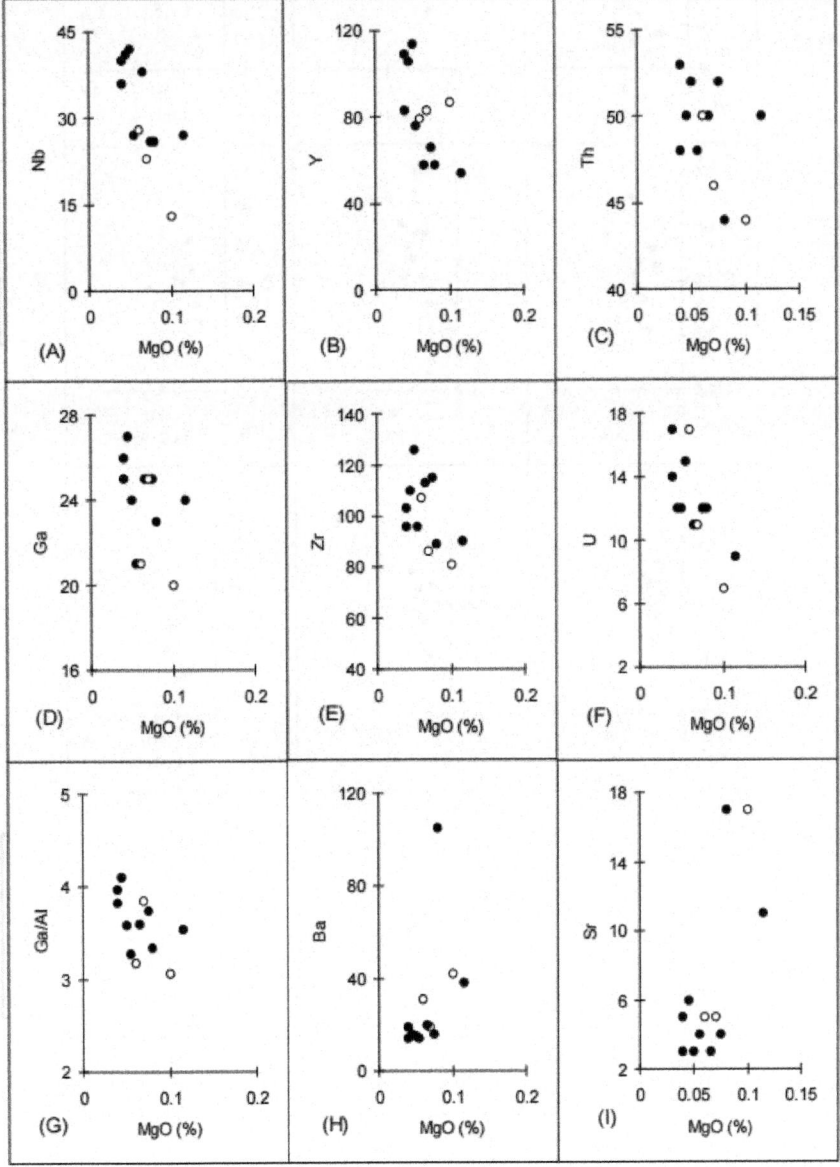

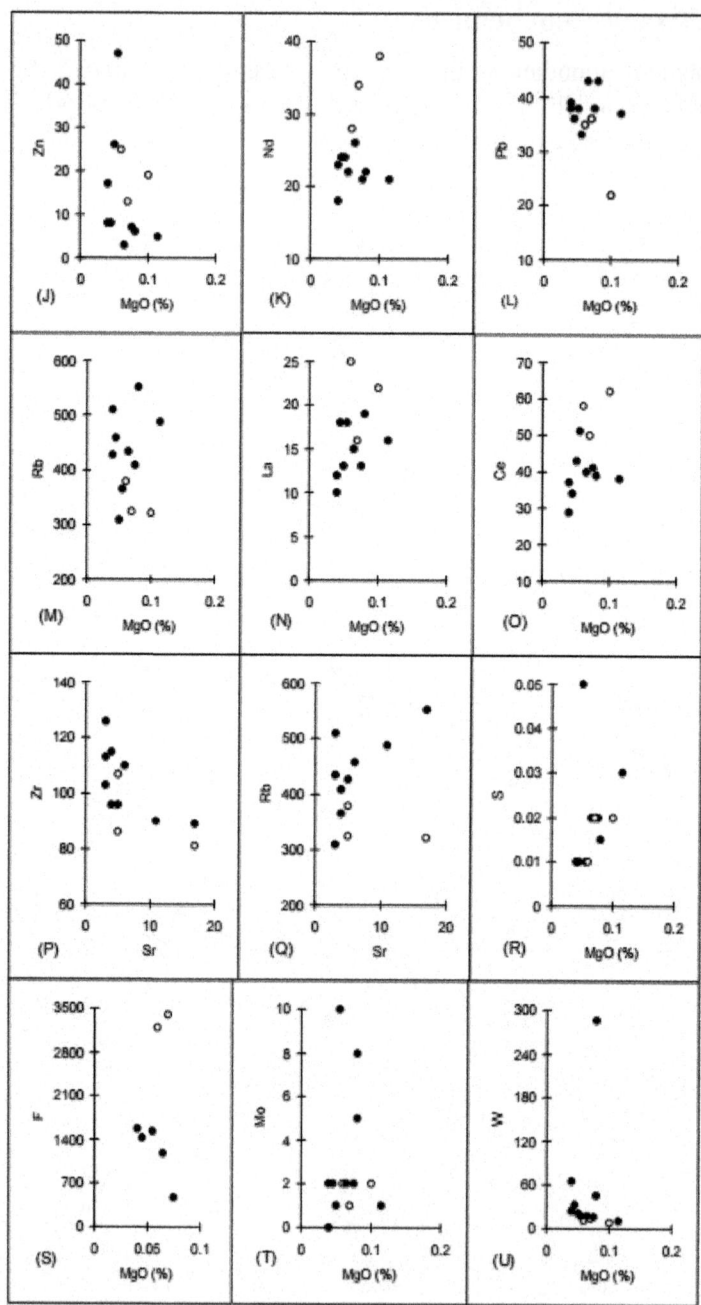

Figure. 5: Chemical variation diagrams for trace elements (in ppm) of the GEG. Open circle = micrographic granite, closed circle = granite porphyry.

However, due to narrow range and low concentrations (near detection limit) of MgO, these trends may not be significant. The important trace element features of the GEG are low concentrations of Sr, Ba, Zr and Zn and high concentrations of Y, Th, U, and Ga relative to average A-type, felsic I- and S-type granites. Also GEG contains high concentrations of F and W, similar to other ore-associated granites (Tischendorf, 1977). High values and wide ranges of Rb/Ba (5 to 43) and Rb/Sr (18 to 170) in microgranite porphyry and micrographic granite indicate crystal fractionation in the magma (Chappell and White, 1992). High Rb/Sr (commonly over 25), very low CaO (

Glen Eden Granite in the Q-Or-Ab system

Average normative $Q:_{Ab}:Or$ in microgranite porphyry is Q_{39}, Ab_{29}, Or_{32} and in micrographic granite is Q_{42}, Ab_{31}, Or_{27}. These approximate the eutectic composition Q_{39}, Ab_{30}, Or_{31} for the calcium-poor granite system at P H O2 =0.5 kbar (Winkler, 1974). Holtz et al. (1992) showed that decreasing H_2O content of the melt causes a rise in liquidus temperatures and a progressive shift of minimum and eutectic compositions toward the Q-Or join at approximately constant normative quartz content. The GEG does not show such shift. Microgranite porphyry and micrographic granite samples plot around the minimum melt composition on the Q-Or-Ab ternary diagram for F-poor Q-Or-Ab-H_2O systems, whereas aplite samples plot toward the Or apex, reflecting the potassic alteration of these samples (Fig. 6). The minimum melt composition of these samples explains the absence of welldefined trends in Harker diagrams. The samples do not have Ab-enriched compositions expected of near-minimum melts in F-rich Q-Or-Ab-H_2O systems (Fig. 6A) (Manning and Pichavant, 1988), but this does not necessarily prove that the GEG melt was F-poor. There is a possibility that the GEG was a minimum melt at P less than 1 kbar but with higher concentrations of fluorine. However, in a calcium-poor granite system, like GEG, with Ab/An >15, as little as 0.5% fluorine in the melt could significantly affect the crystallization processes, since crystallization of fluorite will not occur until the late stages. The pattern of data in the Q-Or-Ab system in the GEG is very similar to that of East Kemptville, Nova Scotia, in which data points plot around the minimum in the F-poor system, suggesting F concentration of less than 1% in the melt, while the effects of F in the various geochemical trends and greisen formation are clear (Richardson et al., 1990). The large amount of F as topaz and fluorite which accompany all the alteration assemblages in the Glen Eden Mo-W-Sn deposit and the presence of primary magmatic fluorite and F-rich biotite in this granite indicate the presence of F in the magma and its effects on the magmatic differentiation of GEG. F-rich granitic rocks typically contain between 0.5 and 2.5 wt% F (Keppler, 1993). F contents measured in granitic rocks should be considered as lower limits for the original F contents of the respective melts as large amounts of F could have been lost by the evolution of a fluid phase (Keppler, 1993). In the Glen Eden prospect, the widespread occurrence of fluorite in all assemblages indicates that a large amount of F has been concentrated in the late-stage fluids.

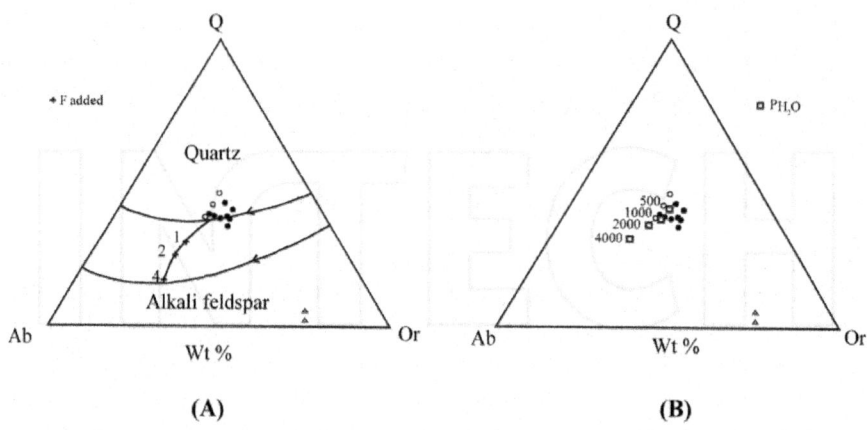

(A)   (B)

Figure. 6: A- The GEG compositions (triangle = aplite, open circle = micrographic granite, closed circle = granite porphyry) compared with liquidus phase relationships in the system Q-Ab-Or at 1 kbar total pressure with excess water (Tuttle and Bowen, 1958) and with F added (in %) under water-saturated conditions (after Manning and Pichavant, 1988). BPosition of minima and eutectics in the system Q-Ab-Or at various P_{H_2O} (Tuttle and Bowen, 1958; Luth, 1976) H$_2$O pressures are given in bars.

Crystallization of GEG has probably occurred between 500 and 1000 bars (Fig. 6B) which suggests high-level emplacement of the GEG. Also, compositional uniformity of analyzed granite samples and their similarity to minimum melt composition at low P_{H_2O} may imply that at least this part of the GEG has crystallized under low P_{H_2O}, since with increasing P_{H_2O} there is more potential for differentiation. This may suggest that these dykes have crystallized after separation and escape of the first episode of aqueous phase. The absence from the GEG of older hydrothermal alterations, which are recognized in the volcanic wall rock, supports this idea and indicates that there was an activation of the still-unsolidified magma chamber which yielded the emplacement of granitic and aplitic dykes within the alteration products. These observations do not imply that water content of the melt was low, since the presence of breccia pipe and the great quantity of hydrothermal veins and assemblages in the central zone reflects a high content of water.

CLASSIFICATION OF THE GLEN EDEN GRANITE

It seems that three factors have influenced the concentrations of major and trace elements in the GEG.

The composition of the protolith: High concentrations of elements such as Th and U, which are high not only in the GEG but also in other I-type leucogranites of the New England area [e.g. Gilgai (Walsh, 1991; Stroud, 1995; Vickery et al., 1997), Kingsgate and

- Mount Jonblee (Plimer, 1973), Stanthorpe (Bampton, 1988), Mole (Brodie, 1983; Stegman, 1983; Kleeman, 1985; Vickery et al., 1997), Dumboy-Gragin (Vickery et al., 1997) and Oban River (Le Messurier, 1983)] most probably reflect the composition of the protolith.

- Post-magmatic hydrothermal alteration: The concentrations of CaO, Sr and possibly Ba are less than those that can be attained by fractional crystallization alone and it seems that destruction of feldspars by hydrothermal solutions can account for their low concentrations. Since P_2O_5 is more immobile than those components mentioned above, it seems that the low concentrations of P_2O_5 are unlikely to be the result of hydrothermal leaching.

- Fractionation in the melt: It seems that fractionation was the most important factor controlling the composition of the GEG. Almost all of the geochemical characteristics of the GEG, such as high SiO_2, Rb, U, Th, Nb, Y, Ga and W and low concentrations of CaO, P_2O_5, Ba, Sr, Zr and Zn and high values and wide ranges of Rb/Ba and Rb/Sr can be explained by various degrees of fractionation of the magma.

Comparison of the GEG with well-known I-, S- and A-type granites is complicated since various investigators have reported different average values for some elements and other features of these granites (e.g. Whalen et al, 1987; Chappell and White, 1992). For example DI of the GEG resembles that of average A-type granite of Whalen et al. (1987) (Table 1), but is more similar to that of average fractionated I-type granite of Chappell and White (1992) (Table 3). The problem of determining I-, S- or A-type affinities of highly felsic granites (such as GEG) has been addressed by several authors (e.g., Whalen et al., 1987; Eby, 1990; Chappell and White, 1992). Aluminium saturation index (ASI; Zen, 1986), molecular Al2O3/(Na$_2$O+K$_2$O+CaO), in the GEG varies between 1 and 1.2. Avila-Salinas (1990) used ASI=1.1 as a boundary between I- and S-type granites. Chappell and White (1992) showed that ASI in S-type granites of Lachlan Fold Belt (LFB), Australia, are always greater than 1 whereas I-type granites generally show ASI1.1. They explained higher ASI in S-types to be a reflection of sedimentary source rocks which contain more clay and so more Al. In contrast, lower ASI in I-types results from lower Al contents in igneous sources. However, they suggested that compositionally very similar felsic granites can be produced from these two quite different source rocks. In such circumstances, the only clue to the nature of a granite protolith might well be isotopic compositions (Chappell and White, 1992). Non-diagnostic values of ASI in the GEG and the very felsic composition of this granite may suggest that it cannot be classified on this criterion alone, and also no conclusion can be made about the source rocks, without isotopic data. Low concentrations of CaO and Sr and resultant higher values of ASI in the GEG may partly reflect slight leaching of these elements by hydrothermal solutions.

The average compositions of microgranite porphyry and micrographic granite are compared, in Table 3, with the average compositions of unfractionated and fractionated felsic I- and S-type granites and A-type granites (data from Chappell and White, 1992). As can be seen, GEG in both major and trace elements is mainly similar to fractionated I- and Atype granites. However, in some elements, GEG shows similarity to other types as well. GEG has very low concentrations of Fe_2O_3 in comparison with others, which indicates the reduced character of this granite. K_2O concentrations of the GEG overlap all types of granites. CaO, Sr and Ba concentrations are more similar to fractionated I-type, but actually they are very low in GEG due to hydrothermal leaching. A-type granites have higher Nb, Y, La, Ce, Sc, Zn, Zr and Ga in comparison to all I- and S-types. However, for Nb and Ga, the fractionated I- and S-type averages move towards the A-type values, relative to unfractionated values, as also does Y for the I-types (Chappell and White, 1992). These changes due to fractionation also increase Rb, U and Sn and decrease Ba and Sr concentrations in fractionated I- and S-types relative to A-types. It seems that increasing Ga and decreasing Al in fractionated granites, especially in fractionated I-types which contain less Al_2O_3, would cause highly fractionated granites to plot in the A-type field in discrimination diagrams of Whalen et al. (1987). Whalen et al. (1987) showed that A-type granites have a high ratio of 10000Ga/Al (>2.6) and they used this ratio for the construction of discrimination diagrams. On these diagrams, high concentrations of Ga and resultant high Ga/Al ratios cause the GEG to plot within the A-type granite field (Fig. 7). However, Whalen et al. (1987) stated that highly fractionated felsic I- and S-type granites can have high Ga/Al ratios and overlap with A-types. They suggested that these fractionated rocks can be distinguished from A-types using Zr+Nb+Ce+Y as a discriminator. Use of this discriminator is based on the principle that at any given degree of fractionation, the A-type granites would contain higher abundances of these elements. On the FeOtotal/MgO versus Zr+Nb+Ce+Y diagram, GEG plots in all fields. This may be due to post-magmatic alteration. On the K_2O+Na_2O/CaO versus Zr+Nb+Ce+Y diagram of Whalen et al. (1987),

FeO$_{total}$/MgO versus SiO$_2$ and 10000Ga/Al versus Zr+Nb+Ce+Y diagrams of Eby (1990) (Fig. 8), who used a higher minimum Ga/Al ratio for A-type granites, GEG samples plot in both 'Fractionated Granite' and A-type granite fields. In the multicationic diagram of Batchelor and Bowden (1985) (Fig. 9), GEG plots in 'Anorogenic' and 'Post-orogenic' granitoids fields which mainly include A-type granites. Based on these observations, a few conclusions can be made.

Table 3: Comparison of the GEG with average compositions of various types of granites (data from Chappell and White, 1992)

	Average GEG (porphyry)	Average GEG (micro-graphic)	Unfractionated I-type	Fractionated I-type	Unfractionated S-type	Fractionated S-type	A-type	Similar type
SiO₂	76.69	77.39	72.90	76.17	71.58	74.40	73.47	FI
TiO₂	0.06	0.07	0.30	0.10	0.42	0.16	0.30	FI
Al₂O₃	12.62	12.39	13.48	12.51	13.83	13.50	12.88	FI, A
Fe₂O₃	0.16	0.13	0.54	0.32	0.45	0.28	0.90	?
FeO	0.50	0.61	1.47	0.71	2.38	1.14	1.63	FI
MnO	0.02	0.03	0.05	0.04	0.05	0.04	0.06	FI, FS
MgO	0.06	0.08	0.66	0.12	1.02	0.27	0.30	FI
CaO	0.30	0.39	1.63	0.61	1.74	0.67	1.06	FI
Na₂O	3.30	3.42	3.27	3.37	2.57	3.06	3.50	UI, FI, A
K₂O	5.05	4.23	4.42	4.92	4.33	4.84	4.62	?
P₂O₅	0.01	0.01	0.09	0.02	0.14	0.18	0.07	FI
FeO*	0.64	0.73	1.96	1.00	2.79	1.39	2.44	FI
Nb	34	21	14	21	12	19	26	FI, A
Zr	104	91	151	116	168	92	322	FS
Y	80	83	38	75	34	28	71	FI, A
Sr	6	9	147	31	114	43	96	FI
Rb	439	341	219	424	221	475	188	FI, FS
Th	50	47	25	47	19	17	24	FI
Pb	38	31	29	35	28	25	27	?
U	13	12	6	16	4	11	5	FS
Ga	24	22	16	19	17	21	22	A
Zn	14	19	38	29	53	46	95	FI
Cu	3	2	6	2	7	3	5	FI, A
Ni	13	4	5	<1	10	2	2	UI, US, FS, A
Ce	39	57	74	79	63	37	130	US, FS
Ba	28	31	488	99	512	150	547	FI
V	3	3	25	3	41	7	9	FI
La	15	21	35	35	28	16	55	FS
Sc	4	6	8	6	10	5	11	FI, FS
Sn	<3	4	7	13	8	23	8	?
Mo	3	2						
W	54	11						
F	1240							
As	4	5						
Nd	22	33						
S (%)	0.02	0.02						
Q	37.00	40.00	31.89	35.87	33.65	35.98	32.06	FI, FS
Or	30.00	25.00	26.12	29.08	25.59	28.61	27.31	?
Ab	28.00	29.00	27.67	28.52	21.75	25.89	29.62	UI, FI, A
An	1.44	1.87	7.50	2.90	7.72	2.15	4.80	FS
Hy	0.85	1.10	3.49	1.25	5.94	2.34	2.61	FI
Mt	0.23	0.19	0.78	0.46	0.65	0.41	1.30	?
Ilm	0.11	0.13	0.57	0.19	0.80	0.30	0.57	FI

Ap	0.02	0.02	0.21	0.05	0.33	0.42	0.17	FI
C	1.20	1.50	0.57	0.58	2.09	2.44	0.36	?
Py	0.04	0.03						
DI	95	94	86	93	81	90	89	FI
ASI	1.10	1.14	1.03	1.04	1.15	1.17	1.02	?
100Mg/Mg+Fe	23	22	54	29	50	35	34	FI
K_2O/Na_2O	1.53	1.24	1.35	1.46	1.68	1.58	1.32	?
Rb/Sr	71	38	1.50	14	2	11	2	FI
Rb/Ba	16	11	0.50	4.30	0.40	3	0.30	FI
10000*Ga/Al	3.66	3.36	2.24	2.87	2.32	2.93	3.22	A

- The GEG is highly fractionated. High Ga concentrations have been considered as a diagnostic feature of A-type granites by many investigators (e.g., Collins et al., 1982; Whalen et al., 1987; Eby, 1990; Haapala and Ramo, 1990; Whalen and Currie, 1990). As stated by Whalen et al. (1987), Sawka et al. (1990) and Chappell and White (1992), high fractionation of I- and S-type granites could enrich Ga in the magma. So high Ga concentration is not exclusively a feature of A-type granites

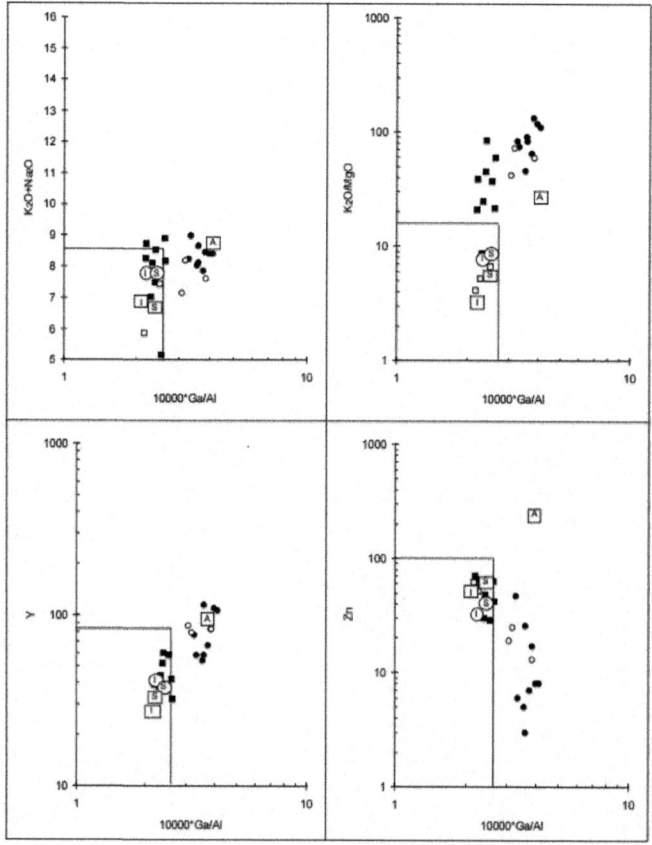

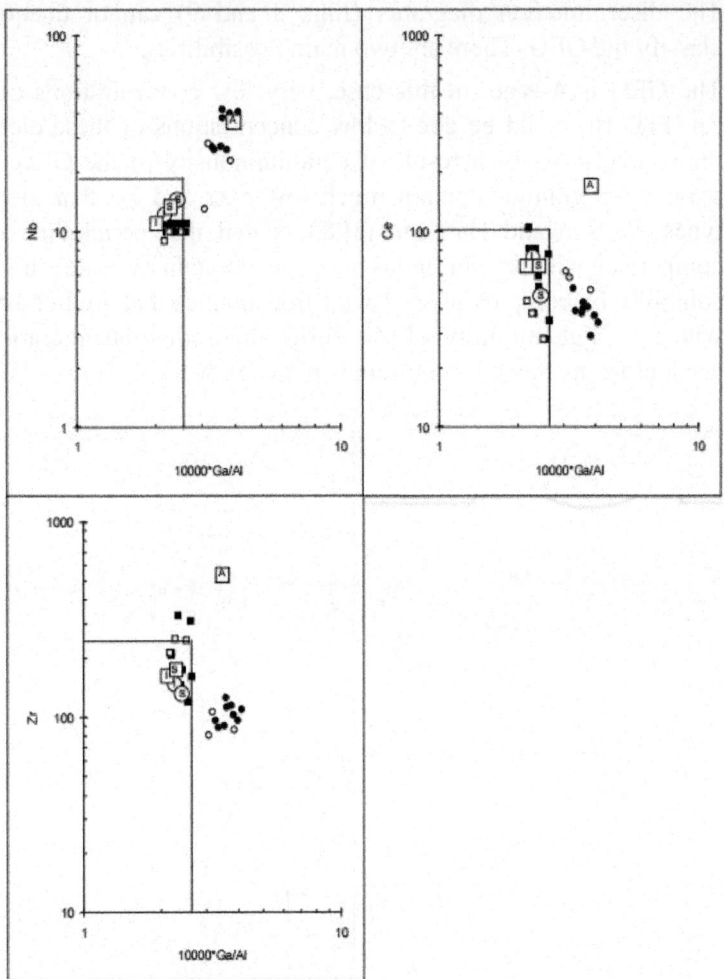

Figure. 7: Data from the GEG and Emmaville Volcanics in discrimination diagrams of Whalen et al. (1987), showing the possible A-type characteristics of this granite. I I-type granite average, S S-type granite average, I felsic I-type granite average, S felsic S-type granite average, A A-type granite average. . Oxides and trace elements are in percent and ppm, respectively. Closed circle = granite porphyry, open circle = micrographic granite, closed square = rhyolite, open square = dacite and rhyodacite.

- The GEG most probably is not S-type, as S-type granites have high concentrations of P_2O_5 which increase with fractionation in this type of granite (Sawka et al., 1990; Chappell and White, 1992). Very low concentrations of P_2O_5 in the GEG and rarity of apatite in hydrothermal assemblages of the Glen Eden prospect indicate a low content of this component in the GEG melt.

- The discrimination diagrams (Figs 8 and 9) cannot unequivocally classify the GEG. There are two main possibilities.

a. The GEG is A-type. In this case, very low concentrations of Zr and Zn (Fig. 10) could be due to low concentrations of these elements in the source rocks or a result of peraluminousity of the GEG, as non-peralkaline granites contain much lower Zr and Zn than peralkaline types. Watson and Harrison (1983) stated that peralkaline melts in comparison with peraluminous ones can maintain extremely high zircon solubility by complexing Zr^{4+} with free alkalies that are not associated with Al. High mobility of Zn during hydrothermal alterations may account for its low concentrations in the GEG.

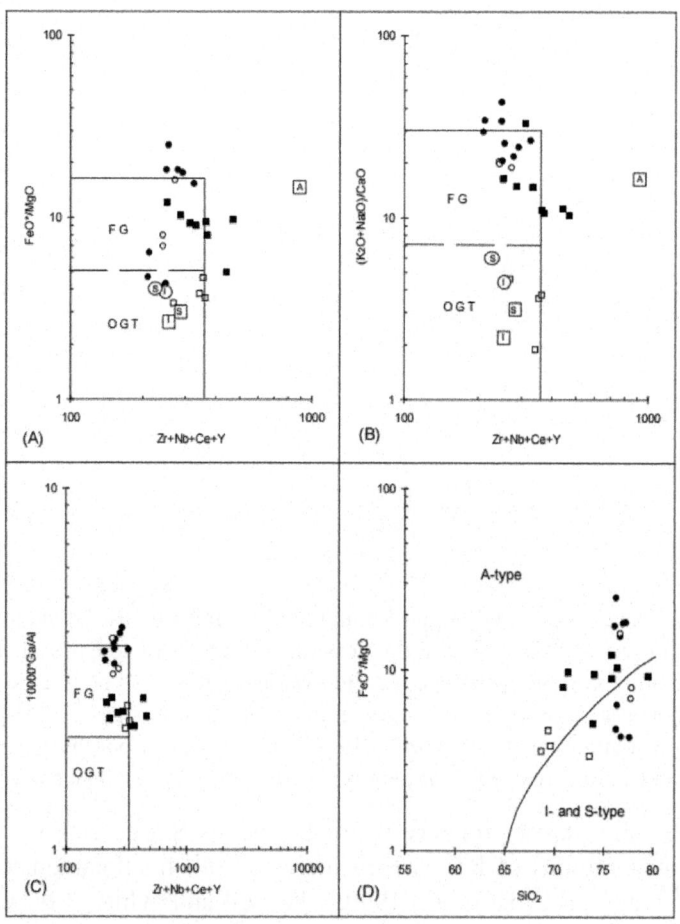

Figure. 8: The GEG and Emmaville Volcanics in discrimination diagrams of Whalen et al. (1987) (A and B) and Eby (1990) (C and D) plot in 'Fractionated

Granites' and 'A-type Granites' fields. FeO*= total FeO, F G = Fractionated felsic Granite, O G T = Unfractionated Granite. Oxides and trace elements are in percent and ppm, respectively. Closed circle = granite porphyry, open circle = micrographic granite, closed square = rhyolite, open square = dacite and rhyodacite..

b.   The GEG is fractionated I-type: It has been found that a large degree of fractional crystallization of I- and even S-type granite magmas can produce a minor amount of evolved magma with high Ga/Al, very low concentrations of Ba and Sr and large variation in Rb/Sr and Rb/Ba ratios (Whalen and Currie, 1990). As can be seen in Table 3, fractionation of I-type granite magmas in the Lachlan Fold Belt increased SiO2, Rb, Pb, Th, U, Nb, Y, Ce, Ga, Sn, DI, Rb/Sr, Rb/Ba and partially Na2O and K2O and decreased TiO_2, Al_2O_3, Fe_2O_3, FeO, MnO, MgO, CaO, P2O5, Ba, Sr, Zr, Sc, V, Ni, Cr, Co, Cu, and Zn. Also, with fractionation, the Fe_2O_3/FeO ratio increases (Fig. 11). So it seems that fractionation of I-type granitic magma could produce the GEG. This is consistent with the field occurrences of the GEG as dykes. These dykes may represent the last stage of fractional crystallization of a major pluton at greater depth.

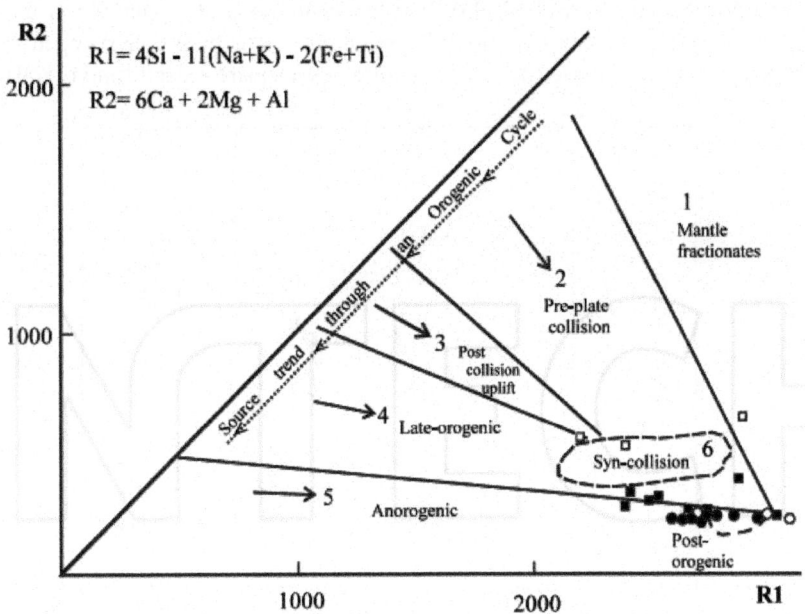

Figure. 9: Data from the GEG and Emmaville Volcanics in the multicationic discrimination diagram for the major granitoids (after Batchelor and Bowden, 1985). Analyzed samples plot mainly in the 'Post Orogenic' and 'Anorogenic' fields. Closed circle = granite porphyry, open circle = micrographic granite, closed square = rhyolite, open square = dacite and rhyodacite.

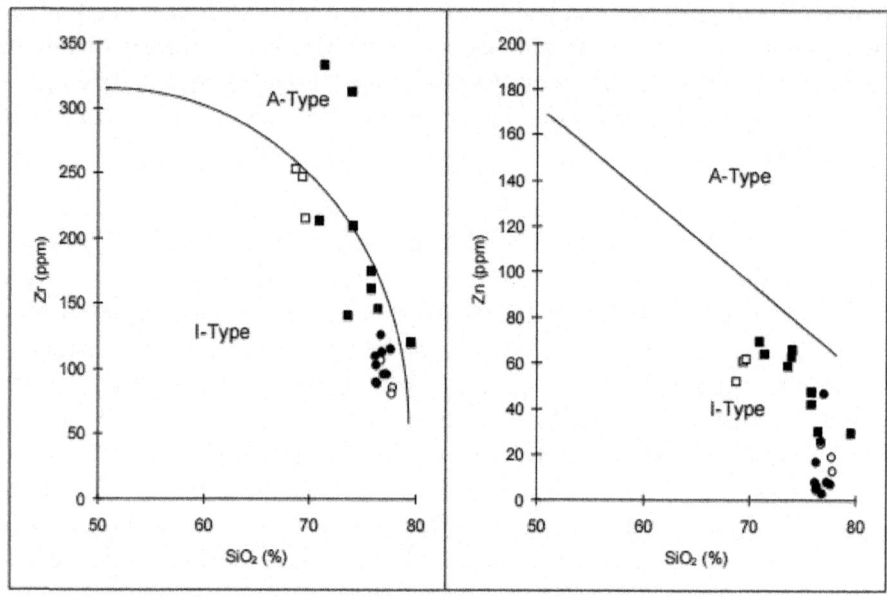

Figure. 10: Data from GEG and Emmaville Volcanics in Zr-SiO2 and Zn-SiO2 diagrams (after Newberry et al., 1990). Low concentrations of Zr and Zn in the GEG cause GEG to plot in the I-type field. Closed circle = granite porphyry, open circle = micrographic granite, closed square = rhyolite, open square = dacite and rhyodacite.

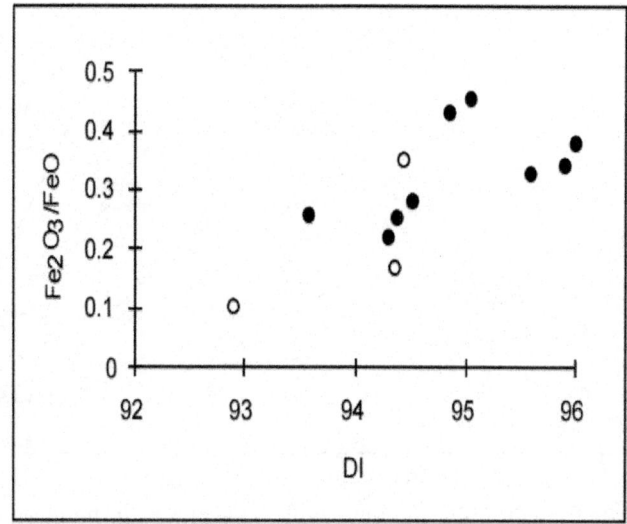

Figure. 11: Plot of DI versus Fe_2O_3/FeO for the GEG, showing increasing Fe_2O_3/FeO with fractionation. Open circle = micrographic granite, closed circle = granite porphyry.

Due to removal of some feldspars and accessory minerals, these dykes show low concentrations of CaO, Sr, Ba, Zr, Zn and high concentrations of U, Th, Nb, Rb, Y, Ga and W. This is consistent with field evidence of other granites associated with Mo-Sn-W ore deposits, wherein fine-grained granites mostly occur as 'carapace' facies formed near the upper contacts of the pluton (e.g., at Krusne hory, Erzgebirge; Stemprok, 1985), or as dykelike products at the apical part of the larger plutons (Tischendorf, 1977; Blevin and Chappell, 1996a). Either way, fine-grained granites as marginal carapace (Plimer, 1987) or as late-stage phases (Stemprok, 1990; Blevin and Chappell, 1995, 1996a) are one of the common features of plutons associated with Sn, W and Mo ore deposits. Most of these plutons show vertical zoning as enrichment in some elements such as W, Be, Sn, F, Cs, Rb and Li at the top with an impoverishment of Ba, Sr, Ni, Cr and V (Tischendorf, 1977). It is noteworthy that the GEG contains high concentrations of Ni, relative to fractionated I- and A-type granite compositions, which are inconsistent with normal fractionation, as Ni concentration decreases with fractionation. Since Ni is relatively immobile, it is unlikely that this element has been added during slight hydrothermal alteration. It is more likely that the high concentrations of Ni and low concentrations of Ce, La, and Sn in the GEG, relative to other fractionated I-type granites, reflect the compositional features of the source.

Regarding the classification of Ishihara (1977), the GEG has formed under reduced conditions (Fig. 12) and belongs to the ilmenite series, since its Fe_2O_3/FeO ratio is <0.5 (Ishihara, 1981). Generally, I-type granites show higher fO_2 than S-types and Chappell and White (1992) consider this feature to be inherited from the source. The GEG shows a wide range of Fe_2O_3/FeO, overlapping with unfractionated and fractionated I- and S-type granites. Fe_2O_3/FeO ratio increases with fractionation in the GEG samples (Table 3, Fig. 11). In summary, the highly fractionated character of the GEG makes classification difficult. The geochemical features and field observations show that GEG could be A-type or fractionated I-type. As stated by Chappell and White (1992), fractionated I-type granites are similar to fractionated A-types and they can be mistaken.

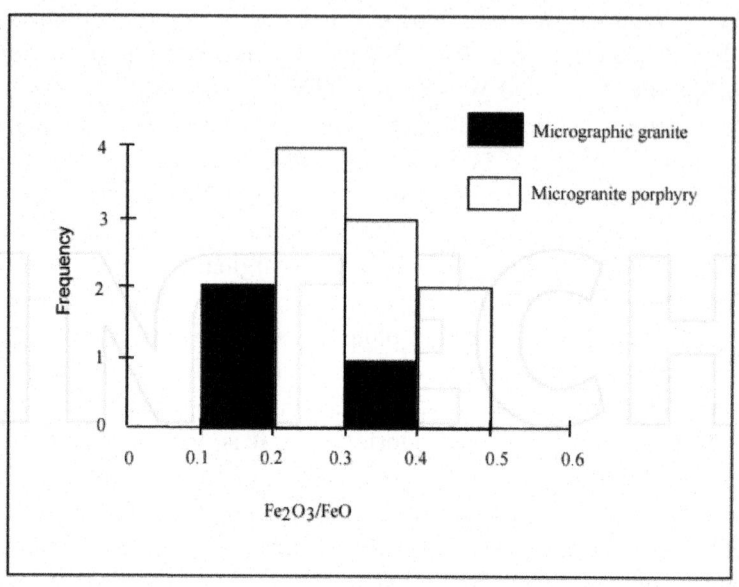

Figure. 12: Histogram of Fe2O3/FeO in the GEG, showing the reduced character of the Glen Eden Granite.

TECTONIC SETTING

On tectonic discrimination diagrams of Pearce et al. (1984), data from the GEG, but not the associated volcanics, mostly plot in the 'Within Plate' field (Fig. 13). This is typical for Atype granites (Pearce et al., 1984; Whalen et al., 1987), but does not mean that the GEG is necessarily A-type (Whalen and Currie, 1990). Although I- and S-type granites mostly plot in the 'Volcanic Arc' field (Whalen et al., 1987), they may also plot in the 'Within Plate' field (Whalen, 1988). It seems that the high fractionation of I-type granites would increase the concentrations of Nb, Y and Rb and would cause these rocks to plot in the 'Within Plate' field. For example, average fractionated I-type granites of the Lachlan Fold Belt (Table 3) plot in the 'Within Plate' field, as does the GEG. There seems to be general agreement that A-type granites were emplaced into tensional (or non-compressive) environments either at the end of an orogenic cycle in continental rift zones or in oceanic basins (Eby, 1990). The Glen Eden Granite plots in 'Post-orogenic' and 'Anorogenic' fields on the multicationic diagram of Batchelor and Bowden (1985) (Fig. 9) and it seems that this granite was emplaced into an unstable active margin. This tectonic setting is similar to that proposed for the A-type Topsails Granite, western Newfoundland (Whalen and Currie, 1990).

SUMMARY

The various features of the GEG can be summarized as follows.

- The compositional features of the GEG are enrichment in SiO_2, Rb, U, Th, Nb, Y, Ga and W and impoverishment in CaO, P_2O_5, Ba, Sr, Zr and Zn and high values and wide range of Rb/Ba and Rb/Sr.

- The GEG, like other mineralization-associated plutons of the New England Batholith (Kleeman, 1978; Blevin and Chappell, 1996a, b; Vickery et al., 1997), is a high-level leucogranite of near minimum melt composition.

- The presence of crenulate quartz layers, micrographic texture and hydrothermal breccia at Glen Eden suggests saturation of magma from water and the presence of fluid-rich environment.

- The highly fractionated character of the GEG does not allow unequivocal classification but it has strong similarities to fractionated I-type and A-type granites.

- The tectonic setting of the GEG, based on geochemical criteria, is 'Within Plate' and possibly it has been emplaced into an unstable active margin.

- Strong fractionation of the granitic magma increased concentration of incompatible elements, including metals such as Sn, W and Mo, in the final melt and magmatic solution. Increasing the pressure of this fluid eventually caused brecciation of the cap rocks and formed a breccia pipe wherein Mo-W-Sn mineralization occurred.

REFERENCES

1.   Augustithis, S.S. 1973. Atlas of the textural pattern of granites, gneisses and associated rock types. Elsevier, Amsterdam, London, New York, 378pp.

2.   Avila-Salinas, W.A. 1990. Tin-bearing granites from the Cordillera Real, Bolivia: a petrological and geochemical review. Geol. Soc. Am., Special Paper, 145-159.

3.   Bailey, J.C. 1977. Fluorine in granitic rocks and melts: a review. Chem. Geol., 19, 1-42. Bampton, M.D. 1988. Alteration and mineralisation of the southern part of the Stanthorpe Adamellite, near Tenterfield, New South Wales. Unpublished BSc (Hons) thesis, University of Sydney, 163pp.

4.   Bankwitz, P. 1978. Remarks concerning the development of the Erzgebirge pluton. In M. Stemprok, L. Burnol and G. Tischendorf (eds), Metallization associated with acid magmatism, 3, 156-167.

5.   Batchelor, R.A. and Bowden, P. 1985. Petrogenetic interpretation of granitoid rock series using multicationic parameters. Chem. Geol., 48, 43-55.

6.   Blevin, P.L. and Chappell, B.W. 1995. Chemistry, origin, and evolution of mineralized granites in the Lachlan Fold Belt, Australia: the metallogeny of I- and S-type granites. Econ. Geol., 90, 1604-1619.

7.   Blevin, P.L. and Chappell, B.W. 1996a. Internal evolution and metallogeny of Permo-Triassic high-K granites in the Tenterfield-Stanthorpe region, southern New England Orogen, Australia. In Mesozoic geology of the eastern Australia plate conference, Geol. Soc. Aust., Abs., 43, 94-100.

8.   Blevin, P.L. and Chappell, B.W. 1996b. Permo-Triassic granite metallogeny of the New England Orogen. In Mesozoic geology of the eastern Australia plate conference, Geol. Soc. Aust., Abs., 43, 101-103.

9.   Brodie, R.S. 1983. Geology and mineralization of the Mole River-Silent Grove area, near Tenterfield, northern New South Wales. Unpublished BSc (Hons) thesis, University of New England, 149pp.

10. Burnham, C.W. 1967. Hydrothermal fluids in the magmatic stage. In H.L. Barnes (ed), Geochemistry of hydrothermal ore deposits, New York, John Wiley, 34-76.

11. Burnham, C.W. 1979. Magmas and hydrothermal fluids. In H.L. Barnes (ed), Geochemistry of hydrothermal ore deposits, 2nd edition, New York, John Wiley, 71-136.

12. Carmichael, I.S.E., Turner, F.J. and Verhoogen, J. 1974. Igneous Petrology. New York. McGraw-Hill Book Co., 739pp.

13. Carten, R.B., Walker, B.M., Geraghty, E.P. and Gunow, A.J. 1988. Comparison of field-based studies of the Henderson porphyry molybdenum deposit, Colorado, with experimental and theoretical models of porphyry systems. In R.P. Taylor and D.F. Strong (eds), Recent advances in the geology of granite-related mineral deposits, Can. Ins. Min. Metall., 39, 1-12.

14. Chappell, B.W. and White, A.J.R. 1992. I- and S-type granites in the Lachlan Fold Belt. Trans. Roy. Soc. Edin.: Earth Sci., 83, 1-26.

15. Collins, W.J., Beams, S.D., White, A.J.R. and Chappell, B.W. 1982. Nature and origin of Atype granites with particular reference to southeastern Australia. Contrib. Mineral. Petrol., 80, 189-200.

16. Collins, W.J., Offler, R., Farrell, T.R. and Landenberger, B. 1993. A revised Late PalaeozoicEarly Mesozoic tectonic history for the southern New England Fold Belt. In P.G. Flood and J.C. Aitchison (eds), New England Orogen, eastern Australia, University of New England, 69-84.

17. Dingwell, D.B. 1985. The structure and properties of fluorine-rich silicate melts: implications for granite petrogenesis. In R.P. Taylor and D.F. Strong (eds), Granite-related mineral deposits geology, petrogenesis and tectonic setting, CIM Conference on Granite-related Mineral Deposits, 72-81.

18. Dingwell, D.B. 1988. The structure and properties of fluorine-rich magmas: a review of experimental studies. In R.P. Taylor and D.F. Strong (eds), Recent advances in the geology of granite-related mineral deposits. Can. Inst. Min. Metall., Special Volume 39, 1-12.

19. Dingwell, D.B., Scarfe, C.M. and Cronin, D.J. 1985. The effect of fluorine on viscosities in the system Na2O-Al2O3-SiO2 - implications for phonolites, trachytes and rhyolites. Am. Mineral., 70, 80-87.

20. Eby, G.N. 1990. The A-type granitoids: a review of their occurrence and chemical characteristics and speculations on their petrogenesis. In A.R. Woolley and M. Ross (eds), Alkaline igneous rocks and carbonatites. Lithos, 26, 115-134.

21. Fenn, P.M. 1979. On the origin of graphic intergrowth [abs.]. Geol. Soc. Am., Abstr. Programs, 11, 424.

22. Flood, P.G. and Aitchison, J.C. 1993a. Understanding New England geology: the comparative approach. In P.G. Flood and J.C. Aitchison (eds), New England Orogen, eastern Australia, University of New England, 1-10.

23. Flood, P.G. and Aitchison, J.C. 1993b. Recent advances in understanding the geological development of the New England Province of the New England Orogen. In P.G. Flood and J.C. Aitchison (eds), New England Orogen, eastern Australia, University of New England, 61-67.

24. Glyuk, D.S. and Anfiligov, V.N. 1973. Phase equilibria in the system granite-H2O-HF at a pressure of 1000 kg/cm2. Geochem. Internat., 10, 313-317.

25. Haapala, I. and Ramo, O.T. 1990. Petrogenesis of the Proterozoic rapakivi granite of Finland. In H.J. Stein and J.L. Hannah (eds), Ore-bearing granite systems ; petrogenesis and mineralizing processes. Geol. Soc. Am., Special Paper, 246, 275-286.

26. Hannah, J.L. and Stein, H.J. 1990. Magmatic and hydrothermal processes in ore bearing systems. In H.J. Stein and J.L. Hannah (eds), Ore-bearing

granite systems; petrogenesis and mineralizing processes. Geol. Soc. Am., Special Paper, 246, 1-10.

27. Hensel, H.D. 1982. The mineralogy, petrology and geochronology of granitoids and associated intrusives from the southern portion of the New England Batholith. Unpublished Ph.D. thesis, University of New England, 273pp.

28. Holtz, F., Pichavant, M., Barbey, P. and Johannes, W. 1992. Effects of H2O on liquidus phase

29. relations in the haplogranite system at 2 and 5 kbar. Am. Mineral., 77, 1223-1241.

30. Ishihara, S. 1977. The magnetite-series and ilmenite-series granitic rocks. Min. Geol., 27, 293-

31. 305.

32. Ishihara, S. 1981. The granitoid series and mineralization. Econ. Geol., 75th Anniversary Volume, 458-484.

33. Keith, J.D. and Shanks, W.C. 1988. Chemical evolution and volatile fugacities of the Pine Grove porphyry molybdenum and ash-flow tuff system, southwestern Utah. In R.P. Taylor and D.F. Strong (eds), Recent advances in the geology of granite-related mineral deposits. Can. Inst. Min. Metall., Special Volume, 39, 402-423.

34. Keppler, H. 1993. Influence of fluorine on the enrichment of high field strength trace elements in granitic rocks. Contrib. Mineral. Petrol., 114, 479-488.

35. Kirkham, R.V. and Sinclair, W.D. 1988. Comb quartz layers in felsic intrusions and their relationship to porphyry deposits. In R.P. Taylor and D.F. Strong (eds), Recent advances in the geology of granite-related mineral deposits. Can. Inst. Min. Metall., Special Volume, 39, 50-71.

36. Kleeman, J.D. 1978. Tin mineralizing granites in New England [abs]. Australian Geology Convention, 3rd, Townsville, August 1978, Abstracts and Programs, 37.

37. Kleeman, J.D. 1982. The anatomy of a tin-mineralizing A-type granite. In P.G. Flood and B. Runnegar (eds), New England Geology., University of New England and AHV Club, 327-334.

38. Kontak, D.J. 1994. Geological and geochemical studies of alteration processes in a fluorinerich environment: the east Kemptville Sn-(Zn-Cu-Ag) deposit, Yarmouth Country, Nova Scotia, Canada. In D.R. Lentz (ed), Alteration and alteration processes associated with ore-forming systems, Geological Association of Canada, Short

39. Course Notes, 11, 261-314.

40. Kovalenko, V.I., Kuz'min, M.I., Antipin, V.S. and Petrov, L.L. 1971. Topaz bearing keratophyre (ongonite), a new variety of subvolcanic igneous vein rock. Doklady Academy Science, U. S. S. R., Earth Science Section, 199, 132-135.

41. Le Maitre, R.W. 1976. The chemical variability of some common igneous rocks, Jour. Petrology, 17, 589-637.

42. Le Messurier, L.A. 1983. The genetic relationships between two alkali granites and associated enclosing I-type plutons within the New England region. Unpublished BSc (Hons) thesis, University of New England, 131pp.

43. London, D. 1987. Internal differentiation of rare-element pegmatites: effects of boron, phosphorus, and fluorine. Geochim. et Cosmochim. Acta, 51, 403-420.

44. Luth, W.C. 1976. Granitic rocks. In D.K. Bailey and R. Macdonald (eds), The evolution of the crystalline rocks, Academic Press, London, 335-417.

45. Manning, D.A.C. 1981. The effect of fluorine in liquidus phase relationships in the system Qz-Ab-Or with excess water at 1 kb. Contrib. Mineral. Petrol., 76, 206-215.

46. Manning, D.A.C. and Pichavant, M. 1988. Volatiles and their bearing on the behavior of metals in granitic systems. In R.P. Taylor and D.F. Strong (eds), Recent Advances in the Geology of Granite-Related Mineral Deposits, Can. Ins. Min. Metall., Special Volume, 39, 13-24.

47. Moore, J.G. and Lockwood, J.P. 1973. Origin of comb layering and orbicular structure, Sierra Nevada Batholith, California. Geol. Soc. Am. Bull., 48, 1-20.

48. Munoz, J.L. and Ludington, S.D. 1974. Fluorine-hydroxyl exchange in biotite. Am. Jour. Sci., 274, 396-413.

49. Murray, C.G. 1988. Tectonic evolution and metallogenesis of the New England Orogen. In J.D. Kleeman (ed), New England Orogen - tectonics and metallogenesis. University of New England, 204-210.

50. Nabelek, P.I. and Russ-Nabelek, C. 1990. The role of fluorine in the petrogenesis of magmatic segregations in the St. Francois volcano-plutonic terrane, southeastern Missouri. In H.J. Stein and J.L. Hannah (eds), Ore-bearing granite systems; petrogenesis and mineralizing processes. Geol. Soc. Am., Special Paper, 246, 71-87.

51. Newberry, R.J., Burns, L.E., Swanson, S.E. and Smith, T.E. 1990.

Comparative petrologic evolution of the Sn and W granites of the Fairbanks-Circle area, interior Alaska. In H.J. Stein and J.L. Hannah (eds), Ore-bearing granite systems; petrogenesis and mineralizing processes. Geol. Soc. Am., Special Paper, 246, 121-142.

52.   Nockolds, S.R. 1954. Average chemical compositions of some igneous rocks. Geol. Soc. Am. Bull., 65, 1007-1032.

53.   Pearce, J.A., Harris, N.B.W. and Tindle, A.J. 1984. Trace elements discrimination diagrams for the tectonic interpretation of granitic rocks. Jour. Petrology, 25, 956-983.

54.   Pichavant, M. and Manning, D. A. C. 1984. Petrogenesis of tourmaline granites and topaz granites; the contribution of experimental data. Physics of the Earth and Planetary Interiors, 35, 31-50.

55.   Plimer, I.R. and Kleeman, J.D. 1985. Mineralization associated with the Mole Granite, Australia. In: High heat production (HHP) granites, hydrothermal circulation and ore genesis. St. Austell, England, Inst. Mining Metallurgy, 563-570.

56.   Plimer, I.R. 1973. The pipe deposits of tungsten-molybdenum-bismuth in eastern Australia. Unpublished PhD. thesis, Macquarie University, 288pp.

57.   Plimer, I.R. 1987. Fundamental parameters for the formation of granite-related tin deposits. Geologische Rundschau, 76, 23-40.

58.   Povilaitis, M.M. 1978. Effect of the conditions of magmatic emplacement on hightemperature postmagmatic ore mineralization. In M. Stemprok, L. Burnol and F. G. Tischendorf, Metallization associated with acid magmatism, 3, 375-384.

59.   Richardson, J.M., Bell, K., Watkinson, D.H. and Blenkinsop, J. 1990. Genesis and fluid evolution of the East Kemptville greisen-hosted tin mine, southwestern Nova Scotia, Canada. In H.J. Stein and J.L. Hannah (eds), Ore-bearing granite systems; petrogenesis and mineralising processes. Geol. Soc. Am., Special Paper, 246, 181-203.

60.   Sawka, W.N., Heizler, M.T., Kistler, R.W. and Chappell, B.W. 1990. Geochemistry of highly fractionated I- and S-type granites from the tin-tungsten provinces of western Tasmania. In H.J. Stein and J.L. Hannah (eds), Ore-bearing granite systems; petrogenesis and mineralizing processes. Geol. Soc. Am., Special Paper, 246, 161- 179.

61.   Schroecke, H. 1973. Grundlagen der magmatogenen lagerstättenbildung. Enke Verlag. Stuttgart, 287pp.

62.   Shaver, S.A. 1984a. Origin of crenulate quartz layers - evidence from

the Hall (Nevada Moly) molybdenum deposit, Nevada [abs]. Geol. Soc. Am., Abstr. Programs, 16, 254-255.

63. Shaver, S.A. 1984b. The Hall (Nevada Moly) molybdenum deposit, Nye Country, Nevada: geology, alteration, mineralization and geochemical dispersion. Unpublished Ph.D. thesis, Stanford Univ., 261pp.

64. Shaw, S.E. and Flood, R.H. 1981. The New England Batholith, eastern Australia: geochemical variations in space and time. Jour. Geoph. Res., 86, 10530-10544.

65. Sheppard, S.M.F. 1977. Identification of the origin of ore-forming solutions by the use of stable isotopes. In: Volcanic processes in ore genesis, Geol. Soc. London, Special. Publication, 7, 25-41.

66. Somarin, A.K. 1999. Mineralogy, geochemistry and genesis of the Glen Eden Mo-W-Sn deposit, New England Batholith, Australia. Unpublished PhD thesis, University of New England, Armidale, Australia, 340pp.

67. Somarin, A.K., and Ashley, P., 2004. Hydrothermal alteration and mineralization of the Glen Eden Mo-W-Sn deposit: a leucogranite-related hydrothermal system, Southern New England Orogen, NSW, Australia. Mineralium Deposita, 39, 282-300.

68. Stegman, C.L. 1983. The Mole Granite and its Sn-W-Mo-base metal mineralization - a study of its southern-central margin. Unpublished BSc (Hons) thesis, University of New England, 177pp.

69. Stemprok, M. 1985. Vertical extent of greisen mineralization in the Krusne hory/Erzgebirge granite pluton of central Europe. In: High heat production (HHP) granites, hydrothermal circulation and ore genesis. St. Austell, England, Inst. Mining Metallurgy, 41-54.

70. Stemprok, M. 1990. Intrusion sequences within ore-bearing granitoid plutons. Geological Jour., 25, 413-417.

71. Stewart, J.P. 1983. Petrology and geochemistry of the intrusives spatially associated with the Logtung W-Mo prospect, south-central Yukon Territory. Unpublished M.Sc. thesis, University of Toronto, 243pp.

72. Strong, D.F. 1988. A review and model for granite-related mineral deposits. In R.P. Taylor and D.F. Strong (eds), Recent advances in the geology of granite-related mineral deposits. Can. Inst. Min. Metall., Special Volume, 39, 424-445.

73. Stroud, W.J. 1995. Inverell 1: 250000 metallogenic map. Taylor, R.P. 1992. Petrological and geochemical characteristics of the Pleasant Ridge zinnwaldite-topaz granite, southern New Brunswick, and comparison with other topaz-bearing felsic rocks. Can. Mineral., 30, 895-921.

74. Tischendorf, G. 1977. Geochemical and petrographic characteristics of silicic magmatic rocks associated with rare-element mineralization. In M. Stemprok, L. Burnol and G. Tischendorf (eds), Metallization associated with acid magmatism, 2, 41-96.

75. Tuttle, O.F. and Bowen, N.L. 1958. Origin of granite in the light of experimental studies in the system $NaAlSi3O8-KAlSi_3O_8-SiO_2-H_2O$. Geol. Soc. Am. Mem., 74.,153pp.

76. Velde, B. and Kushiro, I. 1978. Structure of sodium aluminosilicate melts quenched at high pressure; infrared and aluminum K radiation data. Earth Planet. Sci. Lett., 40, 137- 140.

77. Vickery, N.M., Ashley, P.M. and Fanning, C.M. 1997. Dumboy-Gragin Granite, northeastern New South Wales: age and compositional affinities. In P.M. Ashley and P.G. Flood (eds), Tectonics and Metallogenesis of the New England Orogen, Geological Society of Australia Special Publication, 19, 266-271.

78. Walsh, J. 1991. Two distinctive granitoids from the Copeton region: a mineralogical, geochemical and mineralization study. Unpublished BSc (Hons) thesis, University of New England, 200pp.

79. Watson, E.B. and Harrison, T.M. 1983. Temperature and compositional effects in a variety of crustal magma type. Earth Planet. Sci. Lett., 64, 295-304.

80. Webster, J.D. and Holloway, J.R. 1990. Partitioning of F and Cl between magmatic hydrothermal fluid and highly evolved granitic magmas. In H.J. Stein and J.L. Hannah (eds), Ore-bearing granite systems; petrogenesis and mineralizing processes. Geol. Soc. Am., Special Paper, 246, 21-34.

81. Whalen, J.B. and Currie, K.L. 1990. The Topsails igneous suite, western Newfoundland, fractionation and magma mixing in an "orogenic" A-type granite suite. In H.J. Stein and J.L. Hannah (eds), Ore-bearing granite systems; petrogenesis and mineralizing processes. Geol. Soc. Am., Special Paper, 246, 287-299.

82. Whalen, J.B., Currie, K.L. and Chappell, B.W. 1987. A-type granites; geochemical characteristics, discrimination, and petrogenesis. Contrib. Mineral. Petrol., 95, 407- 419.

83. Whalen, J.B. 1988. Granitic rocks of New Brunswick and Gaspe, Quebec: a transect across the southern Canadian Appalachians. Geol. Assoc. Can. Prog. Abst., 13, A133.

84. White, W.H., Bookstrom, A.A., Kamilli, R.J., Ganster, M.W., Smith, R.P., Ranta, D.E. and Steininger, R.C. 1981. Character and origin of Climax-

type molybdenum deposits. Econ. Geol., 75th Anniversary Volume, 270-316.

85. Winkler, H.G.F. 1974. Petrogenesis of Metamorphic Rocks. Berlin, Springer Verlag. Wyllie, P.J. 1979. Magmas and volatile components. Am. Mineral., 64, 469-500.

86. Zen, E. 1986. Aluminum enrichment in silicate melts by fractional crystallization, some mineralogic and petrographic constraints. Jour. Petrology, 27, 1095-1117.

Chapter 4

INTEGRATION OF SATELLITE IMAGERY, GEOLOGY AND GEOPHYSICAL DATA

Andreas Laake

WesternGeco Cairo Egypt

INTRODUCTION

Satellite imagery is a large scale surface geological mapping too, which offers the unique opportunity to investigate the geological characteristics of remote areas of the earth surface without the need to access the area on the ground. The resolution of the technique is limited by the resolution of the imagery. This chapter explains how geological information can be extracted from satellite imagery and how this information can be merged with geological and geophysical data to build consistent geological models for the surface and subsurface. On the one hand, the interpretation of satellite imagery can generate start models prior to the start of geophysical surveys. On the other hand, geological and geophysical data can calibrate models derived from satellite imagery.

METHODOLOGY

When studying the shape of the earth surface in connection with the rock layers and their deformation by tectonic forces, we often notice a correlation between shapes and structures at the surface and in the subsurface (Short and Blair 1986). This opens the opportunity to map the characteristics of the surface and infer characteristics of the subsurface. We can describe the surface by its shape and by its structure. The surface shape depends on topography, terrain gradient and surface lithology, which we call geomorphological properties. The surface structure is determined by lithological boundaries and fracture zones outcropping at the surface, which we call litho-structural properties. Fracture zones can also be inferred from the characteristics of recent or paleo-drainage (Short and Blair 1986). Geomorphology and litho-structure allow building a static geological model. If information is available about the elevation change with time, then the statics model can be expanded into a dynamic geological model. Figure 1 gives an over of the building blocks for geological model building. We illustrate the methodology at a simple layer-cake geological model which is deformed by a vertical fault (fig. 2). The surface is formed by a soft sandstone layer resting on a hard limestone layer. These two layers form the near-surface. We call the layers between the bottom of the near-surface and the top of the basement subsurface. Prior to the deformation by the fault only the soft sandstone was visible at the surface. The fault has lifted part of the layer package and exposed the near-surface sandstone and limestone layers at the fault plane and made them accessible for mapping by satellites.

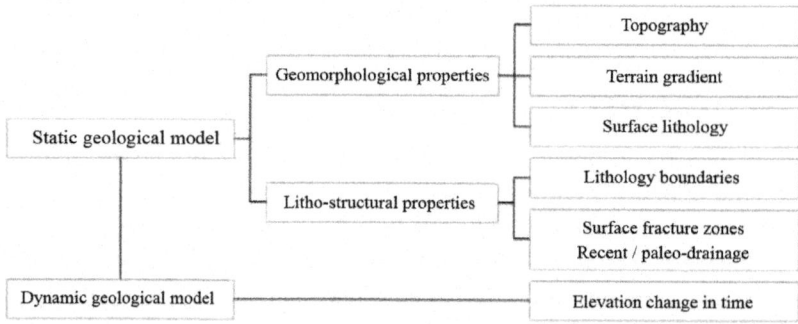

Figure. 1: Building blocks of near-surface and subsurface geological models

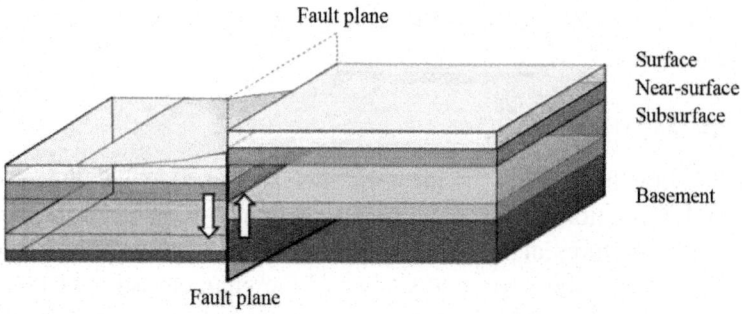

Figure. 2: Correlation of surface shape and subsurface geology

SATELLITE IMAGERY

Earth observation satellites map the physical properties of the earth surface and nearsurface. In the context of geological mapping we distinguish two types of electro-magnetic methods (see figure 3) :

- passive optical methods: use the sunlight as the source and measure the reflectance of the earth surface in the visible and infrared spectral bands. We used Landsat 7 ETM+ and the ASTER instrument from the Terra satellite.

- active microwave radar methods: use a microwave source onboard of the satellite and measure the back-scatter from the earth. We used Radarsat-1 and the radar sensor from the Shuttle Radar Tomographic Mission (SRTM). Details about the acquisition and processing are provided among others by the USGS (2011) and Short (2010). For an introduction into the interpretation of satellite imagery see Sabins (1996). The visible imagery (VIS) covers the colors blue, green and red and provides information about water features, infrastructure and landuse as well as limited information about selected rock types. Infrared imagery is split into three classes : very near infrared (VNIR), which detects specifically vegetation; short wave infrared (SWIR) which is the best option for the discrimination of sedimentary rocks; and finally thermal infrared (TIR). The thermal infrared radiation from the earth surface represents the property of the surface material to convert the solar spectrum into heat radiation. We distinguish between a warm response from dark materials such as non-sedimentary rocks and cool response from ground moisture or voids, where evaporation absorbs energy. In general optical imagery does not penetrate the earth surface.

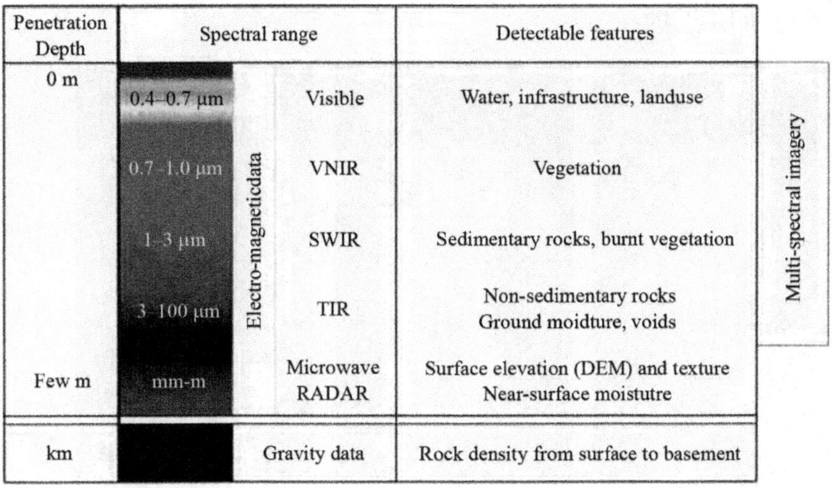

Penetration Depth	Spectral range		Detectable features	
0 m	0.4–0.7 µm	Visible	Water, infrastructure, landuse	
	0.7–1.0 µm	VNIR	Vegetation	Multi-spectral imagery
	1–3 µm	SWIR	Sedimentary rocks, burnt vegetation	
	3–100 µm	TIR	Non-sedimentary rocks / Ground moidture, voids	
Few m	mm-m	Microwave RADAR	Surface elevation (DEM) and texture / Near-surface moistutre	
km		Gravity data	Rock density from surface to basement	

(Electro-magnetic data — spanning Visible through Microwave RADAR rows)

Figure. 3: Spectral overview of electro-magnetic satellite imageryFig. 3. Spectral overview of electro-magnetic satellite imagery

Microwave radar uses electro-magnetic waves in the mm to m range. At hard surfaces microwaves are almost completely back-scattered. Their travel time can be used to determine the distance between the satellite and the surface, which is used to generate digital elevation models (DEM). For soft, non-conductive surfaces the microwaves penetrate into the subsurface; the back-scattered signal is generated from volume scatter in the subsurface. A special application of microwave radar is the estimation of gravity anomalies. Sandwell and Smith (2009) have studied anomalies in the orbits of radar satellites and inferred Bouguer gravity anomalies using a geoid model. These gravity anomalies can be interpreted for the thickness of the sedimentary cover above the crystalline basement. The following examples illustrate the information obtainable from the various satellite imagery sets. Figure 4 shows an example of multi-spectral data from ASTER for each range : visible (fig. 4.a), short wave infrared (fig. 4.b) and thermal infrared bands (fig. 4.c) of the island of Bahrain with a colour scale from blue (low reflection intensities) via green and yellow to red (high intensities). The visible green band penetrates water for a few meters (blue shades). It is relatively insensitive to different rock types (red shades). The short wave infrared band does not penetrate water and therefore shows uniform blue color, whereas it allows the discrimination of the different rock types in the oval anticlinal structure covering most of the island (yellow and red shades). The thermal infrared band shows the thermal properties such as warm response from the built-up areas (dark red). Cool response (greenish areas in the coastal areas) is observed for wet coastal salt flats called sabkha.

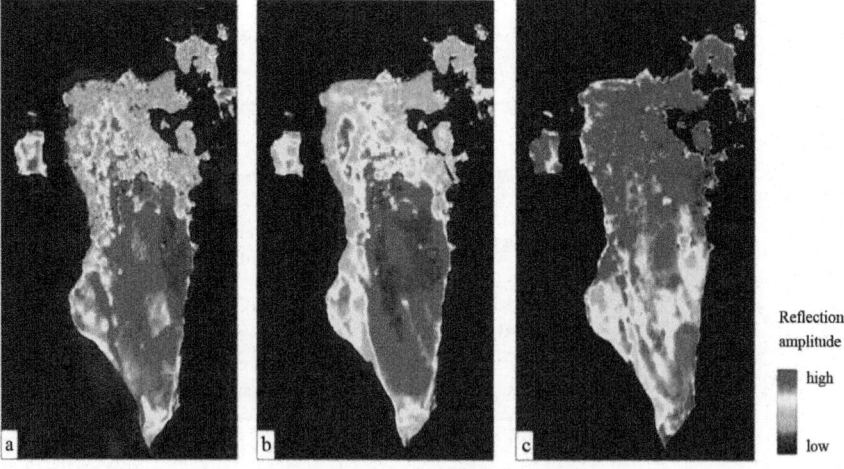

Figure. 4: Examples for optical satellite imagery from ASTER: a – visible blue, b – short wave infrared, c – thermal infrared [after Laake et al. 2006]

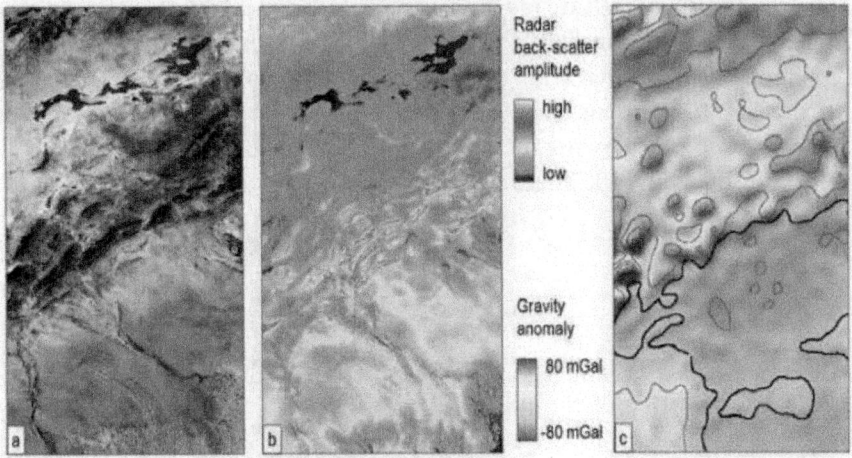

Figure. 5: Examples for optical and radar satellite imagery: a – optical data [Landsat 742 RGB] for comparison, b – radar scatter intensity [Radarsat-1], c – gravity anomaly derived from radar

The examples for microwave radar data show the Atlas area in Algeria. For orientation we have supplied a Landsat 742 RGB image (fig. 5.a), which shows mountain ranges (dark purple), gravel planes (yellow-gray), salt lakes (blue) and vegetation (green). The radar back-scatter intensity from Radarsat-1 (fig. 5.b) is displayed from blue (total absorption in conductive salt brine) via intermediate volume back-scatter (green and yellow) to strong surface back-scatter from hard rock (brown and white). Figure 5.c shows the gravity anomaly inferred from radar satellite imagery. Negative anomalies correspond to low density rocks as from thick sedimentary cover at the foot of the Atlas mountains whereas positive anomalies correlate with dense metamorphic and basement rocks in the Atlas ranges.

EXTRACTION OF INFORMATION FROM SATELLITE IMAGERY

Satellite imagery is provided as sets of digital images, one image for each spectral band. Each image displays the measured values as intensity. The information contained in the satellite imagery can be extracted using either single bands or combinations of bands. Single bands are usually displayed as maps coding measured amplitude as colour. Figure 6 shows radar data from the very dry desert in south-west Egypt : radar data from the shuttle radar tomographic mission (SRTM, Jarvis et al., 2008) interpreted for a DEM using topographic colour coding, i.e. green for low elevations, yellow to brown for high ground (fig. 6.a). Raw radar data from Radarsat-1 reveal high amplitudes from the backscatter at hard sandstone (white and brown colors in fig. 6.b). Low radar back-scatter (fig. 6.c) correspond to volume back-scatter from microwaves penetrating sand sheets. They reveal buried paleo-rivers (blue colors in fig. 6.c) following the interpretation by El-Baz and Robinson (1997) and Robinson et al. (2007).

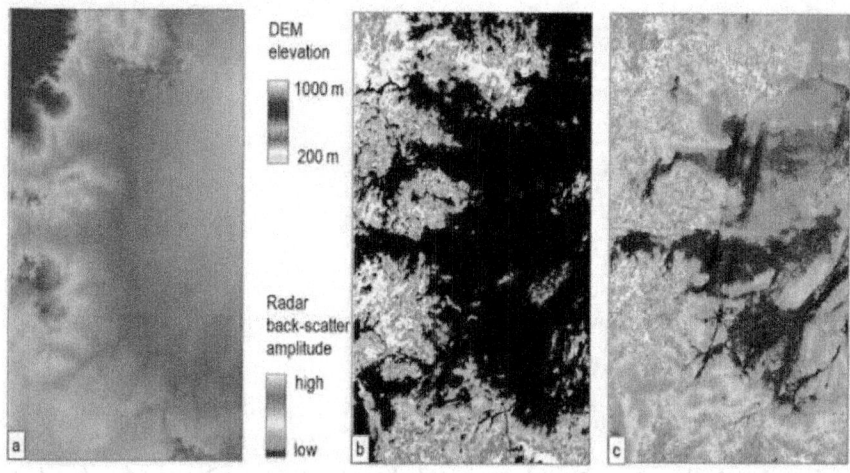

Figure. 6: Single band data examples a – radar based digital elevation model [SRTM DEM], b – high intensity radar data [Radarsat-1], c – low intensity radar data [Radarsat-1]

Dual band images use combinations of two bands from Landsat for the north-western desert in Egypt (fig. 7) thereby enhancing subtle features in the data that would not be imaged by a single band alone. The ratio of the infrared bands 7 and 4 (fig. 7.a) highlights for example clay minerals which fill karst holes (red and cyan) in an otherwise homogeneous limestone plateau (dark blue). The band difference of the very near infrared band 4 and the green band 2 for the same area highlights difference in lithology between the pure limestone (yellow to red) and the more sandy cover (blue tones) towards the top of the image (fig. 7.b). Multi-band images use three or more bands combined in continuous colour or red-greenblue (RGB) images. RGB images provide significantly more shades than single or dual band images: for 8 bit imagery an arithmetic combination of 3 bands provides 256 shades whereas an RGB image offers 16.8 million colors (Guo et al., 2008).

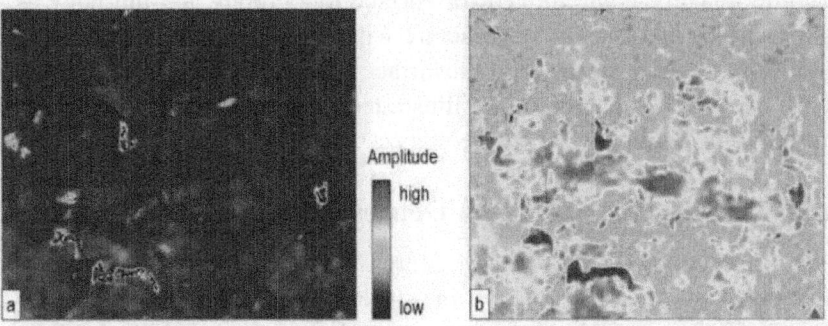

Figure. 7: Dual band data examples: a – band ratio [Landsat 7/4], b – band difference [Landsat 4-2] for the same area

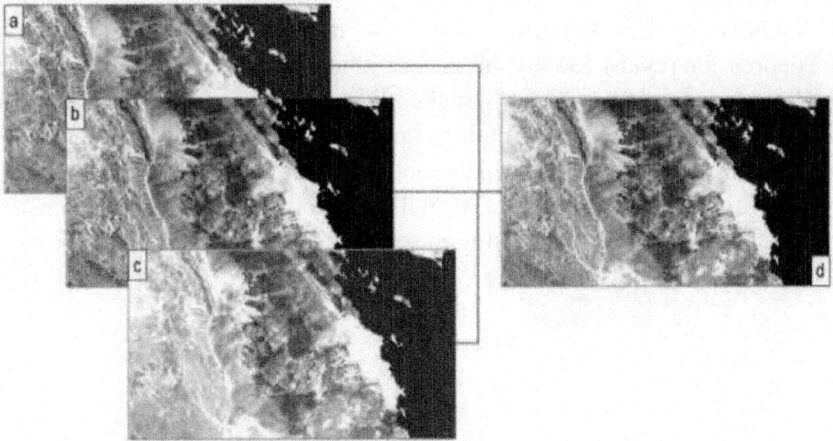

Figure. 8: Multi-band data example : merge of single band data (a to c) into RGB image (d) [Landsat 742 RGB]

To illustrate this we study an example from the southern Red Sea Mountains and the adjacent coastal area in south-east Egypt (fig. 8). The raw input bands (fig. 8 a-c) show only marginally different signatures for the very different rock types, whereas the multi-band RGB image (fig. 8.d)distinguishes clearly between the basement rocks (dark), the Mesozoic clastic sedimentary rocks (yellow tones), coastal carbonates (white) and the sea (black).

INTEGRATION OF SATELLITE IMAGERY AND GEOLOGY

In this section we describe techniques to extract geological information from satellite imagery and how to integrate this information with geological data.

Satellite imagery is interpreted for surface topography and lithology as well as for surface and subsurface structure with the goal of generating geological models for the near-surface and subsurface (Laake and Insley, 2007, and Laake et al., 2008). The techniques are illustrated through a series of case histories starting with simple layer-cake geology.

Layer-cake Geology (Qattara Depression, Egypt)

The surface north of the Qattara depression in Egypt is dominated by flat layering of hard and soft rocks at the surface (fig. 9). The raw DEM (fig. 9.a) shows a platform (yellow), which is located between rough hills (brown tones) and a sharp escarpment towards the Qattara Depression (green to blue). The terrain classification map (fig. 10.c) allocates the different elevations to three classes. The lithological analysis of the multi-spectral image (fig. 9.b) allows a clear separation into rock types : two types of limestone (blue tones), two types of sandstone (yellow to orange) and evaporites of the sabkha at the bottom of the depression (cyan). Ideally the surface lithology interpretation is validated in the field (see fig. 10) using GPS tracked lithological analysis. The combination of terrain classification and lithology suggests that the plane represents the top of the sandstone formation, which continues also below the limestone layers, which form the higher ground. We will use this geological model to estimate statics for seismic data processing in section 5.

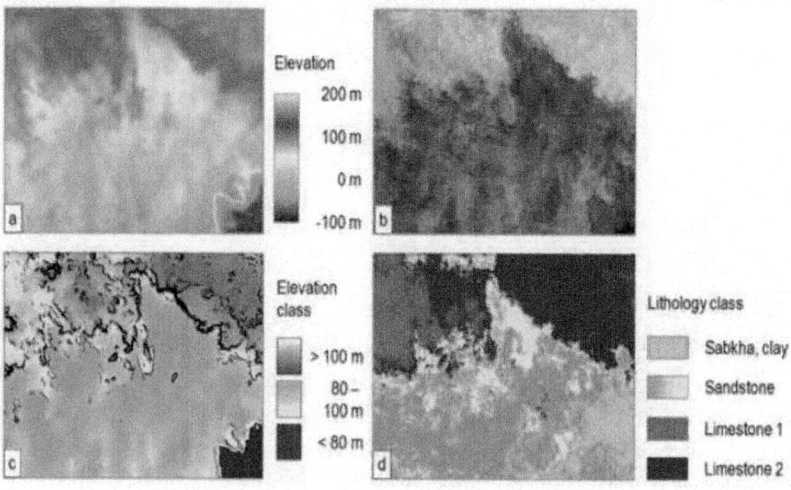

Figure. 9: Layer-cake geology near Qattara Depression, Egypt : a – digital elevation model [SRTM DEM], b – multi-band satellite image [ASTER 631 RGB], c – terrain classification map, d – lithology map [Cutts and Laake 2009]

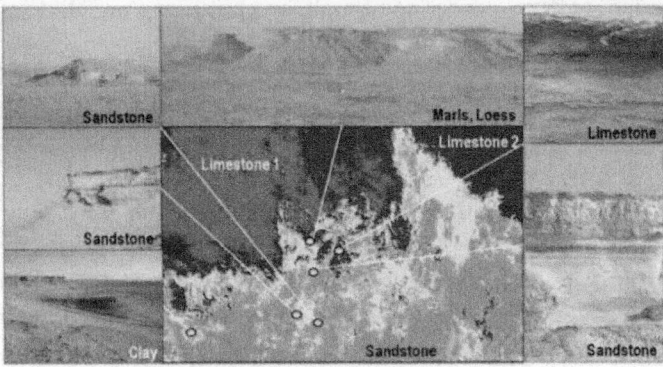

Figure. 10: Validation of lithology map in the field [after Coulson et al. 2009]

Anticline (Awali, Bahrain)

The outline of the island of Bahrain is determined by the topographical and lithological structure of the Awali anticline (fig. 11). We use satellite imagery from the ASTER sensor for the discrimination of clastic and carbonatic rocks (Laake et al., 2006). The continuous colour image from short wave infrared and visible bands (ASTER 631 RGB in fig. 11.a) allows discriminating the coastal farmland in the north (green) from the carbonates of the anticline (purple tints) and coastal sabkha (cyan). The structure of the anticline can be delineated using the difference of visible and short wave infrared data (fig. 11.b), which can be traced even in the built-up area of Manama city in the north. The outer contour of the anticline appears highlighted in the west through the strong signature of the coastal sabkha (dark red). Draping the continuous colour image over the vertically exaggerated DEM generates a strong structural impression, which is useful for structural interpretation (fig. 11.c).

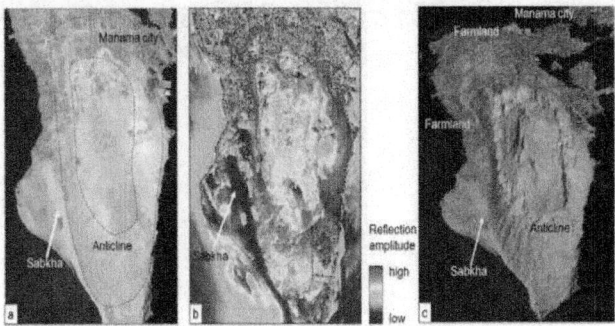

Figure. 11: Anticlinal structure of Awali, Bahrain : a – multi-spectral image [ASTER 631 RGB], geological structure from band difference image [ASTER visible minus

short wave infrared], c – rendering of multi-spectral image on DEM [after Laake et al. 2006]

Mapping of Basins from Gravity and Radar Data (Illizi Basin, Algeria)

The mapping of basins from satellite imagery is the only remote sensing technique which infers deep geological structures. Bouguer gravity anomalies inferred from satellite radar data give an indication of the thickness of the sedimentary cover above the basement. However, this interpretation requires additional information for example from magnetic data to constrain the model. Following the concept of geomorphology, which correlates surface and subsurface geology and litho-structure, surface structural lineaments can be used to infer subsurface structures. In figure 12 we show structural maps from satellite imagery for the Illizi Basin, which is indicated by the black continuous line in each figure. The lithology image (Landsat 742 RGB, fig. mapping 12.a) distinguishes dark paleozoic limestone (blue tones) and Mesozoic clastics (brown tones), which surround recent sand dunes (yellow). The geomorphological map (fig. 12.b) highlights the litho-structural boundaries, which are co-located with topographic ridges. Satellite gravity data (fig. 12.c) reveal a basin (blue). The combination of surface geomorphology and gravity (fig. 12.d) gives an indication of the basin outline.

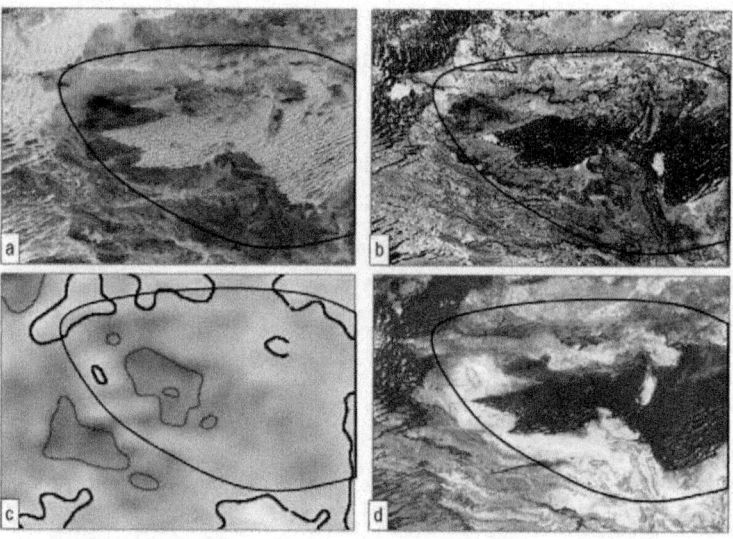

Figure. 12: Basin mapping of Illizi basin, Algeria: a – multi-spectral image (Landsat 742 RGB], b – geomorphological map [SRTM DEM, Landsat 742 RGB], c – gravity anomaly derived from radar, d – overlay of geomorphology on gravity data

Mapping of Fracture Zones Below Sand Cover (Gilf Kebir, Egypt)

When sand dunes cover fracture zones in the near-surface, satellite radar data can map ancient river courses, from which fracture related weak zones in the near-surface can be derived (fig. 13). We start with the outcrops of faults in the hard rock surrounding the sand dune field mapping straight valleys in the hard sand stone. The topography of the study area (fig. 13.a) is composed of the flat Gilf Kebir plateau (brown tones) with gentle slopes (yellow) and a large plane (green to white). This corresponds to a surface lithology (fig. 13.b) of hard sandstones (greenish-brown) and basement rocks (dark brown) as well as belts of sand dunes (white), which are locally discoloured by hematitic iron (purple). The shape of the escarpment and the valleys intersecting the Gilf Kebir plateau reveal fairly straight fault lines (black lines). Below sand neither topographical nor optical satellite imagery can reveal buried fracture zones. Therefore we use satellite radar which can penetrate dry sand for up to 20 m to map the clay contained in buried paleo-river beds. The radar data (fig. 13.c) do not only map the fracture zones in the sandstone of the plateau (brown lineaments), it also shows W-E and SW-NE trends in the paleo-river courses (low radar intensities in blue) which continue from the lineaments in the rock outcrops. The overlay of the paleo-river courses from radar data on the geomorphology delineates the outcropping fracture zones across all terrains (fig. 13.d).

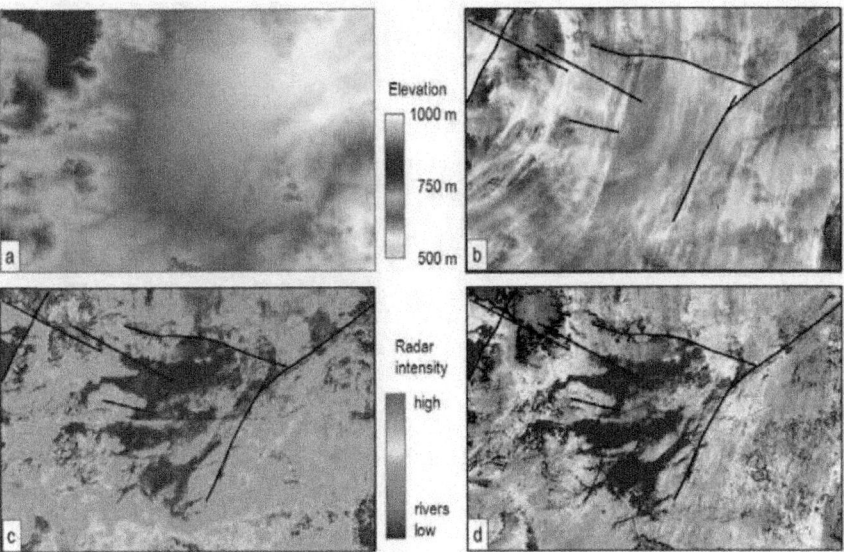

Figure. 13: Structural delineation from radar data close to Gilf Kebir, Egypt: a – DEM, b – multispectral image [Landsat 742 RGB], c – low intensity radar data [Radarsat1], d – merge of radar data with geomorphology. Faults indicated by lines.

Mapping of Glacial Moraine Structures (Pechora Basin, Russia)

The last case study concerns the mapping of post-glacial structures in the Pechora basin in northern Russia using vegetation and water features as indicators; the lithology is not exposed in the study area. The structures studied comprise different types of moraines as well as drainage features in front of the glaciers (Laake 2009, Astakhov et al. 1999). We distinguish terminal and lateral moraines, which are composed of gravel and glacial till, from ground moraines, which are characterized by undulating terrain. Moraine ridges are often drier than the surrounding terrain because the elevated gravels cannot support a shallow water table. In contrast to this, ground moraines deposit more glacial till, which provides the basis for a shallow water table with very wet terrain and lakes Significant differences in the ground moisture attract different plant species, which can be distinguished by their different response in the very near infrared and visible bands of satellite imagery. The topography allows delineating of the lateral and terminal moraines as long as they are still elevated above the surrounding plane (fig. 14.a).

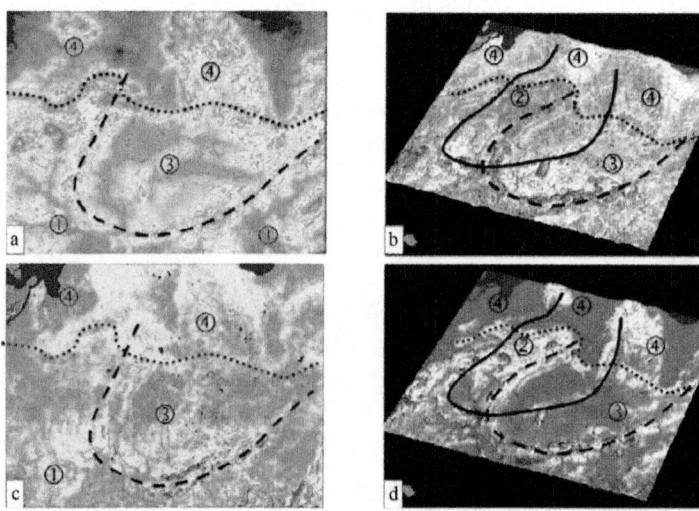

Figure. 14: Identification of glacial structures in Pechora basin, Russia: a – DEM [SRTM DEM], b – rendering of geomorphological map in 3D, c – landuse classification from Landsat, d – interpretation of geomorphology for glaciers and moraines [after Laake 2009]

Rendering the geomorphological map on the DEM in 3D improves the detection of these moraines. The location of lakes, which is indicative of ground moraines, is obtained from the landuse analysis of short wave infrared and visible green data. (fig. 14.c). The combination of all maps yields a clear delineation of the glacial moraines particularly when rendered in 3D (fig. 14.d). In this case four glacial stages can be interpreted (numbers in fig. 14) : an initial stage where the entire area was covered by a thick ice shield (1), which deposited ground moraines over the entire study area. The second (2) and the third (3) phase comprise two distinguishable glaciers, where the glacier from phase two created the partly broken moraine wall in the west. The glacier in phase three covered the centre of the study area thereby partly levelling the lateral moraines from phase two. After melting, this glacier left the huge Lake Komi between the lateral and terminal moraines, which forms an extensive swamp today. The final glacier advance (4) covered only the northern third of the study area, leaving an irregular line of terminal moraines and extensive planes of ground moraines behind.

INTEGRATION OF SATELLITE IMAGERY AND GEOPHYSICS

The geological information extracted from satellite imagery can be used to build geological models from basin scale (several 105 sqkm) to survey scale (few 103 sqkm) well before any geophysical data are acquired on the ground. These models can be used for the estimation of logistic risks for personnel and vehicles as well as for the estimation of the quality of the acquired geophysical data (Laake and Insley (2004a and b, Laake and Cutts 2007, Coulson et al. 2009). In turn, geophysical data can be employed to calibrate geophysical characteristics of near-surface layers inferred from satellite imagery. We distinguish the following types of geophysical surveys :

- Frontier exploration aiming at identification of new basins in large unexplored areas. Satellite imagery can assist defining the outline of potential basins and the definition of scouting surveys.

- Structural imaging focuses on potential structures once the outline and character of a basin has been identified. Satellite imagery can support the design and logistic planning of geophysical surveys on the ground and can provide estimates for the quality of the acquired data before the data are acquired.

- Reservoir characterization targets the most comprehensive study of the subsurface geological structure and fluids for individual reservoirs and therefore requires the best data quality. Satellite imagery can provide detailed models of the surface and nearsurface which provide input to data quality estimation before and during acquisition. For data processing satellite imagery can supply input to processes that correct for noise related to near-surface properties.

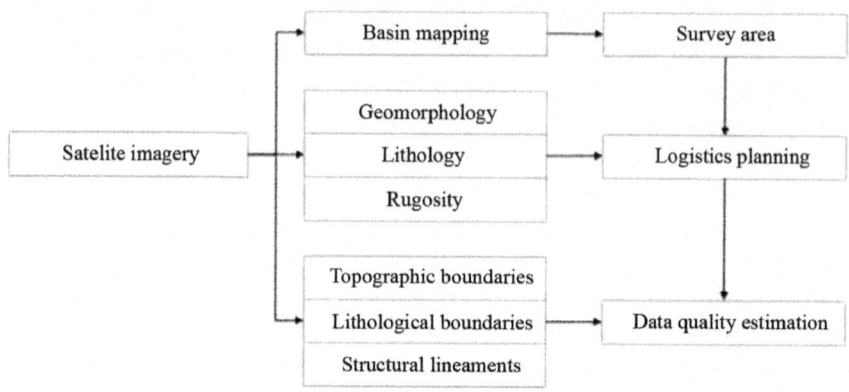

Figure. 15: Workflow for application of satellite imagery to geophysical methods

An outline of the information satellite imagery can provide for geophysical methods and the parameters studied is shown in figure 15. This section focuses on the aspects of survey design and data quality estimation. The design of geophysical surveys requires an understanding of the logistics and data quality aspects of the target area to correctly estimate the effort required to provide the desired quality of the final subsurface geological product. Satellite imagery can provide the information about the surface and near-surface before the start of the data acquisition (Cutts and Laake, 2009a and b). The method uses satellite imagery to generate geomorphological and litho-structural models of the surface and near-surface as described in the previous section to derive logistic planning and data quality estimation maps. Figure 16 shows the result of the method for a survey in the Western Desert of Egypt. DEM and multi-spectral data provide topography (fig. 16.a) and lithology class maps (fig. 16.b). The planning of the logistics requires information about limitations for access and maneuver for personnel and vehicles. The terrain classification provides the locations and steepness of escarpments as well as the surface roughness related to hard limestone. The combination of these terrain attributes defines the logistics risk (fig. 16.c). A major obstacle for data quality for this survey is the scattering of seismic waves at the numerous escarpments resulting from the different weathering of limestone and sandstone. Fig. 16.d shows an estimate map for the scatter risk associated with escarpments.

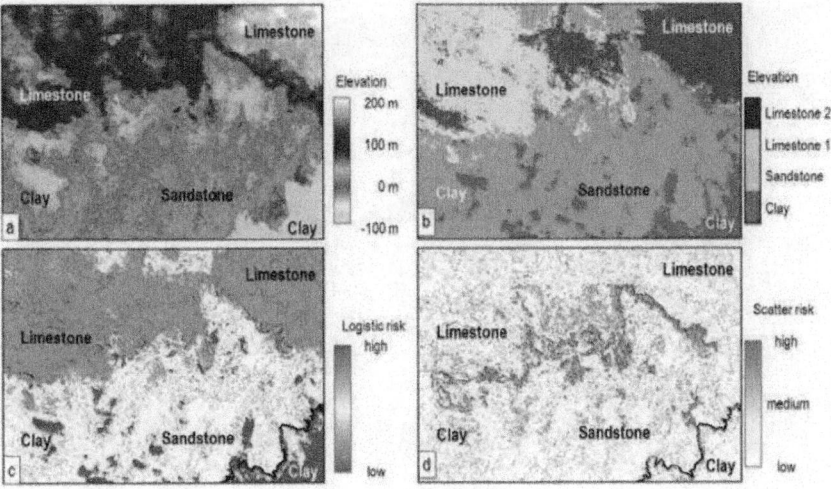

Figure. 16: Survey design based on satellite imagery (Qattara Depression, Egypt) : a – terrain class map based on DEM, b – lithology map, c – logistic risk estimate, d – data quality risk estimate [after Cutts and Laake 2008]

Logistic Planning in Volcanic Terrain

Vegetation may obstruct the assessment of the logistic risk (Laake 2005b). The basalt plateau close to the Payun Volcano in the Andean foothills of Argentina exposes the risk of big basalt blocks obstructing the access to large areas whereas in other parts of the survey basalt grit exposes no maneuver risk at all (fig. 17.a to c). The analysis of the multi-spectral imagery from ASTER for big basalt blocks is challenging because in both cases – big basalt blocks and bush as well as basalt grit and grass – the ratio of basalt and vegetation is very similar. The resolution of the satellite image of 15 m does not allow the direct mapping of the basalt blocks (fig. 17.d). However, the thermal properties of big basalt blocks are sufficiently different from basalt grit, which allows the discrimination of the basalt block size using the thermal infrared bands from ASTER imagery (fig. 17.e), which is supported by the intermediate back scatter intensity recorded by radar data from Radarsat-1 (fig. 17.f).

Planning of Safe Operation in Sand Dunes

In high sand dunes, both logistics for wheeled vehicles and high absorption for seismic waves may impact the survey design severely (Laake and Insley, 2004a). The analysis of the topography can assist in outlining the dunes and characterizing their shape (fig. 18).

Figure. 17: Satellite imagery based logistic planning in Andean foothills, Argentina : a – basalt gravel plane, b – rendering of ASTER image, c – basalt blocks, d – landuse image [(ASTER 631 RGB], e - basalt block map, f – surface rugosity from Radarsat-1 [after Laake 2005b]

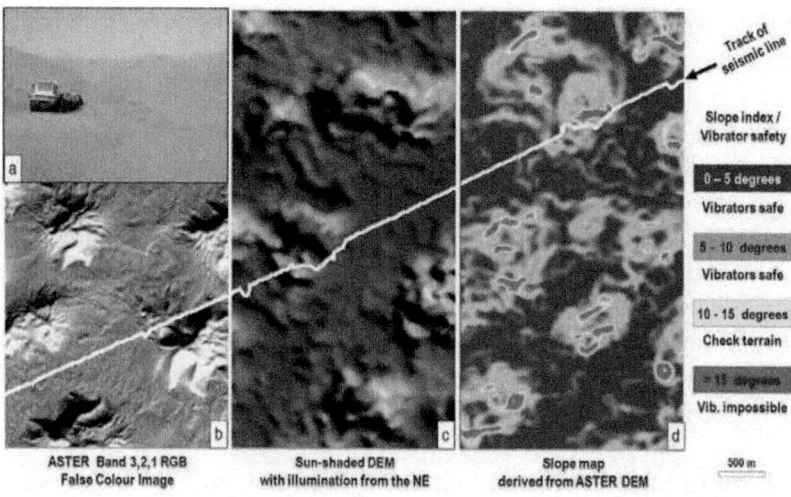

Figure. 18: Planning of safe vibroseis operation in high sand dunes, Berkine basin, Algeria: a – vibrator array in high dunes, b – multi-spectral satellite image [ASTER 321 RGB], c – sunshaded DEM, d – Slope map derived from DEM [ASTER DEM] [Laake and Insley, 2004a]

Digital elevation models derived from radar or stereo optical data (ASTER in our case) provide the slope index, which can be represented as an index for safe operation. In seismic operation surface gradients above 15 degrees are considered unsafe for operation, whereas in areas with 10 to 15 degrees slope an inspection of the terrain would be advised. Figure 18 shows a composite image of seismic line crossing a dune field. The visible band ASTER image (fig. 18.b) gives a photographic impression of the surface, whereas the DEM (fig. 18.c) provides an impression of the elevations. The slope risk map (fig. 18.d) is obtained from the gradient of the DEM.

Prediction of Accessibility and Data Quality in Sabkha

In arid coastal areas often sabkha called salt flats pose a severe threat to personnel and equipment (Cutts and Laake 2009b).

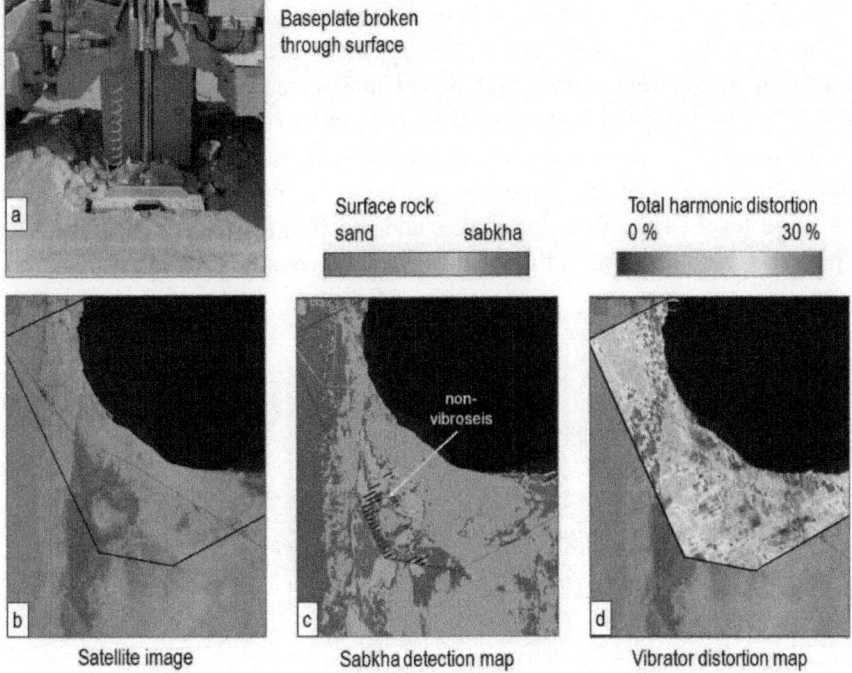

Figure. 19: Prediction of accessibility and data quality of vibroseis operations in sabkha, Sabkha Matti, UAE: a – vibrator baseplate broken through surface crust, b – satellite image [Landsat 321 RGB], c - sabkha detection map, d – map of vibrator total harmonic distortion [after Cutts and Laake 2008]

The salt flats may be subject to seasonal inundation, which may change their accessibility significantly. Analysis of the multi-spectral satellite Landsat imagery can allow the detection of halite minerals found in the surface crusts of sabkha. Thermal infrared imagery provides information about the different thermal properties of wet and dry sabkha. The visible band Landsat image in figure 19.b gives an indication of the location of the wet sabkha. The risk that the sabkha surface would not support heavy vehicles (see fig. 19.a) is directly correlated with presence of sabkha and can be derived from a sabkha detection map (fig. 19.c). In addition to its impact on the logistics sabkha also affects the vibrator data quality. High distortion of the vibrator signal, an undesirable data property, is directly related to the presence of sabkha (fig. 19.d).

Estimation of the Vibrator Baseplate-To-Ground Coupling

Hard rock terrain affects the coupling of the vibrator baseplate to the ground (Laake and Tewekesbury 2005). Good ground coupling is achieved when the entire baseplate is in contact with the ground for example on gravel (fig. 20.c). If the baseplate rests on a big boulder, the coupling surface from the baseplate to the ground may be reduced to a very small area resulting in substantial distortion of the transmitted signal and in severe damage to the baseplate (fig. 20.a). In this case the so-called point loading risk is correlated with hard limestone. The limestone prediction map draped over the DEM is shown in figure 20.b. But even for gravel the ground coupling may be compromised if the force level of the vibrator shaker and the frequency exceed the levels at which the ground supports the weight of the vibrator.

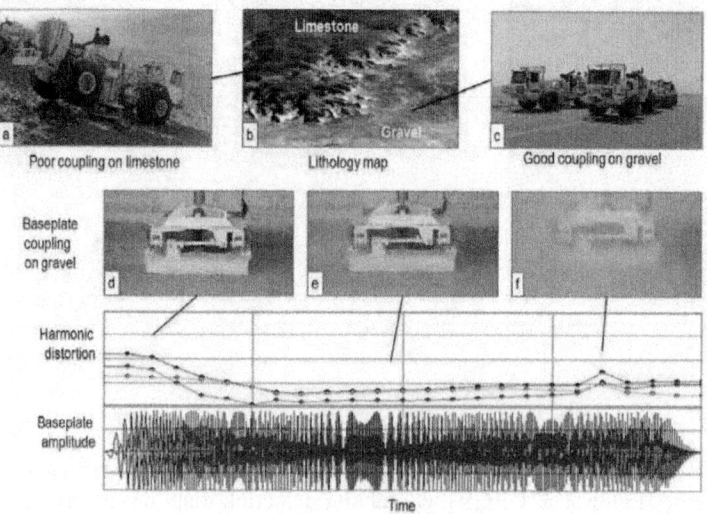

Figure. 20: Prediction of accessibility and data quality for vibroseis operations in hard rocky terrain, Tademait, Algeria: a, c – operations photos, b – virtual 3D lithology map,

d - f – photos from baseplate coupling, bottom – baseplate distortion and amplitude time signal [after Laake and Tewkesbury 2005]

The time signal of the baseplate amplitude (fig. 20 bottom) shows the increasing frequency of the shaker force with time along with measured signal distortion. At three moments of the sweep photos of the shaker were taken. As long as the coupling of the baseplate to the ground is perfect the signal distortion is low (fig. 20.d and e). When the ground surface breaks dust is whirled up by the baseplate motion and the distortion goes up (fig. 20.f). Usually frequencies higher than the frequency at which the baseplate breaks through the surface show high phase distortion and deteriorate the data quality. The high distortion for very low frequencies (corresponding to fig. 20.d) is associated with motions of the entire vibrator and are not considered here.

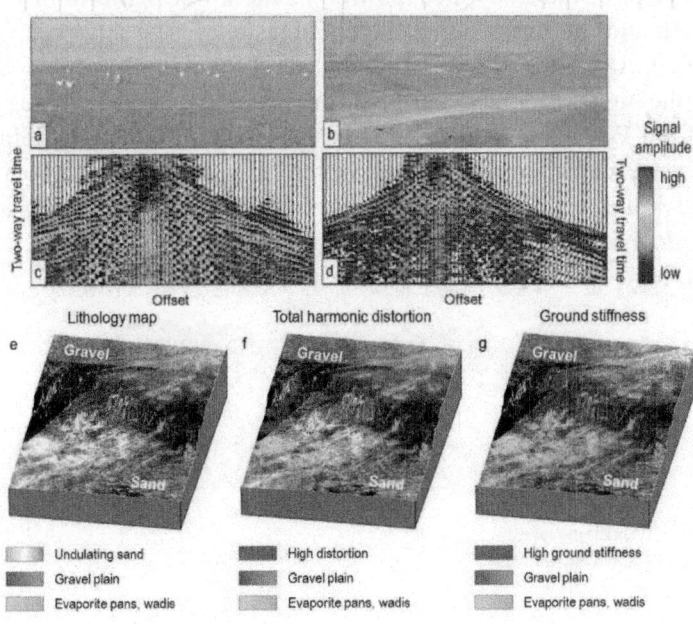

Figure. 21: Correlation of source and receiver data quality with terrain, Kuwait: a – gravel plain, b – undulating sandy surface, c, d – shot record displays, e – surface lithology, f – total harmonic distortion, g – ground stiffness [after Laake 2005a]

Soft terrain may have an impact on both the source coupling as well as on the receiver signal output (Laake 2005a). Due to higher absorption in soft sand, the signal level on sandy surface (fig. 21.b and d) is generally lower than on hard gravel plain (fig. 21.a and c). Satellite imagery can discriminate gravel with shallow evaporite pans on the higher ground from sandy terrain on the lower ground (fig. 21.e). The quality of the source signal also correlates with the terrain character : signals from the sandy terrain show higher distortion (fig. 21.f) and lower ground stiffness (fig. 21.g) than signals from the gravel plain. Interpretation of satellite imagery can provide estimates for the quality of seismic signals obtained from the terrain characterization.

Improvement of Static Corrections

Near-surface geological models can be generated from the interpretation of satellite imagery for geomorphology and lithology (Laake et al. 2008). Using standardized seismic velocities the geological model is converted into an elastic model, from which estimates for statics corrections can be computed (Laake and Zaghloul 2009). The analysis of the DEM delivers the topographic classification into three layers (see fig. 22): the platform (sandstone), which occupies most of the study area, as well as the lower platform of the Qattara Depression and the rough higher layer (limestone). The interpretation of the shortwave infrared data reveals that the platform is composed of sandstone, whereas the higher ground consists of limestone (fig. 22 top). When choosing standardized P-wave velocities of 2200 m/s for sandstone and 3300 m/s for limestone we can compute estimated statics values (fig. 22 bottom). The estimated values are compared against refraction static corrections from a 3D seismic survey (fig. 23).

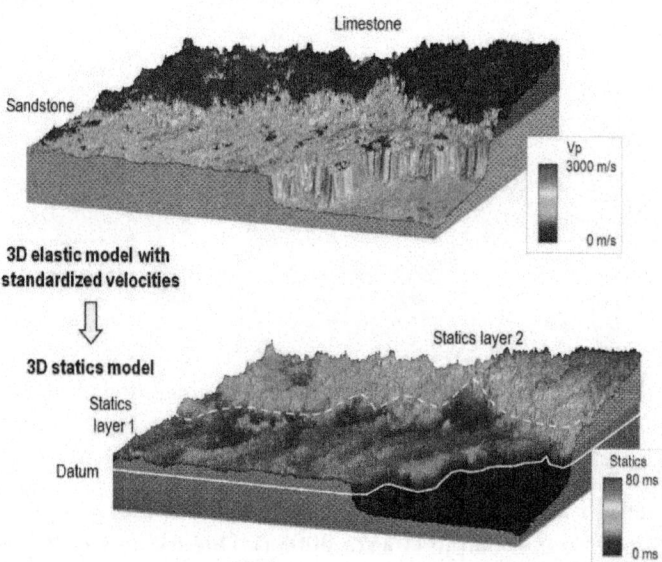

Figure. 22: Estimation of statics corrections from shallow geological model, near Qattara depression, Egypt : top – near-surface geological model from satellite imagery, bottom – 3D statics model [after Laake and Zaghloul 2009]

The static corrections for model 1 (fig. 23.a) use the geological model generated from satellite imagery. The statics contain the lithological details of the sandstone plateau and retain the sharp velocity contrast at the edge of the rough limestone plateau (see fig. 23.c for comparison with the lithology). The statics estimates for model 2 are based on the first break picks from a 3D seismic survey, which are remarkably smoother than the statics from model 1 (fig. 23.d). First break picking involves velocity smoothing over offsets up to 1500 m, which results in a spatial low pass filter. This may result in attenuation of sharp structural lineaments such as fracture zones outcropping at the surface. The difference between the two models (fig. 23.d) reveals another effect of the spatial smoothing: local velocity anomalies may not be corrected properly and might remain in the data as an artefact, which may lead to the generation of structural artefacts in the deeper seismic data.

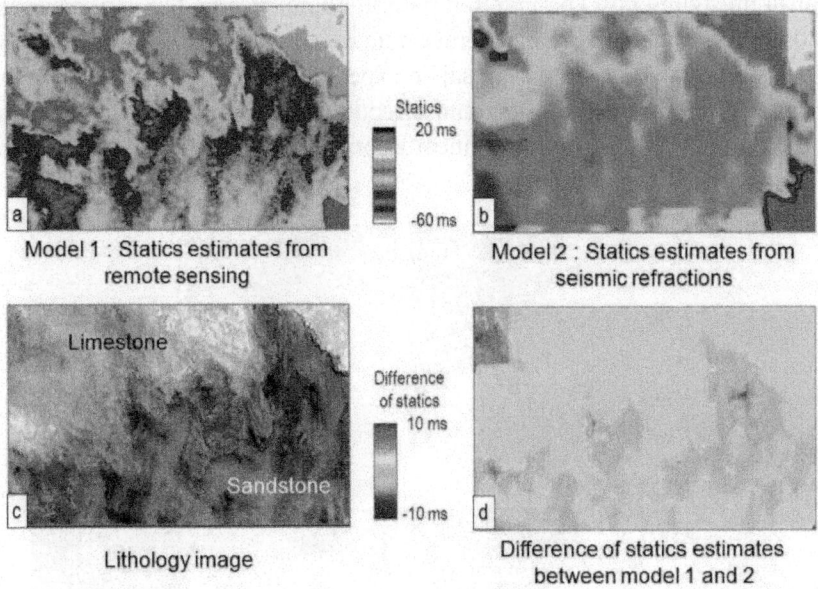

Model 1 : Statics estimates from remote sensing

Model 2 : Statics estimates from seismic refractions

Lithology image

Difference of statics estimates between model 1 and 2

Figure. 23: Comparison of statics from near-surface geological model and from seismic refractions, Qattara depression, Egypt : a – statics corrections from satellite imagery, b – statics from seismic refractions, c – lithology image, d – difference between statics from both models [after Laake and Zaghloul 2009]

Consistent Deep Geological Models from Data Integration

In the final section we will show the benefits of merging interpreted satellite imagery with geological and geophysical data on the western shore of the Gulf of Suez in Egypt by demonstrating the correlation of the geomorphology with the subsurface litho-structure . The processing of multi-spectral imagery from Landsat for lithology and drainage reveals fault structures which are masked by recent wadi deposits (Laake 2010, Laake et al. 2011). The approach follows the idea that tectonic movements deviate the courses of wadis in a characteristic way.

Mapping of Fault Outcrops at Surface from Satellite Imagery

The geological setting of the study area is determined by rift faulting in NW – SE direction and transform faults in the perpendicular direction (Darwish and El-Araby 1993, Alsharhan and Salah 1993). The natural colour Landsat 321 RGB image (fig. 24.a) shows the almost featureless beige gravel plain between the Red Sea Mountains in the west and the Gulf of Suez in the East. The only visible features are SW – NE running wadis delineated through their light sediments. The inverted natural colour image directs the eye to the wadi courses, which reveal several anomalously straight sections (fig. 24.b). When we use all data from Landsat and spectrally enhance the resulting image boundaries along the main tectonic directions are imaged (fig. 24.c): parallel to the coast NW – SE trending linear anomalies point to outcropping parallel fracture zones (dashed lines).

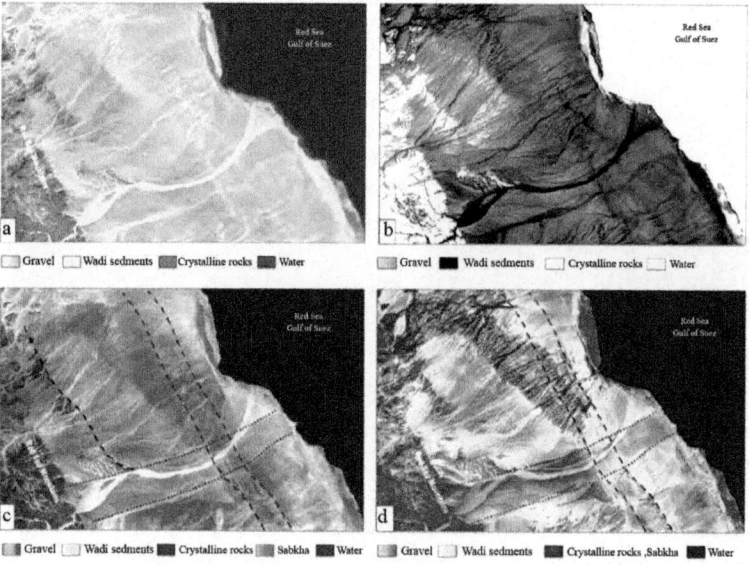

Figure. 24: Surface and near-surface litho-structural mapping from Landsat : a – natural colour image [Landsat 321 RGB], b – inverted natural colour image, c – high-discrimination lithology image, d – drainage map [Landsat bands 6 – 8] [after Laake et al. 2011]

The dominating wadi in the southern part of the study area is confined between two straight SW – NE trending lineaments which point to faults running perpendicular to the rift orientation (dotted lines). This map also reveals differences in the mineral composition of the bedrock of the Red Sea Mountain granites. The erosion fans from these two granites may actually be used as tracers to highlight the structural boundaries on the gravel plain. The lineaments are also highlighted in the drainage map, which results from the ratio of the thermal and pan-chromatic bands (fig. 24.d).

Mapping of Faults in Near-Surface from Shallow Geophysical Data

In the study area, the fault outcrop pattern mapped from satellite imagery is calibrated with surface features and shallow seismic data from a 3D seismic survey. The only tree in the study area has been found at the intersection of two fault outcrops (fig. 25.a) indicated by the arrow in the lithology map (fig. 25.b). The surface structure becomes evident when draping the litho-structural map over the vertically exaggerated DEM (fig. 25.c). Arrows indicate the outcrop of faults at the surface. The correlation with faults in the near-surface is achieved by extracting weak zones of low surface wave velocity from seismic 3D data (fig. 25.d). Surface wave velocity analysis provides an iso-velocity horizon which corresponds to a shallow lithologic horizon. This horizon shows very low velocities along the rift parallel fault outcrop zones mapped by remote sensing. In the direction orthogonal to the rift faults a high-velocity structure below the wadi is revealed but no sharp boundaries.

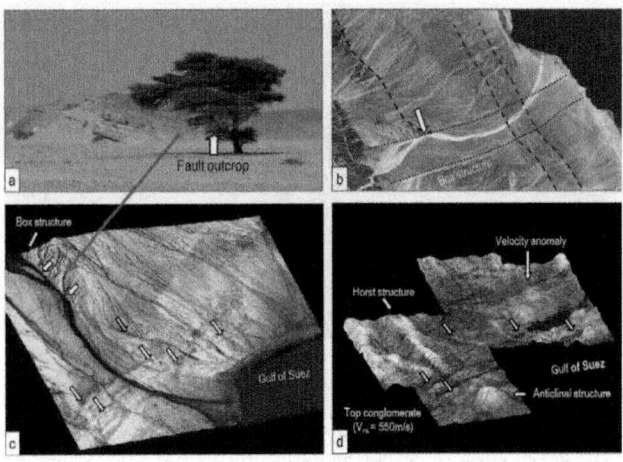

Figure. 25: Calibration of satellite imagery results with near-surface seismic data: a – tree located at a fault intersection, b – high-discrimination lithology map, c – virtual

3D rendering of the litho-structural map – arrows indicate fault outcrops, d – near-surface formation surface obtained from the velocity analysis of seismic surface waves [after Laake et al. 2011]

This may be due to the transform character of these faults which does not provide an impedance contrast that could be detected from the seismic.

Mapping of Faults in the Subsurface Using Surface Tamplates

The characteristics of drainage patterns delineated from satellite imagery may also be used to generate shape templates for the detection and extraction of similar structures in subsurface seismic data (fig. 26). We use the drainage map to define the template for a wadi (fig. 26.a) consisting of the braided stream and the fan delta. In our case the braided stream is confined by the SW – NE trending faults perpendicular to the rift orientation, whereas the fan delta part shows so little elevation change that the fan delta crosses the boundary faults (fig. 26.b).

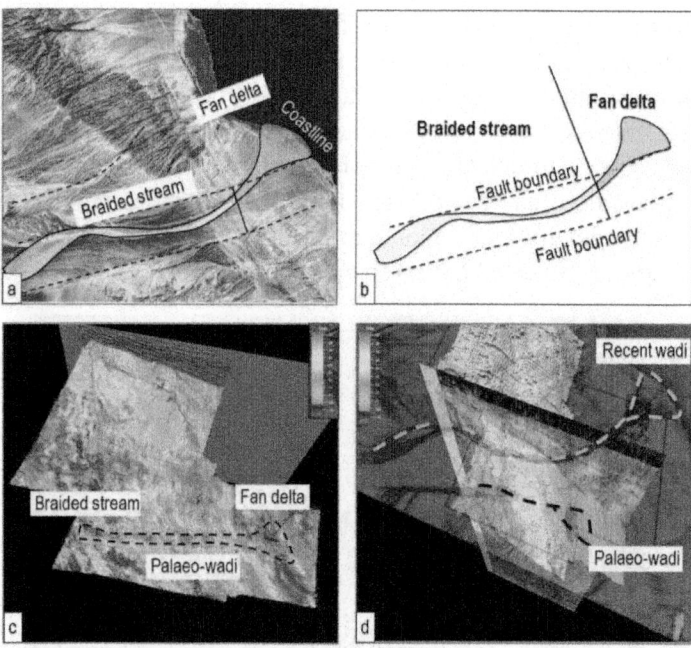

Figure. 26: Satellite based generation of templates for subsurface geobody extraction: a – drainage map, b – extracted wadi pattern, c – detection of similar pattern in shallow seismic data, d – delineation of geobodies in deep seismic data [after Laake et al. 2011]

The shape template is used in the processing of the shallow seismic data to search for the correct seismic attribute suited to map the faults which are hidden in most other seismic attributes. In our case the instantaneous frequency attributes were identified for mapping the SE – NE faults because this attribute highlights locally high spectral amplitudes which were interpreted as resonances within the palao-wadi structures (fig. 26.c). Once the shallow paleo-wadi has been identified the same methodology is applied to the deeper data; a total of four paleo-wadis spanning the entire rift history were mapped (fig. 26.d).

Mapping of Shallow Drilling Risk Related to Faults

Finally the surface and shallow subsurface mapping can be merged into a shallow structural map with indications of shallow drilling risks related to outcropping fault zones (fig. 27). The map basis is a landuse map (Landsat 742 RGB) onto which the near-surface topography is projected. The fault lines delineate areas of shallow drilling risk related to fault induced weak zones, which might lead to collapsing boreholes and/or loss of circulation.

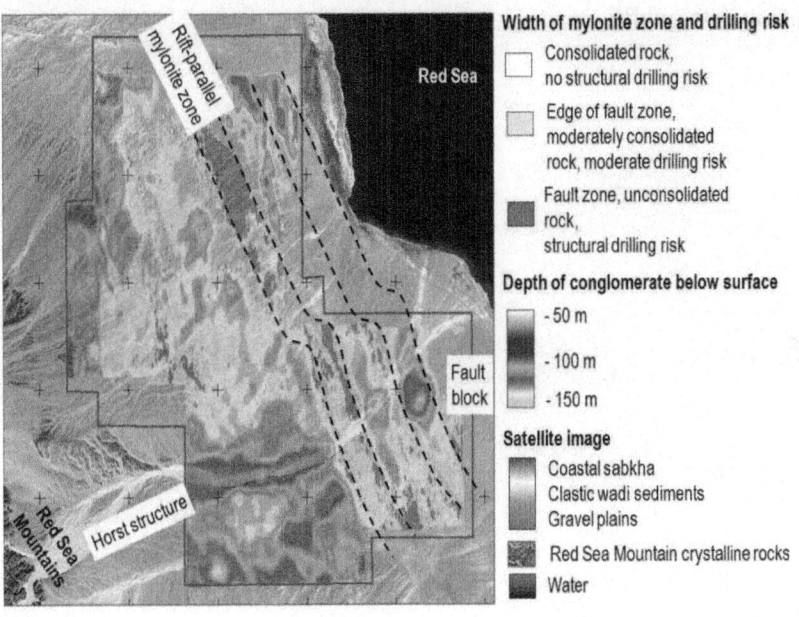

Figure. 27: Shallow structural and drilling risk mapping guided by interpretation of satellite imagery [after Laake et al. 2011]

CONCLUSION

Satellite imagery enables the investigation of the properties of the earth surface in remote areas and over large areas through the mapping of physical properties from satellite based sensors. Optical imagery delivers information about land-use, water features and surface rocks. Microwave radar imagery maps surface roughness when back-scattered at hard surfaces and paleo-drainage when penetrating dry sand cover. Microwave radar distance measurements can be converted into surface topography maps or digital elevation models. The joint interpretation of radar elevation measurements and the geoid shape delivers satellite based gravity anomaly maps, which reflect the thickness of the sedimentary layers above the crystalline basement. The interpretation of satellite imagery can assist in the planning of surface seismic surveys through assessment of logistic and data quality risks early in the planning specifically when exploring in frontier areas. Interpretation of satellite imagery can provide estimates of the source and receiver data quality and static corrections. Consistent geological models can be generated from the interpretation of satellite imagery for geomorphological and lithostructural properties integrated with geological and geophysical data. Fig. 28 and table 1 give an overview of the geological features detectable from satellite imagery and their impact on seismic data quality.

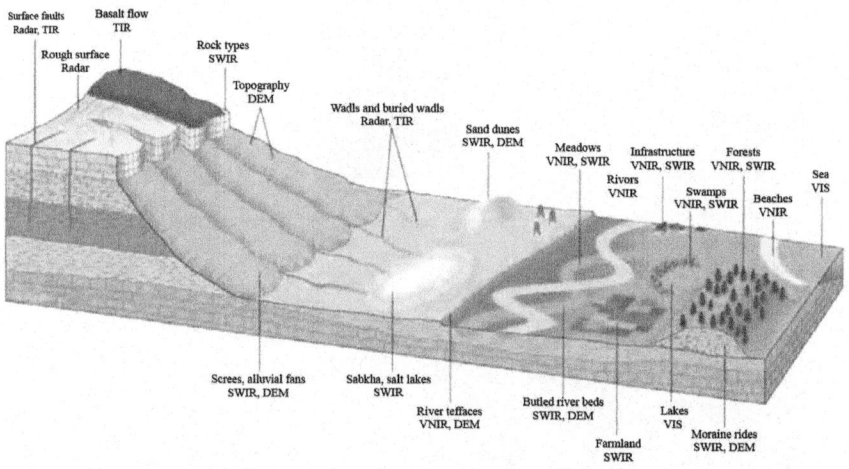

Figure. 28: Characterization of surface and near-surface properties from satellite imagery

Table 1: Detection of surface geological features from satellite imagery and impact on seismic data For satellite imagery codes see section 3

Surface Class	Surface Feature	Satellite Imagery	Impact on Data Quality	Impact on Logistics
Topography and texture	Escarpments, river terraces	DEM, radar	Scattering noise	Severe risk for 15 - 25% slope, no access above
	Rough surfaces	Radar	Poor source / receiver coupling	Severe risk of tire damage for vehicles
	Surface faults	TIR, radar	Scattering noise	Only on escarpments
Land use	Swamps, marshes	VIS–NIR	Resonance; velocity statics	If wet, no access for vibrators and vehicles
	Water features	VIS	Special equipment	No vehicle access
Lithology	Basalt flows	TIR	Poor coupling; strong scattering	Often risk for vibrator and vehicle tires
	Claypans	DEM, NIR	Resonance	No access if wet
	Hard rock outcrops	NIR, SWIR, radar	Poor source / receiver coupling	Limited risk of access for vibrators
	Sabkhas, salt lakes	DEM, SWIR	Resonance; velocity statics	Severe risk for vibrator and vehicle access
	Buried rivers	TIR, radar	Velocity statics	No risk
Geo-morph-ology	Moraine ridges	SWIR, DEM	Low velocity, high attenuation	No risk
	Sand dunes	SWIR, DEM	Elevation statics; strong attenuation	Access for vibrators severely limited

ACKNOWLEDGEMENTS

The author thanks ADNOC, Anadarko, Apache Oil Egypt, Bahrain Oil Company, Dara Petroleum Company, Egypt General Petroleum Corporation, Gaz de France, Kuwait Ministry of Oil, Kuwait Oil Company Ltd., Repsol YPF, Sonatrach, TransGlobe Energy, and WesternGeco for the permission to publish the data. The original data for Landsat 7 ETM+ and MrSID is provided by NASA. The original data for the SRTM DEM is provided by NASA through the CIAT-CSI SRTM website (http://srtm.csi.cgiar.org), the Canadian Space Agency provided Radarsat microwave data. ASTER original data and ASTER GDEM are products of METI and NASA, the original data is property of METI and NASA. Radar based satellite gravity data are provided by the University of California, La Jolla Satellite. The author also thanks Elisabeth DeTemple, Charles Woodward and Nick Moldoveanu for their reviews, Andrew Cutts for the support with GIS and Claudio Strobbia, Mohamed Sheneshen and Larry Velasco for their contributions to the data integration.

REFERENCES

1. Alsharhan, A.S. and Salah, M.G. (1995). Geology and hydrocarbon habitat in rift setting: Northern and central Gulf of Suez, Egypt, Bulletin of Canadian Petroleum Geology, Vol. 43, 156–176.

2. Astakhov, V. I., Svendsen, J. I, Matiouchkov, A., Mangerud, J., Maslenikova, O. and Tveranger, J. (1999), Marginal formations of the last Kara and Barents ice sheets in northern European Russia. Boreas, Vol. 28, pp. 23–45. Oslo.

3. Coulson, S., Grabak, O., Cutts, A., Sweeney, D., Hinsch, R., Schachinger, M., Laake, A., Monk, D. and Towart, J. (2009). Satellite sensing : risk mapping for seismic surveys, Schlumberger Oilfield Review, Winter 2008/2009, pp. 40-51.

4. Cutts, A. and Laake, A. (2009). An Analysis of the Near Surface Using Remote Sensing for the Prediction of Logistics and Data Quality Risk, Tunis 2009 – 4th North

5. African/Mediterranean Petroleum and Geosciences Conference & Exhibition, Tunis, March 2009, paper S30.

6. Darwish, M. and El-Araby, A.M. (1993). Petrography and diagenetic aspects of some siliciclastic hydrocarbon reservoirs in relation to the rifting of the Gulf of Suez. In: Philobbos, E.R. & Purser, B.H. (eds) Geodynamics and Sedimentation of the Red Sea–Gulf of Aden Rift System. Geological Society of Egypt, Special Publication, 1, pp. 155–187.

7. El-Baz, F. and Robinson, C. (1997), Paleo-channels revealed by SIR-C data in the Western Desert of Egypt : implications to sand dune accumulations, 12th International Conference and Workshops on Applied Geologic Remote Sensing, Denver November 1997.

8. Guo, H., Lewis, S. and Marfurt, K.J. (2008). Mapping multiple attributes to three- and fourcomponent colour models – A tutorial. Geophysics, Vol. 73, pp. W7–W19.

9. Jarvis A., H.I. Reuter, A. Nelson, E. Guevara, 2008, Hole-filled seamless SRTM data V4, International entre for Tropical Agriculture (CIAT), available from: http://srtm.csi.cgiar.org.

10. Laake, A. (2005a). Application of Landsat data to seismic exploration – Case study from Kuwait, Kuwait First International Remote Sensing Conference and Exhibition, Kuwait September 2005.

11. Laake, A. (2005b). Remote sensing application for vibroseis data quality estimation in the Neuquen Basin, Argentina, VI Congreso de Exploración

y Desarrollo de Hidrocarburos, Mar del Plata, November 2005. Laake, A. (2009). Hybrid near-surface modeling for seismic property estimation in arctic areas, 71st EAGE Conference & Exhibition, Amsterdam, June 2009, paper T003.

12. Laake, A. (2010). Enhancing the value of remote sensing data through integration with ground based data in 3D, ESA workshop on satellite earth observation for the oil and gas sector, Frascati, September 2010. Laake, A. and Cutts., A. (2007). The role of remote sensing data in near-surface seismic characterization, First Break, Vol. 25, pp. 51 – 55.

13. Laake, A. and Insley, M. (2007). Near-surface characterization from remote sensing data, ENVISAT Symposium 2007, Montreux, April 2007, A457029.

14. Laake, A. and Tewkesbury, A. (2005). Vibroseis data quality estimation from remote sensing data, Proceedings of the 67th Conference and Exhibition, Madrid, June 2005, Expanded abstracts, paper G017.

15. Laake, A. and Zaghloul, A. (2009). Estimation of static corrections from geologic and remotesensing data, The Leading Edge, February 2009, pp. 192-196.

16. Laake, A., Al-Alawi, H. and Gras, R. (2006). Integration of remote sensing data with geology and geophysics – Case study from Bahrain, GEO 2006, Manama, March 2006.

17. Laake, A., and Insley, M. (2004a). Applications of satellite imagery to seismic survey design, The Leading Edge, Vol. 23, No. 10, 1062-1064.

18. Laake, A., and Insley, M. (2004b). Satellite-based seismic technology, World Oil, Vol. 225, No. 9, pp. 27-33.

19. Laake, A., Sheneshen, M., Strobbia, C., Velasco, L. and Cutts, A. (2011). Integration of 4surface/subsurface techniques reveals faults in Gulf of Suez oilfields, Petroleum Geoscience, Vol. 17, 2011, pp. 165–179.

20. Laake, A., Strobbia, C. and Cutts, A. (2008). Integrated approach to 3D near-surface characterization, First Break, Vol. 26, pp. 109-112.

21. Robinson, C.A., Werwer, A., El-Baz, F., El-Shazly, M., Fritch, T. and Kusky, T. (2007), The Nubian aquifer in Southwest Egypt, Hydrogeology Journal, Vol. 15, 2007, pp. 33-45.

22. Sabins, F. (1996). Remote sensing, principle and interpretation (3rd ed.), Freeman, ISBN 0716724421, New York.

23. Sandwell, D.T. and Smith, W.H. (2009). Global marine gravity from retracked Geosat and ERS-1 altimetry: Ridge segmentation versus spreading rate, Journal of Geophysical, Vol. 114, B01411, pp. 1-18.

24. Short, N., (2010), The remote sensing tutorial, NASA 2010, Date of access : 15/06/2011, Available from : http://landsat.gsfc.nasa.gov/about/L7_td.html

25. Short, N.M.Sr., and Blair, R.W.Jr. (eds.) (1986), Geomorphology from Space, NASA 1986, Date of access : 19/09/2007, Available from : http://geoinfo.amu.edu.pl/wpk/geos/GEO_COMPLETE_TOC.html

26. USGS (2011), Landsat 7 science data users handbook, Date of access : 24/06/2011, Available from: http://landsathandbook.gsfc.nasa.gov/pdfs/Landsat7_Handbook.pdf

Chapter 5

FACTORS CONTROLLING THE INCORPORATION OF TRACE METALS TO COASTAL MARINE SEDIMENTS: CASES OF STUDY IN THE GALICIAN RÍAS BAIXAS (NW SPAIN)

Belén Rubio[1], Paula Álvarez-Iglesias[1], Ana M. Bernabeu[1], Iván León[2], Kais J. Mohamed [1], Daniel Rey[1] and Federico Vilas[1]

[1]Universidad de Vigo, Vigo, Pontevedra Spain

[2]Universidad del Atlántico, Barranquilla Colombia

INTRODUCTION

Transitional coastal environments such as the Galician Rías in the Atlantic coast of NW Spain are densely populated areas. Their environmental problems are highlighted by the conflicting interests of different economic sectors: extensive mariculture activities are located in its waters and intertidal zone; shipbuilding, carbuilding, canning and other industries compete with tourism on their shores; and dairy farming is the main agricultural activity in its surrounding hills and hinterland (Vilas et al., 2008). As a result, the management of the coastal zone is highly complex and it is difficult to balance quality of coastal waters with economic activities. For instance, in the Ría de Vigo, the southernmost of the Rías Baixas, wastewater treatment plants were not installed until the 1990s, and in spite of regional environmental legislation (Lei 8/2001), their capacity was still insufficient in 2005 when the European Court of Justice found Spain guilty of failure to fulfill its obligations under the Article 5 of the Council Directive 79/923/EEC on the quality required for shellfish waters (Case C-26/04 ECJ). This case was closed following Spain's submission of a pollution-reducing programme specifically pertinent to shellfish waters; success of this plan will depend critically on the behaviour of the sediments on the ría bottom. Galician Rías experience seasonal upwelling, which increases marine productivity. This promotes the deposition of high organic matter contents in the bottom sediments and contributes to the observed intense sedimentation rates of 1-6 mm yr-1 (Álvarez-Iglesias et al., 2007; Rubio et al., 2001). Current levels of trace metals (Prego & Cobelo, 2003) in sediments of these Rías have caused a significant concern by local and European authorities in the last ten years, especially in relation to the application of the Water Framework Directive (WFD), aimed to ensure that all waters reach "good status" by 2015. Some of these studies (Álvarez-Iglesias et al., 2003; Belzunce-Segarra et al., 2008; Rubio et al., 2000a) showed that the highest concentration of trace metals occurs in the muddiest surficial sediments of the rías, and that their fate and bioavalability depends on the intensity and speed of bacterial-mediated redoxomorphic post-sedimentary processes (Álvarez-Iglesias & Rubio, 2008, 2009; Rubio et al., 2010). This chapter will review the main factors that control the incorporation of metals to the sediments in these environments, with a focus on the forcing factors and their temporal evolution in the recent sedimentary record. This study will also show the critical importance of distinguishing and quantifying the various metal forms by using sequential extractions and by the determination of magnetic properties in order to reach a full understanding of the potential and present environmental impacts of contaminated sediments. Special emphasis will be put in the role of mussel rafts on the diagenetic inmobilization of heavy metals. Finally, the solubility of these metals by aerobic oxidation will be analyzed in some laboratory experiments in order to improve coastal risks prevention and management.

SEDIMENTS AS TRACE METAL SINKS AND SOURCES

Water analyses proposed in the WFD (Directive 2000/60/CE) are the most obvious way to quantify the degree of metal contamination in an area. However, these analyses are not easy because concentrations of metals in solution are very low, contamination can occur during collection and analysis, and sampling needs to be repeated in specific time intervals (weeks, months, and seasons). Moreover, most metals transported in aquatic ecosystems quickly set on the solid material, due to their low solubility (Forstner & Wittman, 1981). Binding of metals in suspension will eventually lead to their incorporation into the sediment (Fig. 1).

Therefore, sediments are a sink of metals with concentrations several orders of magnitude higher than those in the adjacent waters, both interstitial and overlying (Tessier & Campbell, 1988). Nevertheless, sediments are dynamic reservoirs subject to rapidly changing conditions. When the environmental variables change, remobilization of metals can occur. Although there are different mechanisms of metal binding to sediments, adsorbed metals appear to be more readily available, and therefore can be recycled. In these cases, the sediment acts as a source of metals to other biotic and abiotic compartments (Fig. 1). In addition, sediments can be considered as archives of environmental information due to their "memory capacity", so that the sedimentary record allows us to reconstruct the recent historical record of coastal pollution (Álvarez-Iglesias et al., 2007; Rubio et al., 2001, 2010; Valette-Silver, 1993; among others).

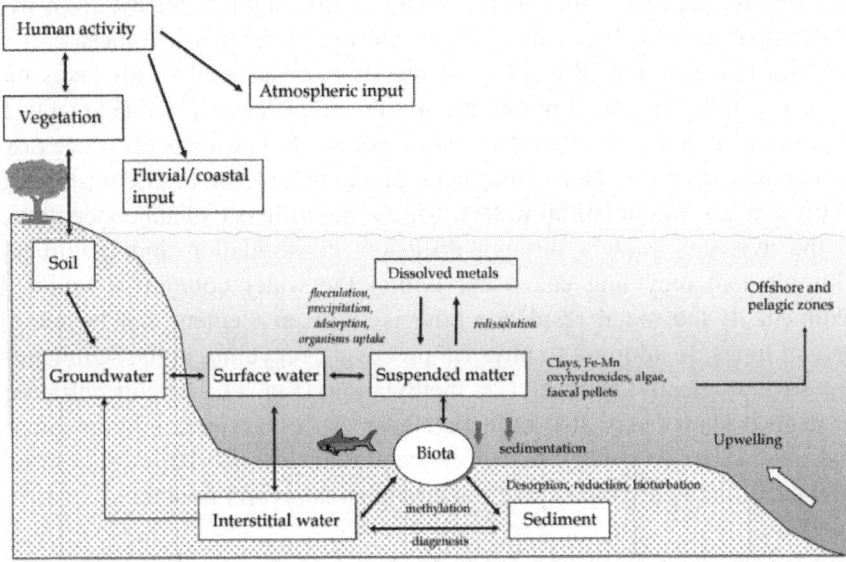

Figure. 1: Schematic representation of metals reservoirs and their interactions in coastal systems.

INCORPORATION OF METALS TO THE SEDIMENT

Human activities have drastically altered the biogeochemical cycles and equilibria of some trace metals. These metals cannot be degraded or destroyed and become stable and persistent contaminants that tend to accumulate in sediments. Metals can be transferred from sediments to benthic organisms and then become a potential risk to human consumers by incorporation through the food web (Soto-Jiménez et al., 2011). The main anthropogenic metals sources are industrial point sources, including present and former mining activities, foundries and smelters, shipbuilding, chemical industries, metallic industries, and diffuse sources such as combustion by-products. Dispersion of metals in the particulate phase is usually small, but relatively volatile metals and those that become attached to air-borne particles can be widely dispersed over very large scales. Trace metals carried in dissolved or particulate forms (e.g., river run-off) enter the normal coastal biogeochemical cycle and are largely retained within near-shore and shelf regions (Fig. 1).

Processes Affecting the Cycles of Metals in Coastal Zones

Trace elements may suffer varying degrees of internal recycling before they are buried in the sediment and/or carried into the ocean (Fig.1). Such recycling may involve processes such as flocculation, precipitation, release from living or dead particulate phases, and subsequent regeneration when these particles undergo redissolution. Recycling of metals in suspended solids takes place by coprecipitation, adsorption, desorption and flocculation (Fig. 1). The suspended matter and deposited sediments are linked through processes of sedimentation and erosion. Diagenetic processes release high concentrations of trace metals to interstitial waters, which can influence metal concentrations in the overlying waters through diffusion, consolidation and bioturbation. This element recycling can occur within the water column or within the sediment. If the metal residence time is short an element can be recycled several times. In addition to physical processes, recycling in the sediment can also be biologically mediated (i.e. methylation) (Fig. 1). The concentration of suspended matter may also influence these processes, especially in estuaries and rías, where suspended matter concentration is much larger than in other systems of the hydrologic cycle, such as most lakes and oceans.

The Interaction Between trace Metals and Aquaculture

In the last decades, marine aquaculture has experienced an important development around the world. Galicia is the second largest producer of mussels in the world after China. They are cultivated in wood frames called mussel rafts. Most of these are concentrated in the Rías Baixas, with more than 3,000 rafts located in Arousa, Pontevedra and Vigo rías. An important environmental impact of these activities is the high amount of particulate matter discharged by mussels from faeces. Although the concentration of heavy metals in these particles is relatively low, the amount of solids is so high that the total accumulation of metals in the sediments may become an important problem. This fact has been mentioned in previous works (Otero et al., 2005; Prego et al., 2006), but it has not been studied in depth. Table 1 compares the accumulation of some trace metals (especially Pb, Ni, and V) in sediments collected below mussel rafts and in adjacent areas in the Ría de Pontevedra. Despite intensive marine aquaculture these results indicate that the differences are not very high, and they seem to be more related to textural differences than to aquaculture activitities. However, there are very significant differences in the elements and ratios of the organic matter characterization (table 1). Sediments collected below mussel rafts areas showed higher contents of total organic carbon (TOC), total N (TN) and total S (TS) than those collected in adjacent areas (table 1). Significant differences were also observed in the mean values for the ratios C/N and S/C, showing that the increase in TOC in mussel rafts areas influences the redoxomorphic organic matter degradation. C/N ratios are, on average, higher than those reported for biodeposits by other authors ([<10; Calvo de Anta, 1999;] Otero et al., 2006). S/C ratios are below the global average in normal marine sediments (Raiswell & Berner, 1986), indicating a moderate stage of diagenetic evolution.

Table 1: Comparison of trace elements concentration obtained by X-ray Fluorescence (XRF) and other sediment parameters (TOC, TN, TS, C/N and S/C) for a group of sediment cores (101 samples) collected below mussel rafts and in adjacent areas (35 samples) in the Ría de Pontevedra.

Trace elements (μg g^{-1})	Mussel rafts areas	Adjacent areas
Sr	865 ± 504	1888 ± 438
Rb	210 ± 31	169 ± 41
Ba	403 ± 73	341 ± 58
Co	13 ± 2	14 ± 1
Cu	20 ± 7	29 ± 20
Zn	77 ± 20	89 ± 32
Ni	34 ± 12	24 ± 4
Pb	18 ± 15	3 ± 8
Cr	65 ± 12	66 ± 11
V	89 ± 29	65 ± 17
Other parameters (%)		
TOC	3.69 ± 1.76	2.42 ± 0.68
TN	0.24 ± 0.13	0.13 ± 0.77
TS	0.95 ± 0.62	0.37 ± 0.16
C/N	19.83 ± 10.18	25.19 ± 13.33
S/C	0.25 ± 0.13	0.16 ± 0.09

THE RÍA ENVIRONMENT - FACTORS CONTROLLING TRACE METAL CONTENTS IN RÍA SEDIMENTS

The Rías Baixas are a characteristic geomorphological coastal feature of the Northwest Iberian Margin consisting of four deep and narrow V-shaped Tertiary river valleys that have been flooded during the last sea-level transgression. Most regional studies in rías have shown that although the hydrodynamic processes are similar to those identified in estuaries, the rías are clearly dominated by the waves while the estuarine circulation is restricted to the innermost areas (Piedracoba et al., 2005; RuizVillarreal et al., 2002; Souto et al., 2003; Vilas et al., 2005). These environments are also characterized by a lesser continental freshwater input, and a higher primary productivity due to seasonal upwelling (Fraga, 1981) in comparison to estuarine environments. In addition, the sediment characteristics and distribution of the Galician Rías Baixas (Vilas et al., 2005) also show significant differences from the facies models of wave- or tidedominated estuaries (Vilas et al., 2010) as we will discuss in the following sections.

Factors and Forcings Controlling Grain-Size Distributions

Wave conditions exert an important control on sediment distribution (Rey et al., 2005; Vilas et al., 2005, 2010). Organic-rich fine-grained sediments accumulate in low-energy areas along the deep central axis, and in protected areas of the inner ría sector with maximum mud percentages near 100% (Fig. 2). Mud accumulation is also promoted by the agglutinating effect of organic matter. As a result, organic matter content is higher in muds, and increases towards the inner ría to values in excess of 10% (Vilas et al., 2005). Sediment composition inside the rías is predominantly siliciclastic, as a result of the granitic and metamorphic rocks of their catchment areas. As an example, figure 2 shows the similarities between quartz distribution and mud contents. On the contrary, biogenic carbonates are predominant in the sand and gravel fractions. Production of these coarse calcareous bioclastic sediments is favoured by upwelling fertilization of the rías. $CaCO_3$ abundance is greatest at the margins of the ría and towards the outer areas (Fig. 2), where wave energy is stronger. In these areas $CaCO_3$ contents can reach values higher than 90%. Many authors have recognized the sediment grain size as a factor directly related to the ability for retaining trace elements (Horowitz & Elrick, 1987). This relationship is clearly observed in the sediments of the rías by the surficial distribution of Pb (Fig. 2) and other trace elements. This is also shown by the strong positive correlations between mud percentage and trace elements concentrations (Fig. 3). This correlation is explained by a combination of physico-chemical factors, since materials with a higher capacity to retain contaminants have smaller particle sizes and therefore also have higher specific surfaces and cation exchange capacities. In addition, the effect of grain size is enhanced by organic matter, which is a complexing agent for some pollutant metals and is concentrated in fine-grained particles (Wangersky, 1986). Note in figure 3 the typical association of Pb and Cu with organic matter and the strong relationship of Co with finer fractions. The diluent effect, expressed as a negative correlation, caused by coarser fractions and/or carbonates, is exemplified by the concentration of Zn vs the percentage of $CaCO_3$ or the concentration of Cu vs the percentage of sand.

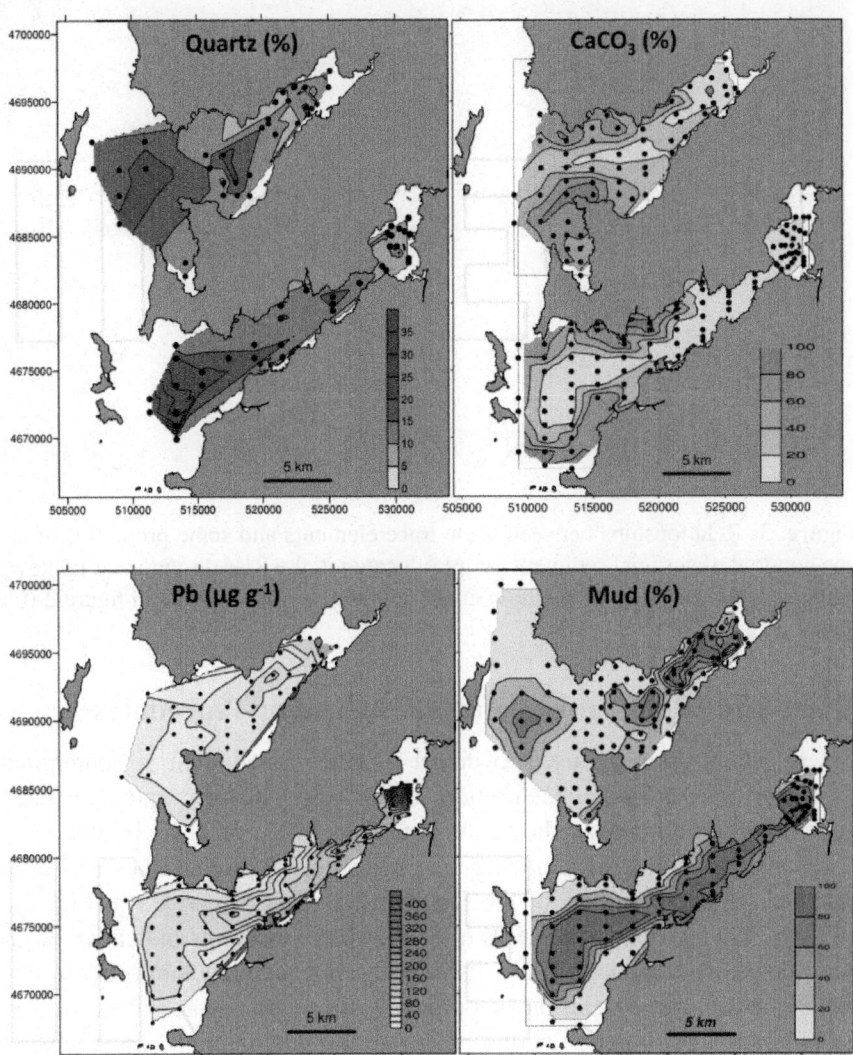

Figure. 2: Distribution maps of quartz, calcium carbonate, lead and mud concentrations in surface sediments of the Rías de Vigo and Pontevedra (NW Spain) measured on more than 100 samples (black dots).

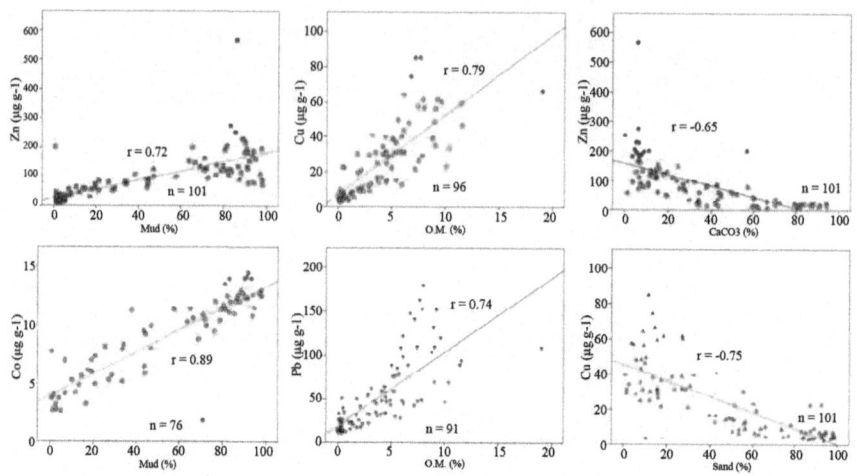

Figure. 3: Relationships between some trace elements and some properties of about one hundred of surficial sediment samples located in the Rías de Vigo and Pontevedra (data from Ría de Vigo from Rubio et al., 2000a). Sample location in figure 2 (black dots).

Grain-Size Effect: Proxies and Normalization Procedures

A very simple method used to detect whether a sediment is contaminated is to map the surface concentration of the target element and try to detect geochemical anomalies (Chester & Voutsinou, 1981) that highlight areas or regions with anomalous contents. For instance, too high values of Pb were detected in the inner part of the Ría de Vigo (San Simón Bay) (Fig. 2). In addition, the distribution patterns of conservative elements indicative of grain size should be compared to the distribution of trace metals in order to detect whether or not these metals are supplied by anthropogenic activities. However, a first approach to determine the presence of contamination is to analyze the relationships between a normalizer element or grain-size proxy (Al, Ti, Rb, among others) and the potential contaminant element. If there is no linear relationship between them, this is usually due to contamination. For example for the relationship between Zn and mud (Fig. 3), those data points that are far from the correlation line are indicative of contamination.

Anthropogenic Evidences on Metal Concentration in ría Sediments

Several indexes (contamination factor, enrichment factor, geoaccumulation index, among others) have been developed to assess the degree of metal contamination in a given area. These indexes compare the metal content of the samples with natural values for each metal. The determination of these so-called background levels is a key factor in assessing the degree of contamination or the anthropogenic effect in a given area. Rubio et al. (2000a) showed how the choice of these values determines the geochemical interpretation of a given area, hence the importance of establishing background values adequately. Also the comparison with Sediment Quality Guidelines (SQGs) that allows calculating the effects range low (ERL), effects range medium (ERM) and probable effect levels according to Long et al. (1995) has been used by several authors (Mucha et al., 2003; Pekey et al., 2004). In the following sections we will review some examples for the sediments of the Rías Baixas.

Background Levels

The background value or "background" of a given trace metal in sediments is the natural content of the metal without human intervention. This value will depend on the geochemistry of the source area sediment. Several possibilities have been set up to establish background values for trace metals (Forstner &Wittmann, 1981):

1. Mean values of metals in the crust (Taylor, 1964) or average shale values (Turekian & Wedepohl, 1961; Wedepohl, 1971, 1991).

2. Values determined by various methods, in the same study area, including:

    a. Selection of presumably clean stations (Barreiro et al., 1988; Subramanian & Mohanachandran, 1990).

    b. Statistical methods, among others, include: multiple regression techniques (Summers et al., 1996a), principal component analysis (Rubio et al., 2000), selection of the first percentile of the cumulative distributions of the concentration of metals (Barreiro et al., 1988), and determination of homogeneous populations based on the analysis of frequency distribution curves (Carral et al., 1995).

3. Analysis of sediment cores deep enough to contain the preindustrial record in the sediment (Angelidis & Aloupi, 1995), which is the best recommended technique for establishing background values for a particular area. For example, Rubio et al. (2000b) proposed background values for the Ría de Vigo from a core about 3 m long, with an approximate age of over 1000 years BP enough to reach preindustrial levels (Table 2). Table 2 gives some examples of background concentrations obtained for several authors for typical trace metals found in the rías compared with global background values. In many cases background values at the

global level can be inadequate for a particular area and it is necessary to obtain background values at local or regional level.

Table 2: Regional background values obtained by different authors for ría sediments, and its comparison to average shale values from Turekian & Wedepohl (1961). Shadowed values are similar between the different authors

Metal	B (1)	C (2)	R(3)	R (4)	A (5)	T (6)
Al	--	--	6.48	6.48	9.82	8.0
Fe	2.69	2.95	3.51	3.51	3.53	4.72
Ti	--	--	0.34	0.34	0.36	0.46
Mn	225	273	244	244	216	850
Zn	100	133	105	105	110	95
Cu	25	22	29	20	21	45
Pb	25	73	51	25	51	20
Cr	43	34	34	55	65	90
Ni	30	32	30	30	33	68
Co	16	12	12	12		19

(1) Barreiro 1991. (2) Carral et al., 1995. (3) Rubio et al., (2000a). (4) Rubio et al. (2000b). (5) Álvarez-Iglesias et al., 2006. (6) Average shale values from Turekian & Wedepohl (1961).

Studies on Sediment Cores in the rías: The Need of Dating with 137Cs and 210Pb

During the last decade, radionuclide dating of sediment cores has been used to establish sources and input rates of pollutants such as trace metals (Lee & Cundy, 2001; Ligero et al., 2002). However they have been very rarely used in sediments for the Galician Rías (ÁlvarezIglesias et al., 2007; Rubio et al., 2001). Among the latest methods to determine these rates the depth distribution of ^{210}Pb and ^{137}Cs specific activities have proven to be valid. ^{137}Cs is a good tracer for erosion and sedimentation because there are no natural sources of this radioisotope that is produced during nuclear fission. Its presence in the environment, therefore, is due to nuclear testing or release from nuclear reactors. The distribution of this radioisotope in a sediment core would reflect variations in their inputs to the environment. Average sedimentation rates are obtained by identifying their maximum inputs in the activity profiles if mixing or radionuclide diffusion has not occurred. ^{210}Pb can be used for dating sediments because it is a natural daughter radionuclide in the decay series of ^{238}U. The decay of ^{226}Ra (half-life 1600 years) in soils and sediments produces the rare gas ^{222}Rn (half-life of 3.8 days) which partially diffuses into the atmosphere or into the water column where it decays to ^{210}Pb (half-life of approximately 22 years). ^{210}Pb becomes absorbed onto particles and finally deposits in the bottom sediments (Allen et al., 1993). The ^{210}Pb method is very useful for dating events that have occurred over the last 100-150 years. It has been successfully applied in the sediments of the Ría de Vigo by Álvarez-Iglesias et al. (2007) to obtain sedimentation rates of about 5 mm yr-1 in intertidal sediments, whereas Rubio et al. (2001) determined values between 1 and 3 mm yr-1 for sediments in inner areas of the Ría de Pontevedra. The analysis of dated sediment cores is tremendously useful because it provides a historical record of natural background levels while it also records the anthropogenic accumulation of metals over the last century.

The Assessment of Metal Pollution

In order to assess metal pollution in sediment cores it is essential to account for grain-size effects first. The two basic procedures for this purpose are to make analytical determinations on a separate grain-size fraction (Ackerman et al., 1983), or use a normalizing factor to correct the results so that regardless of the sediment size distribution, the analytical results can be compared. Some authors disagree with the grain-size separation because they think that some metals are associated with the coarser fractions, either as aggregates or pellets composed of finegrained particles and organic matter, or as grain coatings, that may contain high concentrations of metals. For instance, Rubio et al. (1999) have confirmed the occurrence of pellets and coatings enriched in metals in the sediments of the Rías Baixas. On the contrary, other authors (Araujo et al., 1988; Salomons & Forstner 1984) recommend the use of the fraction smaller than 63 μm in order to minimize grain-size biases on the results of heavy metal content. However, Rubio et al. (1996) concluded that the analysis of this fraction could not compensate for the grain size effect in sediments of the Ría de Pontevedra. For this reason it is always recommended to normalize the metal content to a conservative element such as Al or Rb. In the case of the Galician Rías, the best results for sediments have been obtained with aluminium (Nombela et al. 1994; Marcet Miramontes et al., 1997, Rubio et al., 2000a, 2001). An example for a sediment core from San Simón Bay (inner Ría de Vigo) is shown in figure 4. The similarity of the profiles of the absolute metal concentrations and the Al-normalized results confirm that the increase of Cu, Pb and Zn concentration towards the top of the core is not due to textural effects but anthropogenic inputs. Therefore, the effectiveness of standardization is itself a way to detect metal contamination.

Normalized enrichment factors (EF) (Zoller et al., 1974) are also a useful tool where $EF = (M/Al)^{sample}/(M/Al)^{background}$. EFs for Zn, Pb and Cu in core SS3 are shown in figure 4. These results show high contamination for Pb, and moderate for Cu and Zn in the upper part of the core, whereas for the core bottom contamination is moderate for Pb and absent, for Cu and Zn.

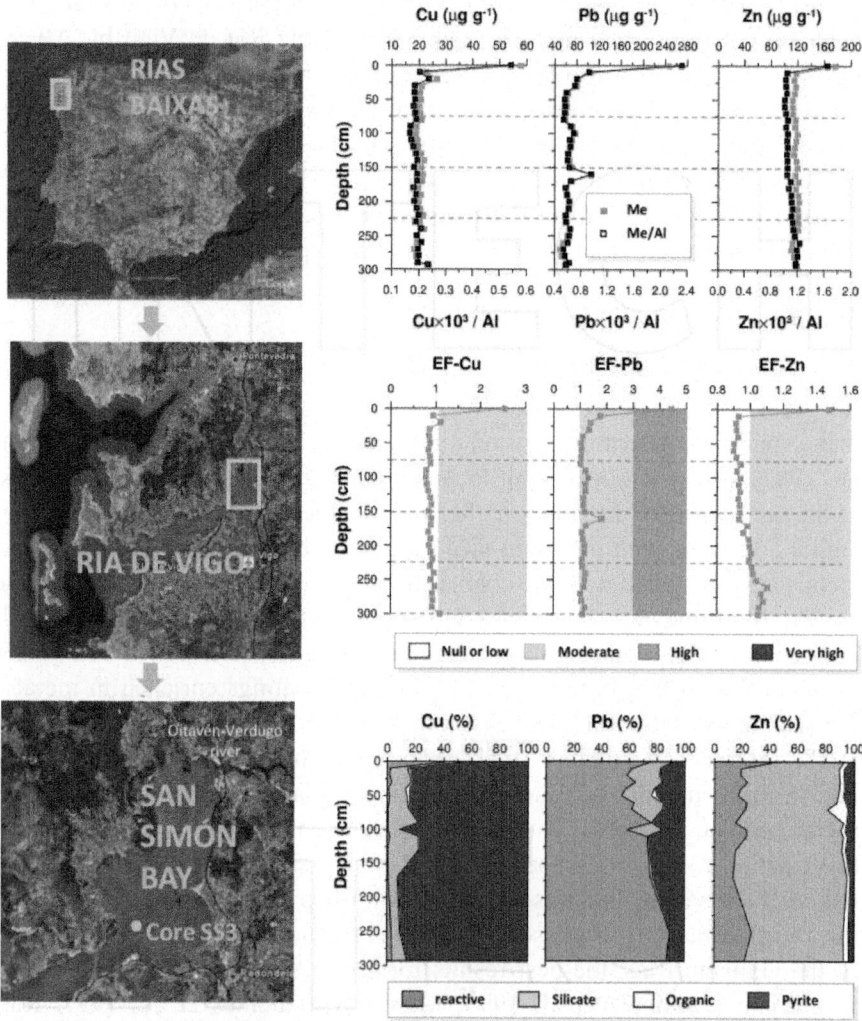

Figure. 4: Left: Location of the core SS3 in the inner area of Ría de Vigo (San Simón Bay). Right: Top, concentrations of Cu, Pb and Zn (orange; data from Álvarez-Iglesias et al., 2006) in a core from San Simón Bay and corresponding metal/Al ratios (black). Middle, enrichment factors (EF) for the same elements and classification of the level of contamination. Bottom, depth distribution of reactive, organic, silicate and pyrite fractions of Cu, Pb and Zn obtained from sequential extractions according to the procedure of Huerta-Díaz & Morse (1990).

The need to Carry out Sequential Extractions

The total amount of metals in the sediment is unrepresentative of the potential toxicity of the metal. To assess the toxicity appropriately it is essential to know the chemical forms in which a metal is presented, i.e. speciation. The chemical form (as dissolved, adsorbed, bound or precipitated) of an element will not only regulate its degree of toxicity but also its availability. Total concentrations are still used very frequently in studies of contamination due to its easy measurement and reproducibility, in spite of the fact that the type of contaminant and the form in which it appears (soluble, exchangeable, bound, adsorbed, occluded, etc.) will decisively influence the pollution effect. For this reason, sequential extractions are usually performed and several operationally defined fractions obtained, which depend on the ability of the chemical extractant to remove certain components. These extractions allow us to determine the chemical forms in which each element is found in the sediment. However, very few studies on trace metals in the Galician Rías have considered the forms adopted by different metals. One example of a sequential extraction following the method of HuertaDíaz & Morse (1990) is shown in figure 4 for inner Ría de Vigo sediments (core SS3). Here we distinguish operationally defined reactive, organic, pyrite and silicate-bound fractions for several trace elements. Pb appeared mostly in the reactive fraction (average, 68.5%), Cu in the pyrite fraction (81.0% on average) and Zn in the silicate-bound fraction (68.7% on average), being the organic fraction very low in all cases (usually lower than 4%). In terms of toxicity these results show that the most problematic trace elements are Pb, because it is found in more biovailable forms, and Cu, because it is found in oxidizable forms. Zn toxicity will mostly come from its reactive fraction These detailed interpretations confirms the interest of the determination of chemical forms when contamination is suspected in a target area.

The Magnetic Properties as a Proxy for Trace Metals in Sediments

The measurement of trace elements in sediments is very laborious and expensive and, therefore, the use of fast and economic alternative techniques is desirable. Environmental magnetism –the use of magnetic properties for environmental applications- can be used to estimate contamination levels and assess possible patterns of dispersion of contaminants. Some authors have shown that certain magnetic properties such as magnetic susceptibility (χ) or the isothermal remanent magnetization (IRM) show significant positive correlations with the concentrations of trace metals in the fine-grained fraction of sediments (Chan et al., 1998, 2001; Scoullos & Oldfield, 1984; Spassov et al., 2004), whereas other researchers (Petrovsky et al., 1998) have reported the contrary. In some studies both behaviours are observed depending on the element considered (Berry & Plater 1998; Georgeaud et al., 1997). A positive association is explained by these authors in terms of the preferent absorption of the metals by the clay fraction and Fe oxides, whereas a negative correlation is sometimes explained in terms of diversity of sources of contamination or due to diagenetical processes.

Some studies in the Rías Baixas pointed out that the distribution of magnetic susceptibility in surficial sediments could be explained mostly by the textural and hydrodynamic interplay (Rey et al., 2000, 2005; Mohamed et al., 2011). The increase in diamagnetic carbonate content toward the ría margins, where coarse-grained material accumulates, results in generally low susceptibility values. The highest susceptibilities lie along the central axis, where the clay content is high and carbonate bioclasts are scarce; and also toward the outer sector of the ría, where oxygenation is more intense and formation of authigenic Fe oxides and oxyhydroxides is favoured. The analysis of the susceptibility of the mud fraction (χmud) that was correlated with trace metals and other properties of the sediments (Fig. 5) revealed a strong negative correlation of susceptibility with Pb. The organic matter content is also correlated with the distribution of elements like Pb, as it is shown in figure 3. Magnetic susceptibility gradually decreases toward the inner part of the central axis because the organic matter decomposition causes reducing conditions and the establishment of an anoxic/sulphidic environment where the magnetic oxides and oxyhydroxides that carry out the susceptibility signal in the outer part of the ría are dissolved. Therefore, low magnetic susceptibility values in sediments of the rías can be a good indicator of reducing conditions, related to organic-rich fine-grained sediments in lowenergy environments where trace metals tend to accumulate.

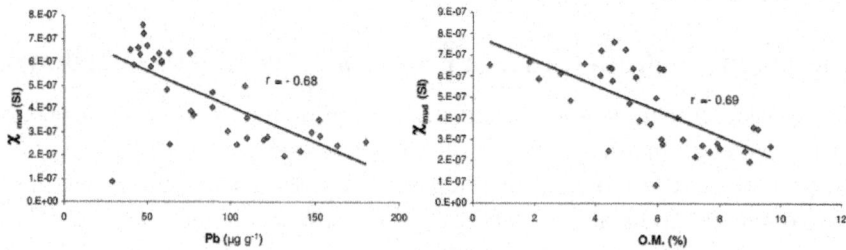

Figure. 5: Relationships between magnetic suspectibility of the mud fraction (χmud) vs Pb concentration and vs organic matter (O.M.) in surficial sediments from the Rías de Pontevedra and Vigo (modified from López-Rodríguez et al., 1999).

Early Diagenesis of Metals in ría Sediments

The early diagenetic reactions that control the formation of authigenic minerals are driven by the oxidation of organic matter, initially by aerobic respiration and subsequently by a series of reactions controlled by anaerobic bacteria, such as reduction of Fe and Mn oxides, reduction of nitrates and sulfates and methanogenesis (Canfield et al., 1993). These reactions release products (e.g., HCO_3., H_{S^-}, Fe^{+2}, Mn^{+2}) to the sediment pore waters, which will precipitate forming new minerals when the saturation is reached (Gaillard et al., 1989). These processes occur ideally sequentially starting with oxic, suboxic, sulfidic and finally methanogenic reactions (Berner, 1981). This diagenetic sequence of events can be evaluated from the analysis of pore-waters and the mineral concentration of typical diagenetic mineral phases in sediment cores, or by using sequential extractions in the sediments.

Diagenetic Zonation in ría Environments: The Hydrodynamic Role

Previous studies in the Rías Baixas allowed the definition of a diagenetic zonation model in these environments by using a combination of geochemical sequential extractions and magnetic properties (Mohamed et al., 2011; Rey et al., 2005; Rubio et al., 2001, 2010). In particular, speciation data of redox sensitive elements such as Fe and Mn are indicative of the different reducing conditions in sediments. Magnetic properties are useful to identify the magnetic minerals and their concentration, which can be used as proxies for the different diagenetic environments. Figure 6 shows the deepening of the redox boundaries from inner to outer ría. The oxic zone expands as it gets deeper toward the outer ría, in a similar way as the suboxic, anoxic and methanic zones. The observed redoxcline deepening can also be related to the different depths at which shallow gas fields have been described in the Ría de Vigo sediments (García-Gil et al., 2002; Kitidis et al., 2006; Iglesias and García-Gil, 2007). This spatial trend can be explained by several factors: 1) A progressive change in the hydrodynamic conditions along the ría, 2) The different origin (marine or terrestrial) of the organic matter and their aging in the water column. Regarding hydrodynamics the outermost ría areas are affected by severe storms in winter that remobilize and oxygenate the top sediments due to wave action. This process buffers sulphate reduction and contributes to the formation of authigenic iron oxides by precipitation of dissolved iron diffusing from underlying anoxic layers (Rubio et al., 2001). This process also seems to contribute to the gradual depletion of organic matter in finegrained sediments observed toward the outer areas of the ría mouth. As for the organic matter characteristics, the decrease in terrestrial sedimentary organic matter toward the outer-ría (Álvarez-Iglesias 2006; Andrade et al. 2011), in addition to the longer aging of organic matter in deeper waters of the outer-ría areas compared to the innerría, could contribute to explain the mentioned diagenetic zonation.

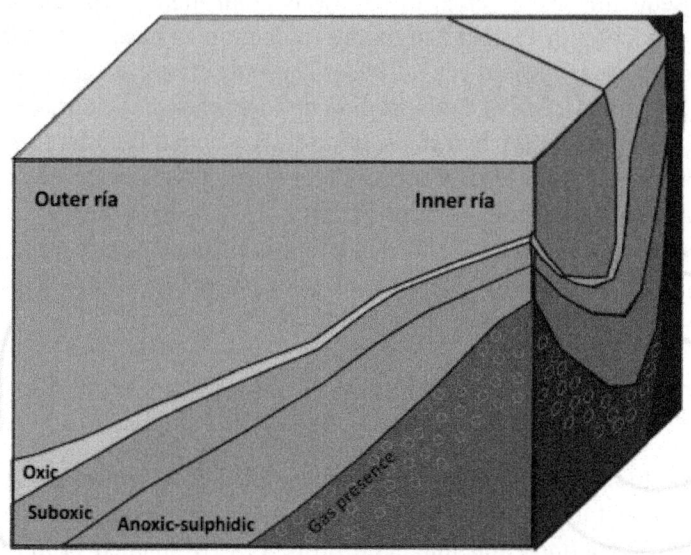

Figure. 6: Block diagram illustrating the variation in depth of the diagenetic zones in the different sectors of the ría.

Diagenetic Mobilization of Trace Metals

Influence of mussel rafts In the last fifty years mussel culture in the Rías Baixas has caused significant changes in the sediment due to the large amounts of detritus originated by these filter feeders, which are deposited mostly as pellets enriched in organic matter on the ría bottoms. Each mussel raft produces approximately 190 kg day^{-1} of dry biodeposit that contains between 31 and 32 kg day^{-1} of organic matter (Cabanas et al., 1979). In addition, sedimentation rate is increased significantly in areas below mussel rafts, with values that range between 5 and 15 mm yr^{-1} (Tenore & González, 1975; Cabanas et al., 1979). This elevated sediment accumulation together with its high concentration of organic matter has led to a change in the physicochemical properties of sediments towards more anoxic environments (Cabanas et al., 1982; León et al., 2004). In these sediments, anoxic degradation of organic matter is responsible for the early diagenesis of sedimentary Fe sulfides that eventually are transformed into pyrite ($FeS2$), which is thermodynamically the more stable compound (Luther, 1991; Morse & Luther, 1999). The study of diagenesis and organic matter degradation can provide very important information about retention and/or mobility of contaminants such as trace metals. Some authors considered that formation of insoluble sulfides under reducing conditions would immobilize and trap trace metals such as Cu, Zn and Pb. On the contrary, other studies (Álvarez-Iglesias & Rubio, 2008, 2009; Rae & Allen, 1993; Rubio et al., 2010; Varekamp, 1991) indicated that these elements can be mobilized or relocated during the degradation of organic matter. It is also important to distinguish between the fraction of the elements incorporated in detrital phases and the fraction which may be available in response to changes in redox conditions, such as variations in the chemical conditions of the bottom water or interstitial water. The two main approaches to make this separation are: 1) The use of statistical techniques of separation of these phases (Calvert, 1976; Dymond, 1981). 2) The application of chemical treatments to remove certain phases or fractions of elements (Huerta- Díaz & Morse, 1990; Tessier et al., 1979; Ure et al., 1993, among others). As we have seen in section 5.4 the latter approach, sequential extractions, are a key tool to assessing the bioavailability of a particular metal. The availability of trace metals in the sediment depends on the fractions to which they are associated to (carbonates, organic matter, sulfides, silicates, oxyhydroxides of Fe and Mn). When conditions are favorable for the formation of pyrite, metals can co-precipitate with it, and pyrite becomes an important metal sink. If environmental conditions change (i.e. oxidation of sediments) metals can be released and pyrite becomes a source.

Solubility of Heavy Metals During Controlled Aerobic Oxidation of Anoxic Sediments:Some Laboratory Experiments

As we have seen in previous sections the concentration levels of certain metal and metalloids in the sediments of the Galician Rías Baixas have shown an increasing trend in the last decades. It is likely that a transfer of these elements to the water column may occur during remobilization of sediments caused by natural events or anthropogenic activities. The inner areas of the rías are exposed to activities that remobilize the sediment such as intense maritime traffic or dredging and cleaning operations. Selected samples of surficial sediments from inner and middle ría sediments of Ría de Pontevedra were subjected to an aerobic oxidation procedure to determine the concentration of some elements (Fe, Mn, Cu, Cr, Pb and Hg) released from the sediment to the aqueous phase. The experiment was done over five days and measurements of pH and total metal concentrations were made both in water and in sediment samples. Metal concentrations were lower in the sediments during aerobic oxidation due to their release to the aqueous phase. The net release of metals was higher in sediments form the inner sector than those from the middle sector of the Ría de Pontevedra (Fig. 9), with the exception of Cu. The high standard deviation of Fe and Mn in the inner sector samples is mainly due to the high redox sensitivity of these two metals and their high abundance as sulphides, as we have mentioned concering the DOP values, which are rapidly oxidized causing the release of these metals to the aqueous phase. The concentrations of these metals together with those of Cu, Cr and Zn increased significantly in the aqueous phase after the experiment. This demonstrates that remobilization of marine sediments tends to increase the mobility and availability of those trace metals.

Metal concentrations in the aqueous phase varied between elements (Fig. 9). Hg and Pb concentrations were below the detection limits in all cases. Cr and Zn concentrations were in general quite low and remained almost constant over time. In contrast, Fe and Mn were released very rapidly although their concentrations decreased sharply to reach undetectable limits, because they precipitated as oxides and oxyhydroxides. Finally, the release of Cu increased with time for most of the samples, with a maximum concentration of total dissolved Cu of 8.9 mg L-1. This concentration is higher than the toxicity threshold for organisms of the Galician Rías reported by other authors (Beiras & Albentosa, 2004).

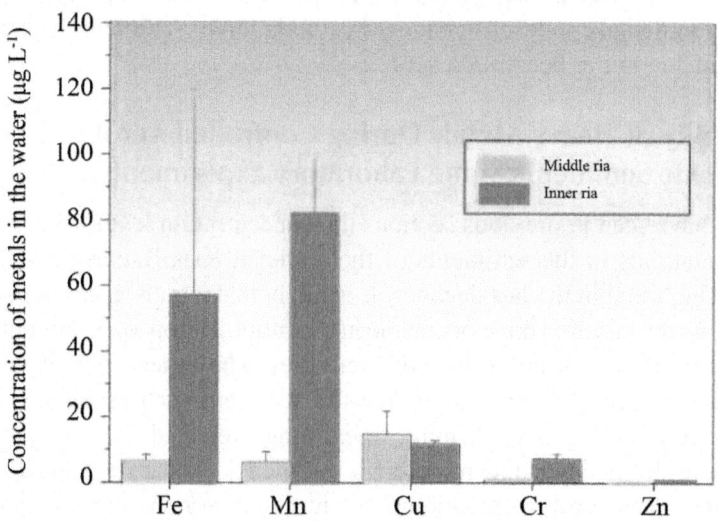

Figure. 7: Mean concentrations of trace metals in water after oxidation of sediments from inner and middle areas of Ría de Pontevedra. Sampling location in figure 7.

CONCLUSION

The main factors controlling the incorporation of metals to the sediments in transitional coastal environments like the Galician Rías Baixas in the NW Spain have been revised in this chapter. It is essential to understand the behaviour of trace metals in the sediments from the ría bottoms in order to improve coastal risks prevention and management, as well as to reach a good status in water quality as one of the great challenges for the European Union in the new millennium. In the rías, wave conditions exert an important control on sediment distribution and in the subsequent diagenetic evolution of the sediments, and thus on trace metal accumulation and immobilization. A strong positive correlation is found between fine–grained sediments and trace metals. Organic matter enhances the grain-size effect on metal concentration, especially in areas influenced by mussel culture. The procedures for normalizing and minimizing grain-size effects have also been revised in order to distinguish natural from anthropogenic metal signals in the sediments.

Inner ría sediments are highly contaminated by Pb, and moderately by Cu and Zn, especially in the most recent sedimentary record. Some examples of point-source Hg contamination have also been shown for the Ría de Pontevedra. The importance of distinguishing and quantifying the various metal forms by using sequential extractions have also been demonstrated with several examples for sediment cores, highlighting the role of the diagenetic processes in the inmobilization and/or relocation of trace metals. A characteristic diagenetic zonation in ría environments is attributed to the local water depth, the distribution of wave energy and the subsequent sediment grain-size distribution. The diagenetic processes have also been emphasized by the results of the magnetic properties, showing that low magnetic susceptibility values in sediments of the rías can be a good indicator of reducing conditions. In such conditions, trace metals are mostly concentrated in sulfide fractions. The degrees of pyritization of iron and trace elements can be valid indicators of the redox status and heavy metal risk, respectively. Experimental aerobic oxidation results have show that the sediments of inner sectors of the rías show a higher release of metals to the aqueous phase than those of the middle sector. However, from an environmental point of view, Cu is the only metal released in quantities that may be toxic for the organisms in the area.

ACKNOWLEDGMENT

This work was supported by the Spanish Ministry of Science and Technology through projects CTM2007-61227/MAR, GCL2010-16688 and IPT-310000-2010-17, by the IUGSUNESCO through project IGCP-526 and by the Xunta de Galicia through projects 09MMA012312PR and 10MMA312022PR.

REFERENCES

1. Ackerman, F., Bergmann, H. & Schleichert, U. (1983). Monitoring of heavy metals in coastal and estuarine sediments. A question of grain size: <20 μm versus <60 μm.

2. Environmental and Technological Letters, Vol. 4, pp. 317-328. Allen, J.R.L., Rae, J.E., Longworth, G., Hasler, S.E. & Ivanovich, M. (1993). A comparison of the 210Pb dating technique with three other independent dating methods in an oxic estuarine salt-marsh sequence. Estuaries and Coasts, Vol. 16, pp. 670-677.

3. Álvarez-Iglesias, P. (2006). El registro sedimentario reciente de la Ensenada de San Simón (Ría de Vigo, Noroeste de España): interacción entre procesos naturales y actividades antropogénicas. Ph.D. Thesis, Universidad de Vigo, 356 p.

4. Álvarez-Iglesias, P., Quintana, B., Rubio, B. & Pérez-Arlucea, M. (2007). Sedimentation rates and trace metal input history in intertidal sediments derived from 210Pb and 137Cs chronology. Journal of Environmental Radioactivity, Vol. 98, pp. 229-250.

5. Álvarez-Iglesias, P. & Rubio, B. (2008). The degree of trace metal pyritization in subtidal sediments of a mariculture area: application to the assessment of toxic risk. Marine Pollution Bulletin, Vol. 56, pp. 973–983.

6. Álvarez-Iglesias, P. & Rubio, B. (2009). Redox status and heavy metal risk in intertidal sediments in NW Spain as inferred from the degrees of pyritization of iron and trace elements. Marine Pollution Bulletin, Vol. 58, pp. 542-551.

7. Álvarez-Iglesias, P., Rubio, B. & Pérez-Arlucea, M. (2006). Reliability of subtidal sediments as "geochemical recorders" of pollution input: San Simón Bay (Ría de Vigo, NW Spain). Estuarine, Coastal and Shelf Science, Vol. 70, pp. 507-521.

8. Álvarez-Iglesias, P., Rubio, B. & Vilas, F. (2003). Pollution in intertidal sediments of San Simón Bay (Inner Ría de Vigo, NW of Spain): total heavy metal concentrations and speciation. Marine Pollution Bulletin, Vol. 46, pp. 491-506.

9. Andrade, A., Rubio, B., Rey, D., Álvarez-Iglesias, P., Bernabeu, A.M. & Vilas, F. (2011). Paleoclimatic changes in the northwestern Iberia península during the last 3000 years as inferred from diagenetic proxies in the sedimentary record of the Ría de Muros. Climate Research, Vol. 48, pp. 247-259

10. Angelidis, M.O. & Aloupi, M. (1995). Metals in sediments of Rhodes

Harbour, Greece. Marine Pollution Bulletin, Vol. 31, pp. 273-276.

11. Araujo, M.F., Bernard, P. C. & Van Grieken, R. E. (1988). Heavy metal contamination in sediments from the Belgian Coast and Sheldt estuary. Marine Pollution Bulletin, Vol. 19, pp. 269-273.

12. Barreiro Lozano, R., Carballeira Ocaña, A. & Real Rodríguez, C. (1988). Metales pesados en los sedimentos de cinco sistemas de ría (Ferrol, Burgo, Arousa, Pontevedra y Vigo). Thalassas, Vol. 6, pp. 61-70.

13. Barreiro, R. (1991). Estudio de metales pesados en medio y organismos de un ecosistema de ría (Pontedeume, A Coruña). Ph.D. Thesis, Universidad de Santiago de Compostela, 227 p.

14. Beiras, R. & Albentosa, M. (2004). Inhibition of embryo development of the commercial bivalves Ruditapes decussatus and Mytilus galloprovincialis by trace metals; implications for the implementation of seawater quality criteria. Aquaculture, Vol. 230, pp. 205-213.

15. Belzunce Segarra, M.J., Prego, R., Wilson, M.J., Bacon, J. & Santos-Echeandía, J. (2008). Metal speciation in surface sediments of the Vigo Ria (NW Iberian Peninsula). Scientia Marina, Vol. 72, pp. 119-126.

16. Berner, R.A. (1970). Sedimentary pyrite formation. American Journal of Science, Vol. 268, pp. 1-23.

17. Berner, R. A. (1981). A new geochemical classification of sedimentary environments. Journal of Sedimentary Petrology, Vol. 51, No.2, pp. 359-365.

18. Berry, A. & Plater, J. (1998). Rates of tidal sedimentation from records of industrial pollution and environmental magnetism: the Tees estuary, North-East England. Water, Air and Soil Pollution, Vol. 106, pp. 463-478.

19. Cabanas, J., González, J. & Iglesias, M. (1982). Physico-chemical conditions in winter in the Ría de Pontevedra (NW Spain) and their influences on contamination. International Council for the Exploration of the Sea-Conseil International pour l'Exploration de la Mer (ICES CIEM), Vol. E53, pp. 15.

20. Cabanas, J., Mariño, J., Pérez, A. & Román, G. (1979). Estudio del mejillón y de su epifauna en los cultivos flotantes de la ría de Arousa. III. Observaciones previas sobre la retención de partículas y la biodeposición de una batea. Boletin del Instituto Español de Oceanografía, Vol. 268, pp. 45-50.

21. Calvert, S.E. (1976). The mineralogy and geochemistry of nearshore sediments, In: Chemical Oceanography, J.P. Riley & R. Chester (Eds), vol. 6, 187-280, Academic Press, London.

22. Calvo de Anta, R., Quintas Mosteiro, Y. & Macías Vázquez, F. (1999). Caracterización de materiales para la recuperación de suelos degradados. I: Sedimentos biogénicos de las Rías de Galicia. Edafología, Vol. 6, pp. 47-58.

23. Canfield, D. E. Thamdrup, B. & Hansen, J. W. (1993). The anaerobic degradation of organic matter in Danish coastal sediments: iron reduction, manganese reduction and sulfate reduction. Geochimica et Cosmochimica Acta, Vol. 57, pp. 3867-3883.

24. Carral, E., Villares, R., Puente, X. & Carballeira, A. (1995). Influence of watershed lithology on heavy metal levels in estuarine sediments and organisms in Galicia (North-west Spain). Marine Pollution Bulletin, Vol. 30, pp. 604-608.

25. Chan, L.S., Ng, S.L., Davis, A.M., Yim, W.W.S. & Yeung, C.H. (2001). Magnetic properties and heavy-metal contents of contaminated seabed sediments of Penny's Bay, Hong Kong. Marine Pollution Bulletin, Vol. 42, pp. 569-583.

26. Chan, L.S., Yeung, C.H., Yim, W.S.-W. & Or, O.L. (1998). Correlation between magnetic susceptibility and distribution of heavy metals in contaminated sea-floor sediments of Hong Kong Harbour. Environmental Geology, Vol. 36, pp. 77-86.

27. Chester, R. & Voutsinou, F. G. (1981). The initial assessment of trace metal pollution in coastal sediments. Marine Pollution Bulletin, Vol. 12, pp. 84-91.

28. Dymond, J. (1981). Geochemistry of Nazca plate surface sediments: An evaluation of hydrothermal, biogenic, detrital and hydrogenous sources. Geological Society of America Memoir, Vol. 154, pp. 133-172.

29. EC (European Communities), 2000. Directive 2000/60/EC of the European Parliament and of the Council of 23 October 2000 establishing a framework for Community action in the field of water policy. Official Journal of the European Communities, L327/1, 22 December 2000.

30. ECJ (European Court of Justice), 2005. Case C-26/04 ECJ, Commission v Spain ECJ 15- 12- 2005, ECRI-11059.

31. EEC (European Economic Community), 1979. Council Directive 79/923/EEC of 30 October 1979 on the Quality required of Shellfish Waters. Official Publications of the European Communities, OJ L281, 10 November 1979.

32. Figueiras, F., Niell, F. & Mouriño, C. (1986). Nutrientes y oxígeno en la Ría de Pontevedra (NO de España). Investigaciones Pesqueras, Vol. 50, pp. 97-115.

33. Förstner, U. & Wittmann, G. T. (1981). Metal pollution in the aquatic environment. SpringerVerlag, London, 486 p.

34. Fraga, F. (1981). Upwelling off the Galician Coast, NE Spain. In: Coastal Upwelling, Coastal Estuarine Studies, F.A. Richards (Ed.), 1, 176-182, American Geophysical Union, Washington, DC.

35. Gaillard, J.F., Pauwells, H. & Michard, G. (1989). Chemical diagenesis in coastal marine sediments. Oceanologica Acta, Vol. 12, No.3, pp. 175-187.

36. García-Gil, S., Vilas, F. & García-García, A. (2002). Shallow gas features in incised-valley fills (Ria de Vigo, NW Spain): a case study. Continental Shelf Research, Vol. 22, pp. 2303- 2315.

37. Georgeaud, V.M., Rochette, P., Ambrosi, J.P., Vandamme, D. & Williamson, D. (1997). Relationship between heavy metals and magnetic properties in a large polluted catchment: The Etang de Berre (South of France). Physics and Chemistry of the Earth, vol. 22, pp. 211-214.

38. Horowitz, A.J. & Elrick, K.A. (1987). The relation of stream sediment surface area, grain size, and composition of trace element chemistry. Applied Geochemistry, Vol. 2, pp. 437-451.

39. Huerta-Díaz, M.A. & Morse, J. (1990). A quantitative method for determination of trace metal concentrations in sedimentary pyrite. Marine Chemistry, Vol. 29, pp. 119-144.

40. Iglesias, J. & García-Gil. S. (2007). High-resolution mapping of shallow gas accumulations and gas seeps in San Simón Bay (Ría de Vigo, NW Spain). Geo-Marine Letters, Vol. 27, pp. 103-114.

41. Kitidis, V., Tizzard, L., Uher, G., Judd, A.G., Upstill-Goddard, R., Head, I.M., Gray, N.D., Taylor, G., Durán, R., Diez, R., Iglesias, J. & García-Gil, S. (2006). The biogeochemical cycling of methane in Ría de Vigo, NW Spain. Journal of Marine Systems, Vol. 66, pp. 258-271.

42. Lee, S.V. & Cundy, A.B. (2001). Heavy metal contamination and mixing processes in sediments from the Humber Estuary, Eastern England. Estuarine, Coastal and Shelf Science, Vol. 53, pp. 619-636.

43. Lei 8/2001, of August 2nd, de protección da calidade das augas das rías de Galicia e de ordenación do servicio público de depuración de augas residuais urbanas. Diario Oficial de Galicia (DOG).

44. León, I., Méndez, G. & Rubio, B. (2004). Geochemical phases of Fe and degree of pyritization in sediments from Ría de Pontevedra (NW Spain): Implications of mussel raft culture. Ciencias Marinas, Vol. 30, pp. 585-602.

45. Ligero, R.A., Barrera, M., Casas-Ruiz, M., Sales, D. & López-Aguayo, F. (2002). Dating of marine sediments and time evolution of heavy metal concentrations in the Bay of Cádiz, Spain. Environmental Pollution, Vol. 118, pp. 97-108.

46. Long, E.R., MacDonald, D.D., Smith, S.L. & Calder, F.D. (1995). Incidence of adverse biological effects within ranges of chemical concentrations in marine and estuarine sediments. Environmental Management, Vol. 19, pp. 81-97.

47. López-Rodríguez, N., Rey, D., Rubio, B., Pazos, O. & Vilas, F. (1999). Variaciones de la susceptibilidad magnética en los sedimentos de la Ría de Vigo (Galicia). Implicaciones para la dinámica sedimentaria y contaminación antropogénica de la zona. Thalassas, Vol. 15, pp. 85-94.

48. Luther, G.W. III. (1991). Pyrite synthesis via polysulfide compounds. Geochimica et Cosmochimica Acta, Vol. 55, pp. 2839-2849.

49. Marcet Miramontes, P., Andrade Couce, M. L., & Montero Vilariño, M. J. (1997). Contenido y enriquecimiento de metales en sedimentos de la Ría de Vigo (España). Thalassas, Vol. 13, pp. 87-97.

50. Mohamed, K.J., Rey, D., Rubio, B., Dekkers, M., Roberts, A.P. & Vilas, F. (2011). Onshoreoffshore

51. gradient in reductive early diagenesis in coastal marine sediments of the Ría de Vigo, Northwest Iberian Peninsula. Continental Shelf Research, Vol. 31, No.5, pp. 433-447.

52. Morse, J. W. & Luther, G. W., III (1999). Chemical influence on trace metal-sulfide interactions in anoxic sediments. Geochimica et Cosmochimica Acta, Vol. 63, No.19/20, pp. 3373-3378.

53. Mucha, A.P., Vasconcelos, M.T.S.D. & Bordalo, A.A. (2003). Macrobenthic community in the Douro estuary: relations with trace metals and natural sediment characteristics. Environmental Pollution, Vol. 121, pp. 169-180.

54. Nombela, M. A., Vilas, F., García- Gil, S., García- Gil, E., Alejo, I., Rubio, B. & Pazos, O. (1994). Metales pesados en el registro sedimentario reciente en la Ensenada de San Simón, parte interna de la Ría de Vigo (Galicia, España). Gaia, Vol. 8, pp. 149-156.

55. Otero, X.L., Calvo de Anta, R.M. & Macías, F. (2006). Sulphur partitioning in sediments and biodeposits below mussel rafts in the Ría de Arousa (Galicia, NW Spain). Marine Environmental Research, Vol. 61, pp. 305-325.

56. Otero, X. L., Vidal-Torrado, P., Calvo de Anta, R. M. & Macías, F. (2005).

Trace elements in biodeposits and sediments from mussel culture in the Ría de Arousa (Galicia, NW Spain). Environmental Pollution, Vol. 136, pp. 119-134.

57. Otero, X.L., Sánchez, J.M. & Macías, F. (2000). Bioaccumulation of heavy metals in thionic fluvisols by a marine polychaete: the role of metal sulfides. Journal of Environmental Quality, Vol. 29, pp. 1133–1141.

58. Pekey, H., Karakaş, D., Ayberk, S., Tolun, L., & Bakoğlu, M. (2004). Ecological risk assessment using trace elements from surface sediments of İzmit Bay (Northeastern Marmara Sea) Turkey. Marine Pollution Bulletin, Vol. 48, pp. 946–953.

59. Petrovsky, E., Kapicka, A., Zapletal, K., Sebestova, E., Spanila, T., Dekkers, M.J., & Rochette, P. (1998). Correlation between magnetic parameters and chemical composition of lake sediments from northern Bohemia-preliminary study. Physics and Chemistry of the Earth, Vol. 23, pp. 1123-1126.

60. Piedracoba, S., Souto, C., Gilcoto, M. & Pardo, P.C. (2005). Hydrography and dynamics of the Ría de Ribadeo (NW Spain), a wave driven estuary. Estuarine Coastal and Shelf Science, Vol. 65, pp. 726-738.

61. Prego, R. & Cobelo-Garcia, A. (2003). Twentieth century overview of heavy metals in the Galician Rias (NW Iberian Peninsula). Environmental Pollution, Vol. 121, pp. 425– 452.

62. Prego, R., Otxotorena, U. & Cobelo-García, A. (2006) Presence of Cr, Cu, Fe and Pb in sediments underlying mussel-culture rafts (Arosa and Vigo rias, NW Spain). Are they metal-contaminated areas? Ciencias Marinas, Vol. 32, No.2B, pp. 339-349.

63. Rae, J.E. & Allen, J.R.L. (1993). The significance of organic matter degradation in the interpretation of historical pollution trends in depth profiles of estuarine sediment. Estuaries, Vol. 16, No.3B, pp. 678-682.

64. Raiswell, R. & Berner, R.A. (1986). Pyrite and organic matter in Phanerozoic normal marine shales. Geochimica et Cosmochimica Acta, Vol. 50, pp. 1967-1976.

65. Raiswell, R., Buckley, F., Berner, R. & Anderson, T. (1988). Degree of pyritization of iron as a paleoenvironmental indicator of bottom-water oxygenation. Journal of Sedimentary Petrology, Vol. 58, pp. 812-819.

66. Rey, D., López-Rodríguez, N., Rubio, B., Vilas, F., Mohamed, K., Pazos, O. & Bógalo, M.F. (2000). Magnetic properties of estuarine-like sediments. The study case of the Galician Rías. Journal of Iberian Geology, Vol. 26, pp. 151-170.

67. Rey, D., Mohamed, K., Bernabeu, A., Rubio, B. & Vilas, F. (2005). Early diagenesis of magnetic minerals in marine transitional environments: geochemical signatures of hydrodynamic forcing. Marine Geology, Vol. 215, pp. 215–236.

68. Rubio, B., Álvarez-Iglesias, P. & Vilas, F. (2010). Diagenesis and anthropogenesis of metals in the recent Holocene sedimentary record of the Ría de Vigo (NW Spain). Marine Pollution Bulletin, Vol. 60, pp. 1122-1129.

69. Rubio, B., León, I., Álvarez-Iglesias, P. & Vilas, F. (2008). Aerobic oxidation of suboxicanoxic sediments: implications for metal remobilization and release. Geotemas, Vol. 10, pp. 651-654.

70. Rubio, B., Gago, L., Vilas, F., Nombela, M.A., García-Gil, S., Alejo, I. & Pazos, O. (1996). Interpretación de tendencias históricas de contaminación por metales pesados en testigos de sedimentos de la Ría de Pontevedra. Thalassas, Vol. 12, pp. 137-152.

71. Rubio, B., Nombela M.A. & Vilas F. (2000a). Geochemistry of major and trace elements in sediments of the Ría de Vigo (NW Spain): An assessment of metal pollution. Marine Pollution Bulletin, Vol. 40, pp. 968-980.

72. Rubio, B., Nombela M.A. & Vilas F. (2000b). La contaminación por metales pesados en las Rías Baixas gallegas: nuevos valores de fondo para la Ría de Vigo (NO de España). Journal of Iberian Geology, Vol. 26, pp. 121-149.

73. Rubio, B., Pye, K., Rae, J. & Rey, D. (2001). Sedimentological characteristics, heavy metal distribution and magnetic properties in subtidal sediments, Ría de Pontevedra, NW Spain. Sedimentology, Vol. 48 No.6, pp. 1277-1296.

74. Rubio, B., Rey, D., Pye, K., Nombela, M.A. & Vilas, F. (1999). Aplicación de imágenes de electrones retrodispersados en microscopía electrónica de barrido a sedimentos litorales. Thalassas, Vol. 15, pp. 71-84.

75. Ruiz-Villarreal, M., Montero, P., Taboada, J.J., Prego, R., Leitão, P.C. & Pérez-Villar, V. (2002). Hydrodynamic model study of the Ria de Pontevedra under estuarine conditions. Estuarine Coastal and Shelf Science, Vol. 54, pp. 101-113.

76. Salomons, W. & Förstner, U. (1984). Metals in the hydrocycle. Springer-Verlag, Berlin, 349 p. Scoullos, M. & Oldfield, F. (1984). Mineral magnetic studies for a pollution monitoring of marine and estuarine sediments. VIIes Journal Etudes Pollution, CIESM/COI/PNUE, Lucerne.

77. Soto-Jiménez, M.F., Arellano-Fiore, C., Rocha-Velarde, R., Jara-Marini, M.E., RuelasInzunza, J. & Páez-Osuna, F. (2011). Trophic transfer of lead through a model marine four-level food chain: Tetraselmis suecica, Artemia franciscana, Litopenaeus vannamei, and Haemulon scudderi. Archives of Environmental Contamination and Toxicology, Vol. 61, No2, pp. 280-291

78. Souto, C., Gilcoto, M., Fariña-Busto, L. & Pérez, F.F. (2003). Modeling the residual circulation of a coastal embayment affected by wind-driven upwelling: Circulation of the Ría de Vigo (NW Spain). Journal of Geophysical Research, Vol. 108, No.C11, pp. 3340-3356.

79. Spassov, S., Egli, R., Heller, F., Nourgaliev, D.K. & Hannam, J. (2004). Magnetic quantification of urban pollution sources in atmospheric particulate matter. Geophysical Journal International, Vol. 159, pp. 555-564.

80. Subramanian, V. & Mohanachandran, G. (1990). Heavy metals distribution and enrichment in the sediments of Sourthern East Coast of India. Marine Pollution Bulletin, Vol. 21, pp. 324-330.

81. Summers, J.K., Wade, T.L., & Engle, V.D. (1996). Normalization of metal concentrations in estuarine sediments from the Gulf of Mexico. Estuaries, Vol. 19, No.3, pp. 581-594.

82. Taylor, S.R. (1964). Abundance of chemical elements in the continental crust: a new table. Geochimica et Cosmochimica Acta, Vol. 28, pp. 1273-1285.

83. Tenore, K. & González, N. (1975). Food chain patterns in the Ría de Arosa, Spain: An area of intense mussel aquaculture, In: 10th European Symposium of Marine Biology, G. Persoone & E. Jaspers (Eds.), Vol. 2, pp. 601-619, Ostend, Belgium, September 17-23, 1975, Universa Press, Wetteren.

84. Tessier, A. & Campbell, P.G.C. (1988). Partitioning of trace metals in sediments, In: Metal Speciation: Theory, Analysis and Application, J.R. Kramer & H.E. Allen (Eds.), 183-199, Lewis Publishers, Inc.

85. Tessier, A., Campbell P.G.C. & Bisson, M. (1979). Sequential extraction procedure for the speciation of particulate trace metals. Analytical Chemistry, Vol. 51, pp. 844-851.

86. Turekian, K.K. & Wedepohl, K.H. (1961). Distribution of the elements in some major units of the Earth›s Crust. Geological Society American Bulletin, Vol. 72, pp. 175-192.

87. Ure, A.M., Quevauviller, P., Muntau, H. & Griepink, B. (1993). Speciation of heavy metals in soils and sediments: an account of the improvement and

harmonization of extraction techniques undertaken under the auspices of the BCR of the Commission of the European Communities. International Journal of Environmental Analytical Chemistry, Vol. 51, pp. 135–151.

88. Valette-Silver, N.J. (1993). The use of sediment cores to reconstruct historical trends in contamination of estuarine and coastal sediments. Estuaries, Vol. 16, pp. 577-588.

89. Varekamp, J.C. (1991). Trace element geochemistry and pollution history of mud flats and marsh sediments from the Connecticut River estuary. Journal of Coastal Research, Vol. SI11, pp. 105-124.

90. Vilas, F., Nombela, M.A., García- Gil, E., García Gil, S., Alejo, I., Rubio, B. & Pazos, O. (1995). Cartografía de sedimentos submarinos, Ría de Vigo. E: 1:50000. Ed. Xunta de Galicia, 40 p.

91. Vilas, F., Bernabeu, A.M., Rubio, B., & Rey, D. (2010). Estuarios, rías y llanuras de marea, In: Sedimentología. Del proceso físico a la cuenca sedimentaria, A. Arche (Ed.), 619-673, CSIC, ISBN 978-84-00-09148-3, Madrid, Spain.

92. Vilas, F., Bernabeu, A.M. & Mendez, G. (2005). Sediment distribution pattern in the Rias Baixas (NW Spain): main facies and hydrodynamic dependence. Journal of Marine Systems, Vol. 54, pp. 261–276.

93. Vilas, F., Rey, D., Rubio, B., Bernabeu, A.M., Méndez, G., Durán, R. & Mohamed, K. (2008). Los fondos de la Ría de Vigo: composición, distribución y origen del sedimento, In: Una aproximación integral al ecosistema marino de la Ría de Vigo, A. González-Garcés, F. Vilas & X.A. Álvarez (Eds.), 17-50, Instituto de Estudios Vigueses, Vigo, Spain.

94. Wangersky, P.J. (1986). Biological control of trace metal residence time and speciation: A review and synthesis. Marine Chemistry, Vol. 18, pp. 269-297.

95. Wedepohl, K.H. (1971). Environmental influences on the chemical composition of shales and clays, In: Physics and chemistry of the Earth, L.H. Ahrens, F. Press, S.K. Runcorn & H.C. Urey, (Eds.), 307-331, Oxford, Pergamon.

96. Wedepohl, K.H. (1991). The composition of the upper Earth›s crust and the natural cycles of selected metals: metals in natural raw materials; natural resources. In: Metals and Their Compounds in the Natural Environment, E. Merian (Ed.), 3-17, Weinheim, VCH. Zoller, W.H., Gladney, E.S., Gordon, G.E. & Bors, J.J. (1974). Emissions of trace elements from coal fired power plants, In: Trace Substances in Environmental Health, D.D. Hemphill (Ed.), Vol. 8., 167-172, Rolla, University of Missouri, Columbia.

Chapter 6

FLUORESCENTLY LABELED PHOSPHOLIPIDS – NEW CLASS OF MATERIALS FOR CHEMICAL SENSORS FOR ENVIRONMENTAL MONITORING

George R. Ivanov[1], Georgi Georgiev[2] and Zdravko Lalchev[2]

[1]Department of Physics, Faculty of Hydraulic Engineering, University of Architecture, Civil Engineering and Geodesy & Advanced Technologies Ltd

[2]Department of Biochemistry, Faculty of Biology, Sofia University, Sofia, Bulgaria

INTRODUCTION

Reliable environmental monitoring strongly depends on the quality of chemical and biochemical sensors. There are still some unsolved problems especially when higher selectivity is required. In this chapter we propose a new class of materials – fluorescently labeled phospholipids, which can be used as chemical and biochemical sensors. We focus our attention on the most promising compound - head labeled with nitrobenzoxadiazole (NBD) phosphatidyl ethanolamines. We were the first to study these compounds in one component layers. Three new phenomena were discovered for this material that can be used for successful sensor applications. In our research we use the Langmuir and Blodgett method for investigation of organic monolayers at the air-water interface and for thin film deposition. It can also independently be used for environmental monitoring, e.g. water purity monitoring.

The Langmuir and Blodgett method - Use of the Langmuir film method for measuring the quality of water in natural basins

Probably the most promising method for the creation of supramolecular architectures in a well controlled manner is the method of Langmuir and Blodgett. This method is schematically described on Fig. 1. A trough, usually manufactured from well cleanable and inert material Teflon® (polytetrafluorethylene) is filled with ultra pure water. The organic substance to be investigated and deposited is spread from a solution. Molecules of the substance should be with the proper hydrophilic–hydrophobic balance so they remain at the air-water interface and do not penetrate the water. These molecules consist of a hydrophilic head group, which is attracted to the water and a hydrophobic tail (most often – hydrocarbon groups) which is repelled by the water. Some time is allowed for the solvent to evaporate until something like a 2D gas of the investigated molecules remains at the airwater interface. This is called Langmuir film. After this a compression of the organic monolayer with a barrier is started.

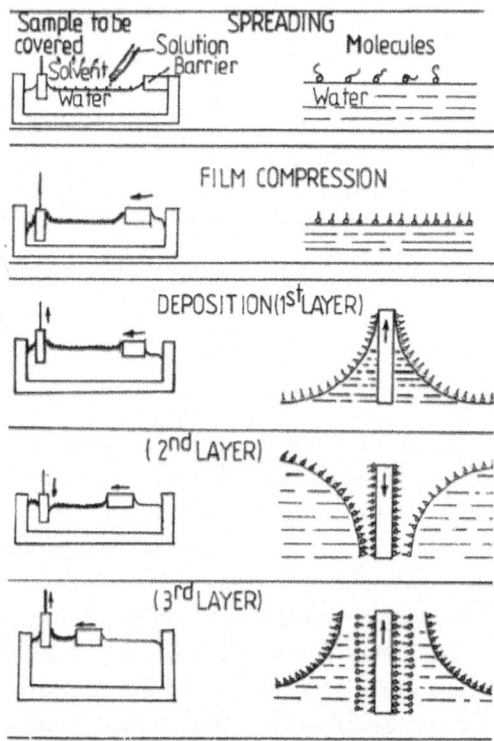

Figure. 1: The Langmuir and Blodgett method for investigation of organic monolayers at the air-water interface (Langmuir films) and for thin film deposition.

The surface pressure is constantly measured by a surface tensiometer (not shown on Fig. 1). A surface pressure – mean area per molecule isotherm can be measured. Additionally, the trough can be integrated with another instrument, e.g. fluorescence microscope and additional data can be gathered. At any point the compression of the monolayer can be stopped, the regime of constant surface pressure maintenance can be switched on, and a deposition on a solid substrate can be started. If we have a hydrophilic substrate which attracts the molecules' heads and it is immersed in the water before the monolayer spread then on the first movement up the first monolayer is deposited. Now the substrate becomes hydrophobic because the hydrophobic molecular tails are on the surface and on a downward movement of the substrate a second layer is deposited. Again the substrate surface becomes hydrophilic and on a subsequent movement upwards a 3rd layer is deposited. And the layer by layer deposition can continue. This deposition method has the following advantages compared to alternative methods of thin film deposition like spin coating, vacuum evaporation and self-assembling

• This is a discreet method of deposition, a complete layer after complete layer are deposited. This gives the possibility for a very precise control of the film thickness. Phospholipid molecules are with height of around 3 nm but also the tail chain length can be varied so a thickness control to 0,1 nm accuracy can be achieved.

• The molecules are well oriented. This is very important for some applications, e.g. nonlinear optics.

• The molecules are prearranged on the water surface before the deposition process. Thus the surface density of defects is much smaller.

• This is the most suitable method for molecular architecture. Different layers can be from different molecules. Inside the layer mixture of different molecules can be used. There is a possibility for interface reactions (e.g. CdSe nanoparticles incorporated in the lipid matrix can be prepared in this way). Also absorption from the water subphase of e.g. proteins is possible.

The development of rapid, economical and sensitive techniques for characterization of the purity of natural and drinking water represents leading ecological problem. The surface properties of natural waters (sampled from rivers, lakes and gulfs) are already successfully used for evaluation of the ecology standard and purity of water basins. Recently Pogorzelski et al. (e.g. Pogorzelski and Kogut, 2003 and references there in) proposed a Langmuir monolayer based technique which by measuring the surface pressure-area isotherm of the samples collected from a range of natural water basins, yields the so called "structural signatures" of water, which adequately predicted the quality and the purity of the basin. Major advantage of the Langmuir monolayer technique is that it combines ease of use, high sensitivity and possibility for rapid application with much lower price in comparison with the most commonly used chromatography techniques. The "structural signatures" of samples of natural water result from the generalized scaling procedures applied to the surface pressure-area isotherms of the natural films. They appear to reflect in a quantitative and sensitive way the film composition, film solubility and the miscibility of its components, the kinetic mobility of surfactant molecules, and the compound's surface concentration. It is suggested that certain classes of film-forming components or "end-members" may dominate the static and dynamic surface properties. Variation in the surface rheological parameters of source-specific surfactants is postulated to reflect organic matter dynamics in natural waters. The reported results demonstrate that natural films are complex mixtures of biopolymeric molecules covering a wide range of solubilities, surface activities and molecular masses with a complex interfacial architecture. The natural water's (sea, lakes, rivers) surface microlayer plays an important role in airwater interactions. A certain fraction of dissolved organic matter in the water basins has surface-active properties and makes up a very reactive part of the organic matter (Druffel and Bauer, 2000). According to their surface-active properties, these substances accumulate at water interfaces thereby influencing gas, mass, momentum and energy transfer between the so modified interfaces. The intensity of the film-effect depends strongly on film surface concentration, composition, and viscoelastic properties of the surface microlayer films. Processes taking place in the water body bulk (biological event, organic matter transformation or degradation, anthropogenic effluents, etc.) are sources of surface-active substances. Surfactants are concentrated at the air–water interface by numerous physical processes including diffusion, turbulent mixing, bubble and particle transport, and convergent circulations driven by wind, tidal forces, and internal waves.

The composition of the natural water's surface films is largely undefined, although significant enrichments of many specific classes of compounds in the surface microlayer have been demonstrated (for review, see Hunter and Liss, 1981). Natural sea/river/lake films mostly resemble layers composed of proteins, polysaccharides, humic-type materials and long chain alkanoic acid esters (Van Vleet and Williams, 1983). The generally accepted view is that the ubiquitous background of degraded biopolymeric and heterogeopolymeric material in the bulk waters has the potential to generate measurable surface films even in oligotrophic waters. Specific inputs of fresh bioexudates and biopolymeric material from local events are superimposed on this background signal. The emphasis in the published studies (Pogorzelski, 2001; Pogorzelski and Kogut, 2001 a, b; 2003 a, b) has been on the multicomponent character of natural surfactant films and the consequent complexities involved in any attempt to predict the interfacial viscoelastic properties (playing a crucial role in modeling of physical systems with surface filmmediated interfaces) due to the diverse chemical composition of such films. A complete compositional or structural description of naturally occurring surfactants is not currently feasible. Instead of analyzing the chemical composition, it should be possible to scale microlayer film surface pressure–area isotherms in terms of the structural parameters, reflecting the natural film morphology, and resulting from the generalized physical formalisms adopted to multicomponent surfactant films. Particularly efficient approach to scale the surface pressure (π) - area (A) isotherms of the microlayer films adsorbed on the surface of natural waters proved to be the fitting of the isotherms by the virial equation of state as proposed by Barger and Means (1985):

$$\pi A = C_0 + C_1 \pi + C_2 \pi^2 \tag{1}$$

where C_0, C_1, C_2 are virial coefficients, and A is the film area (in cm^2). As demonstrated by Pogorzelski, 2001; Pogorzelski and Kogut, 2001 a, b, 2003 a, b C_1 can be interpreted as the limiting specific area occupied by the molecules in the film, and C_0 can be assumed equal to XnkT in the limiting case when π approaches zero:

$$C_0 = XnkT \tag{2}$$

where the parameter X is related to the interaction forces between molecules in the monolayer, n is the number of molecules in the unknown film, k is the Boltzmann constant, T is the temperature in degrees Kelvin. The limiting specific molecular area Alim (in nm^2) can be expressed as (Frew and Nelson, 1992):

where the parameter X is related to the interaction forces between molecules in the monolayer, n is the number of molecules in the unknown film, k is the Boltzmann constant, T is the temperature in degrees Kelvin. The limiting specific molecular area A_{lim} (in nm^2) can be expressed as (Frew and Nelson, 1992):

$$A_{lim} = C_1 n^{-1} \times 10^{14} \tag{3}$$

Since the area covered with a film of a pure substance at a constant value of k is directly proportional to the mass m on the surface, it is possible to extend this computation to all the natural films (Barger and Means, 1985). Similarly, fitting procedures can be applied to quantitatively analyze the hysteresis of natural water's surface films when subjected to cyclic area compression/expansion, and also to describe the sample's surface pressure–temperature isochores.

Thus it is possible to avoid the expensive, time consuming and cumbersome analysis of the chemical composition of natural waters, and instead to characterize the sample's quality by the introduction of sensitive and much easier to obtain physicochemical "structural signatures" of the natural microlayer films. The structural state of natural water films, which can be incorporated with such source-specific markers of both biogenic and anthropogenic origin, can be assessed through the quantification of the parameters variability. They can be useful for tracking organic matter dynamics, as already established factors like the carbon to nitrogen C/N ratio (Bock and Frew, 1993), used in microlayer film studies, for instance. The main expectation of such studies is that variation in the surface rheological parameters of natural biosurfactant films manifested at the air–water interface could be followed to trace and map surface-active sourcespecific compounds spatial-seasonal-temporal evolutions. Compared to the evaluation of the ecological quality of natural waters, the characterization of the purity of drinking water poses higher challenges as it requires precise identification of even trace amount of detrimental ingredients. Some compounds, like the ions of heavy metals or membrane-active molecules, can have profound detrimental effect on the consumers' health even in very low doses that can not be detected by direct measurement of the sample's surface pressure. More precision quantitative measuring techniques are needed. For the purpose of such demanding measurements we further advanced the monolayer technique by introducing the use of fluorescently labeled LB solid supported phospholipid films. This is because fluorescence in some molecules is highly sensitive to even most delicate environmental changes (like slight changes in ionic strength, presence of quenchers in trace concentrations, etc.) LB films from these materials have high potential to be used as sensitive, selective and fast chemical sensors. In this new class of compounds for sensor applications - fluorescently labeled lipids, fluorescence intensity and lifetime are strongly influenced by minimal amounts of tested substances. In the following chapter we look in greater detail to the most promising fluorescence label in this class of compounds – the NitroBenzoxaDiazole (NBD) label. Then, results from our research of NBD labeled phospholipids at the airwater interface, as LB film on solid support, and molecular modeling are presented and discussed in view of sensor applications.

PREVIOUS RESEARCH OF NBD FLUORESCENTLY LABELED LIPIDS

Synthesis, where to the polar head (to the amino group) of egg phosphatidylethanolamine (PE) covalently is bound the NBD chromophore was first described by Monti et al. (1978). Due to the use of egg phosphatidylethanolamine (PE) tail length varies. Solution of NBD-PE in ethanol shows absorption maxima at about 330 nm and 460 nm, and the fluorescence maximum is at 525 nm. Fluorescent intensity in ethanol is proportional to the concentration in the range of 1 ng/ml to about 3 µg/ml. This article studied the dependence of the intensity of absorption and fluorescence of

NBD-PE to the change in dielectric constant of the solvent used. The observed strong sensitivity of the spectral characteristics of NBD-PE to the polarity of its surrounding makes this molecule an excellent indicator of conformational changes in the membrane. This article notes that small amounts of non-ionic detergent can lead to increase in fluorescence intensity and peak position change. Without problems is the incorporation of NBD phospholipid molecules in liposomes and biological membranes. For the NBD chromophores the angle between absorption and emission dipole is about 25 ° (Thompson et al., 1984) and therefore the real environment of the chromophores may be different for absorption and emission. Overview of the spectral characteristics of NBD was made by Suzuki and Hiratsuka (1988). There are a large number of papers in which NBD labeled lipids are used especially as a small percentage additive in the biomembrane studies. Here we will review only the work related to the chemical sensor applications of these molecules. The presence of large paramagnetic metal ions can be monitored by the fluorescence quenching of the NBD chromophore. Morris et al., 1985 used cobalt ions to quench the fluorescence of NBD-PE incorporated in phospholipid liposomes. Large paramagnetic ions such as Co^{2+} efficiently quench the fluorescence. The mechanism that is suggested is of lateral diffusion of Co-lipid complex followed by collisional quenching with NBD-PE. The addition of the chelator EDTA restores the initial fluorescence to 90%. EDTA quenches itself about 10% of the fluorescence. Fluorescence is quenched in the outer layer of the liposomes within milliseconds after the addition of cobalt ions, then, if possible, it penetrates the inner layer. For small monolayer liposomes the process is 10-20 times slower, but in all cases completed in the first few seconds. This technique is used also for measuring the surface potential of the membrane. Another paramagnetic ion copper Cu^{2+} is also used for NBD fluorescence quenching (Rajarathnam et al., 1989). Chattopadhyay and London (1987) proposed a method for measuring the position of NBD chromophores in the biomembrane by quenching its fluorescence by spin-labeled in a different position phospholipids. A comparison of the fluorescence intensity is made when two located in different depths quenchers are used. Results show that the greatest distance from the center of the bilayer is for NBD chromophores in the molecules of the Dipalmitoyl-NBD-PE – 1,42 nm. This means that due to its strong hydrophilicity the NBD chromophore is folded to the hydrocarbon tails and is positioned on the border tail - head, which is 1,5 nm from the center of the bilayer. For 6-NBD-PC this distance is 1,22 nm, for 12-NBD-PC, this distance is 1,26 nm, i.e. the tail in which is the NBD chromophore is folded and goes to the water surface. In this paper is calculated the critical distance Rc, below which the fluorescence of NBD is effectively quenched by the spin-label – 1,2 nm. Calculations show that if fluorescence is quenched due to presence of acceptor this distance is 10% larger.

Another important characteristic of the NBD chromophore that can be used in sensor applications is the dependence of its fluorescence lifetime on the polarity of the surrounding media. In general, reducing the polarity of the environment increases the lifetime. Lifetime of dilauroyl and dimiristoyl-NBD-PE in liposomes of egg lecithin is 6-8 ns (Arvinte et al., 1986). Detailed analysis of the fluorescence lifetime characteristics of NBD-aminohexane acid (NBD-NH(CH$_2$)$_5$C0$_2$H) at low concentrations in solvents of different polarity and donor hydrogen connection strengths was conducted by Lin and Struve, 1991. This substance has aminoalkane side chain similar to chains in which NBD chromophore is conected to phospholipids and the results are comparable. The conclusions are that the line shift of absorption and luminescence is due to the polarity of the solvent, while the drop in luminescence intensity due to non radiation transitions is much more affected by the hydrogen connection strengths. Fluorescence lifetimes in aprotic solvents is from 7,37 ns in DMSO to 10,6 ns in ethyl acetate, but are shorter in alcohols (5,65 ns in methanol). Extremely fast is the NBD luminescence in water - 0,933 ns. Low quantum yield in water is explained by anomalously short lifetime of non radiation transitions combined with radiation transitions which are with 3 times longer lifetime than those in other solvents. Oida et al., 1993 developed the so-called Fluorescence Lifetime Imaging Microscopy (flimscopy) which uses DP-NBD-PE and rhodamine labeled lipids. Fluorescent transduction of changes in the structure of the lipid membranes shows properties necessary for biosensor applications. When connected with the substrate a single membrane associated "receptor" protein may affect a significant number of surrounding molecules via electrostatic interactions, spatial interactions, interface changes in ionic strength or pH. The result is that: 1) perturbation of the lipid layer that is caused by the interaction receptor - ligand can be qualitatively related to the degree of connectivity, and 2) have amplified the original signal after the interaction of biomolecules. Placing a "receptor" protein in the phospholipids layer, which simulate the biological membrane, and provides improved stability against denaturation of the protein, gives biosensors with improved operational life span. Mixed lipid monolayers containing small amounts of DP-NBD-PE were shown to be able to convert changes in pH due to the hydrolytic enzyme activity at the membrane interface. This conversion scheme is used to determine the acetylcholine by acetylcholinesteraze (Brennan et al., 1990) and urea by ureaze (Brennan et al., 1992, 1993). In these studies a small concentration (about 1 mol %) of DP-NBD-PE and the respective enzyme are added in the phospholipid membrane. Changes in interface pH caused by hydrolytic enzyme reaction, lead to a change in the ionization of acidic phospholipid heads. This causes a change in the forces of electrostatic repulsion between neighboring heads. Structural changes in the membrane lead to an analytical signal in the form of change of fluorescence intensity due to fluorescence selfquenching of NBD-group caused by local increase in concentration. A comparison of different fluorophores connected to the same position of a protein showed that NBD-group gives the highest sensitivity (typically 4 times better than the next fluorophore (see Brennan et al., 2000 and references therein). From the viewpoint of sensor applications of DP-NBD-PE important is the optimization of: a) the concentration of DP-NBD-PE molecules in the membrane, and b) the composition and structure of the phospholipids membrane. This is done by the Krull's group in Toronto (Brown et al., 1994 and Shrive et al., 1995). The results are applicable to both LB film layers and liposomes. Fluorescent measurements were performed on liposomes because the fluorescence signal from LB monolayers is weak and leads to significant errors. For the optimization process a model was developed for the fluorescence selfquenching of DPNBD-PE. It considers the probability for static quenching by the formation of emissionless traps consisting of pairs of statistical DP-NBD-PE molecules which are at critical distance Rc. The model also considers the dynamic quenching due to Förster transfer of energy from DPNBD-PE monomers to the traps. Assumptions in this model are: 1) statistical traps are formed according to two-dimensional equation of Perrin; 2) all DP-NBD-PE molecules that do not participate in the traps are uniformly distributed throughout the monolayer; 3) there is no diffusion during the lifetime of the excited state, 4) energy can move between and among fluorophores and traps, but once traps are reached energy immediately and without emission decreases; 5) passing of energy in more than one DP-NBD-PE molecule before reaching the trap is negligible. It is estimated that the distance at which the efficiency of Förster transfer of energy becomes 50% R0 = 2,55 nm and that Rc = 0,94 nm. The optimum concentration of DP-NBD-PE molecules is one in which the theoretical expression undergoes a maximum change, i.e. the second derivative of the expression to the change in concentration is calculated. According to theoretical calculations, the optimal concentrations were 0,027 and 0,073 DP-NBD-PE molecules per nm2. These values were the same within the experimental error when comparing results of three different types of liposome compositions.

Optimization of composition and structure of membrane phospholipids showed the need for structural heterogeneity in the membrane at microscopic and not at molecular level in order to produce significant changes in fluorescence intensity. In membranes without heterogeneity the signal change is only 5-6%. Heterogeneity is achieved by the mixing of dipalmitoyl phosphatidyl choline with dipalmitoyl phosphatidic acid at a ratio of 7:3. At surface pressure of 30 mN/m, which is considered the liposome pressure, this mixture gives domain structure as observed in Langmuir films by fluorescence microscopy. The resulting changes in the average fluorescence intensity on pH change in this case reaches 60%. The mechanism of response of the membrane is shown to depend on the surface potential (Nikolelis et al., 1992) and is the result of changes in the ionic double layer and the rearrangement of the lipid heads and tails. This indicates that the mechanism of response in these biosensors is much more complicated than changing the distance between the heads. Moreover, the choice of phospholipid for these biosensors must be based on constraints coming from the ionic strength and pH, imposed on the activity of immobilized, chemically selective protein as enzyme activity is highly dependent on pH (Brennan et al., 1994).

INVESTIGATIONS OF MONOLAYERS AT THE AIR-WATER INTERFACE

On Fig. 2 are shown the isotherms of head labeled Dipalmitoyl-NBD-PE (DP-NBD-PE, chemical formula is in Fig. 14) at three different temperatures and in the presence of cobalt ions in the water at 20° C. Along with these measurements the monolayer was studied with fluorescence microscopy. The results for 20° C are published and discussed in detail elsewhere (Ivanov, G.R. (1992)). This was the first time that fluorescence self quenching in organic monolayers at the air-water interface (Langmuir films) was described. Here on Fig. 3 for the first time we publish the fluorescence microscopy data at 5° C. At room temperature the average area per molecule in the liquid phase is 1,4 nm2, and in the solid phase - 0,45 nm2. Adding $CoCl_2$ in the water increases the surface area of the molecule in the solid phase to 0,67 nm^2. The addition of $CaCl_2$ (not shown) leads to a smaller increase in the area in the solid phase - 0,56 nm2.

The shape of the solid domains is due to an interplay of several forces: the growth kinetics which at these compression speeds is negligible; the edge energy at liquid phase – solid phase interface which is minimal for circular domains; and the electrostatic repulsion between the similarly oriented dipoles of the molecules which is minimized when the molecules are further apart.

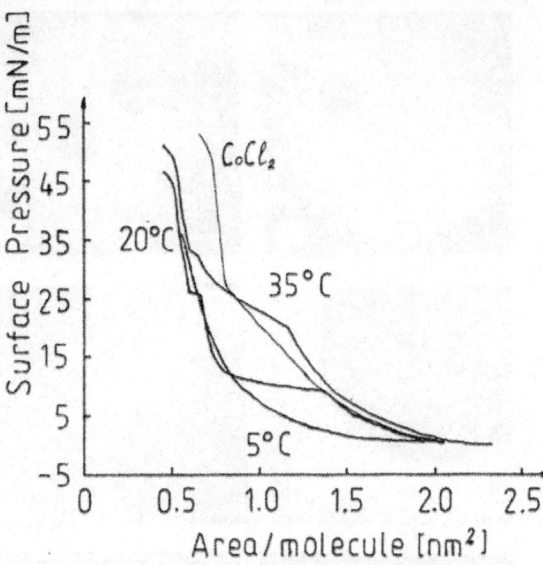

Figure. 2: Isotherms of monolayers from DP-NBD-PE at 5° C, 20° C, 35° C and at 20° C with the presence CoCl$_2$ in the water.

Due to the last force the domains repulse each other at low surface pressures and when the area of the solid domain is increased at higher pressures the domains obtain the dendridic shape which increases the distance between molecules. The fluorescence microscopy data at 5° C reveals also something that is not well observed at higher temperatures. The solid domains grow in size largely due to the attachment of smaller solid domains from the second population of solid domains.

The presented data here is for single component monolayers composed only from the fluorescently labeled in the head phospholipid DP-NBD-PE. The fact that we are able to observe the picture of phase coexistence with an excellent contrast is due to the fluorescence self quenching of this molecule in the solid phase when the distance between the molecules becomes much smaller and this allows for non radiation transfer of energy between them. This new phenomenon can be used with great success in sensor applications.

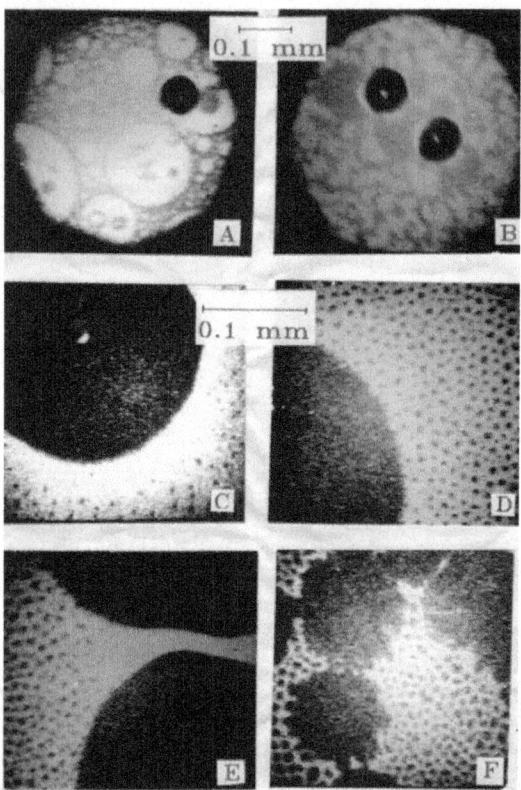

Figure. 3: Fluorescence microscopy of monolayers at the air-water interface from DP-NBD-PE at 5° C at different surface pressures Π: (A) Coexistence of liquid and solid phase at zero pressure; (B) the same at Π = 0,2 mN/m; (C) the occurrence of a second population of solid phase (the small black dots), which is repelled from the large solid domains Π = 6 mN/m; (D) the small solid phase domains overcome the repulsion of the big domains and begin to attach to them at Π = 9 mN/m; (E) large domains close the distance between them at Π = 12 mN/m; (F) domains of the solid phase obtain the dendridic shape and begin to merge with each other at Π = 15 mN/m.

If due to interactions of the sensor with the substance to be detected some conformational changes in the DP-NBD-PE molecules arise, this will lead to a strong measurable change in the fluorescence intensity. So this provides a second mechanism for component detection apart from the already discussed influence of the fluorescence peak maximum, intensity and lifetime on the polarity of the surrounding medium. On Fig. 4. is shown the equilibrium spreading pressure measurement of DP-NBD-PE at 20° C. In the first few seconds after placing some crystals from the material on the water surface the surface pressure increases insignificantly, then within a few seconds it increases by more than 15 mN/m. Then within a minute it reaches its equilibrium value of 19,6 mN/m. This value is quite high and indicates that at room temperature the majority of studies described in this work were conducted under equilibrium conditions. This is not quite so with much of the work conducted with LB films. For example the most widely studied arachidic acid has an ESP of 0 at room temperature indicating that the molecules are in metastable state when deposited.

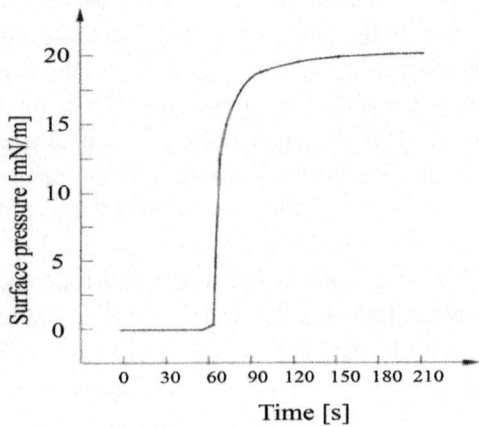

Figure. 4: Measurement of the equilibrium spreading pressure (ESP) of DP-NBD-PE at 20° C.

INVESTIGATIONS OF DEPOSITED ON SOLID SUPPORT THIN FILMS USING THE LB METHOD

For the possible sensor applications of DP-NBD-PE molecule it is important to obtain quality deposition on solid support. Deposition of multilayer structures of phospholipids by the LB method is complicated. Usually when immersed for the second layer deposition the first one is thrown off back into the water probably due to the presence of residual water between the layer and the substrate. A similar phenomenon was observed in the case of DP-NBD-PE. A typical way to overcome this problem is to use extremely low deposition speed. So no water is entrapped and relatively high quality multilayer structures are obtained. The problem is that very few commercial instruments have such low deposition speeds and these speeds are not suitable for industrial applications. Therefore we proposed a new method of obtaining multilayer structures. In it, after each immersion the bilayer was blown with heated air for several minutes at 55°C, which is below the melting temperature of the monolayer. The deposition results are compared in Table. 1. The quality of the film, or more precisely the amount of transferred substance, is judged by two criteria. On one hand this is the transfer ratio (Tr), which ideally is 1. But its determination has large errors due to difficulties in maintaining a constant surface pressure, causing the barrier to move back and forth without much correlation with the deposited layer. Far more accurate method is the measurement of the optical absorption of the film. In it the area of the line of maximum absorption of DP-NBD-PE at 465 nm is integrated. Usually, in order to improve the LB film quality metal ions are added in the water subphase. With fatty acids good results are obtained when divalent ions of heavy metals such as Cd2+ are added but in our case this did not lead to good results. For phospholipids it is usually recommended the use of univalent metal ions. They bind to the negatively charged phosphate head of the phospholipid and neutralize the electrostatic repulsion between neighboring layers. With our molecules best results were obtained with the use of NaCl. Table. 1 shows that the use of thermal treatment of the film increases the transfer rate which means that more substance is deposited. Peeling off the film on the down substrate movement is greatly reduced, although almost always the transfer ratio on the down movement is less than the coefficient in the upward movement of the substrate. The results of optical absorption also confirm that more substance is deposited when heat treatment is used.

To examine the effect of thermal heat treatment on the morphology of the resulting LB films comparative optical microscopy studies of samples G8 and G9 were performed. They are deposited at the optimum surface pressure of 35 mN/m and in the presence of NaCl in the water subphase. The only difference is that in G9 each bilayer was heat treated. Results from the dark field and phase-contrast microscopies clearly showed that considerably more substance is deposited when heat treatment is used and the density of defects is significantly lower. However, a significant number of defects and distinct domain structures with dimensions of tens microns can be seen. Another way of assessing the quality of the deposition is to measure the mass of the deposited substance. This is done when the LB film is deposited directly on a quartz crystal resonator. We used a resonator operating at 10 MHz frequency.

Table 1: Transfer ratios (Tr) and integral absorption of LB films from DP-NBD-PE. H means heat treated, DW means distilled water.

Deposition conditions			Transfer ratios (Tr) for the corresponding layer												Integral Absorption
No	Π (mN/m)	Subphase	Tr_1	Tr_2	Tr_3	Tr_4	Tr_5	Tr_6	Tr_7	Tr_8	Tr_9	Tr_{10}	Tr_{11}	Tr average	(a. u.)
G1	20	D.W./H.	3,2	0,4	2,1	-0,3	1,9	-0,5	1,7	-0,5	1,5	-0,6	1,5	1,40	0,45
G2	31	D.W./H.	2,7	0,8	1,3	0,7	1,3	1,0	1,1	1,0	0,6	1,1	0,3	1,21	3,16
G3	40	D.W./H.	2,1	0,1	0,7	0,4	1,1	0,7	1,0	0,6	0,7	0,7	0,7	0,70	2,56
G4	43	D.W./H.	1,1	0,5	0,5	0,3	0,6	0,4	0,7	0,6	0,7	0,4	0,7	0,80	2,02
G5	47	D.W.	1,3	1,4	1,1	0,9	1,1	0,9	1,9	0,9	1,3	0,9	0,9	1,67	-
G6	31	CdCl₂/H.	3,2	-4,8	4,8	-11,8	11,8	-11,0	11,7	-13,2	10,6	-10,0	9,2	0,35	0,74
G7	31	KCl/H.	4,0	1,9	2,1	1,1	2,3	1,4	3,1	1,3	2,1	1,0	2,3	1,80	0,82
G8	35	NaCl	2,3	-5,7	5,8	-5,6	5,4	-4,4	5,5	-3,7	4,0	-3,7	4,1	0,36	0,76
G9	35	NaCl/H.	2,7	0,4	3,4	0,4	3,1	0,6	2,5	0,5	2,7	1,3	3,4	2,29	2,59

It should be noted that sensitivity depends on the square of the frequency of the resonator. This is one of the most promising methods for creating gas and biochemical sensors which directly measures the mass of the substance to be detected. The accuracy of the measured frequency shift is below 0,1 Hz and the sensitivity of the method can be seen. It was possible to observe the water evaporation from the layer. In the middle of the resonator is evaporated a gold heater which can be used in chemical sensor applications for desorption of the absorbed studied substance. On the same resonator were sequentially deposited 21 layers, then frequency was measured, then 12 more layers were deposited and the frequency was measured, and finally 6 more layers were deposited and frequency measured (Fig. 5). In the case of an ideal deposition the mass (evaluated from the frequency change) should lie on a straight line. In our case this occurs with a deviation of around 10% indicating a high quality deposition. During this deposition every bilayer was heat treated.

To get an idea of the effect of heat treatment at molecular level polarization Fourier transformed infrared spectroscopy in attenuated total reflection mode of multilayer LB films deposited at 35 mN/m was conducted. The results are shown in Fig. 6. The bottom curve shows the spectrum of film obtained at very low deposition speed. The middle curve shows the same film heated for several minutes at 55 ° C. The upper curve shows the spectrum of film obtained by high speed deposition and heat treatment of each bilayer during the deposition. The most important change is the significant broadening of the absorption lines at 1244 cm⁻¹ (which is a mixture of lines Y_w CH₂ and Y_w P=0) and especially at 1738 cm⁻¹ (which corresponds to the Ys C=0) when heat treatment is performed. Particularly noticeable is this broadening when each bilayer is heated. The broadening of these lines indicates greater spread in the orientation of the corresponding parts of the DP-NBD-PE molecule. This is an expected result when heat treatment is performed.

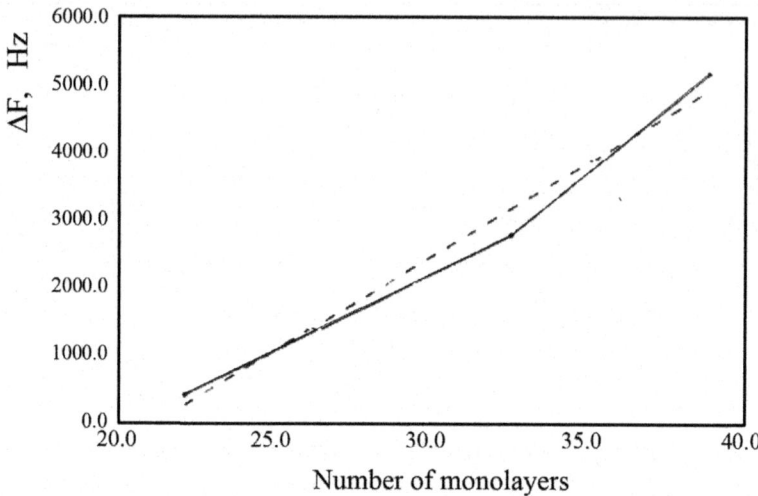

Figure. 5: Increase of the mass of 21, 33 and 39 LB monolayers from DP-NBD-PE measured with quartz resonator.

Additional information about the molecular arrangement and orientation of the chromophore head may be obtained from polarized absorption spectroscopy in the visible region. Simple NBD-derivatives have three main lines of absorption in the visible and near UV region - at around 420 nm, at 306-360 nm, and 225 nm (Lancet and Pecht, 1977). The first line corresponds to the line of 460 nm for NBD-labeled lipids and is due to intramolecular charge transfer (Paprica et al., 1993), which is accompanied by a large (~ 4 Debye) change in dipole moment (Mukherjee et al., 1994). The line absorption at 306-360 nm (for NBD-lipids ~ 335 nm) corresponds to a transition $\pi^* \leftarrow \pi$. Absorption spectra of NBD labeled lipids are shown in Fig. 7. The chloroform solution of a tail NBD labeled dipalmitoyl phosphatidylcholine has a maximum absorption at 475 nm. DP-NBD-PE in chloroform solution has a maximum at 457 nm. When deposited as LB film it has absorption maximum at 460 nm and a pronounced shoulder at 492 nm. This shoulder is probably due to Jaggregation of molecules in which the optical dipoles are arranged like a brick wall (Czikklely et al., 1970 a, b). J-aggregates are characterized by red shift of the absorption spectrum by about 30 nm and therefore this is the most likely interpretation. The presence of a large percentage of J-aggregates in the condensed phase in which deposition was carried out, may explain the fluorescence quenching in the solid phase

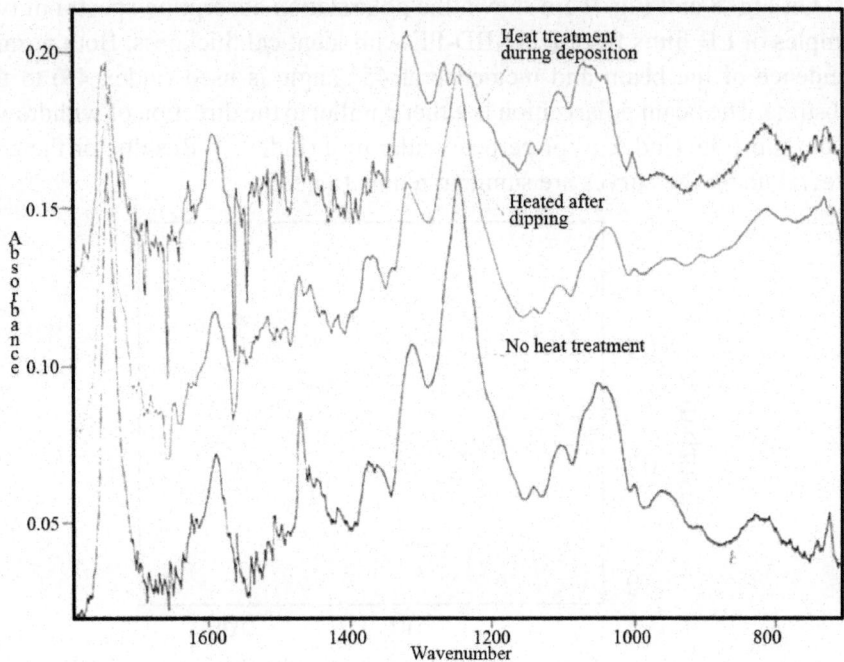

Figure. 6: The influence of heat treatment on multilayer LB films from DP-NBD-PE on the infrared spectra.

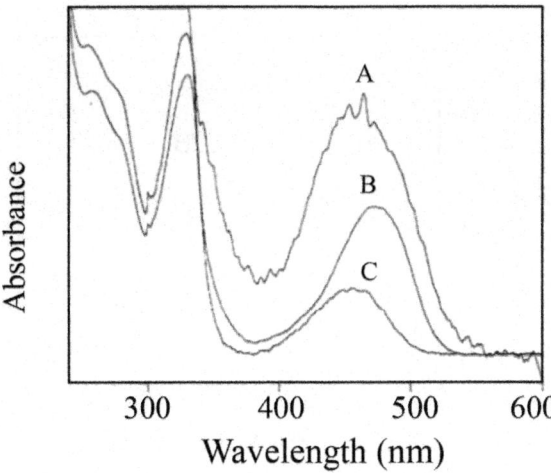

Figure. 7: UV-VIS absorption spectra of: A – LB film from DP-NBD-PE deposited at 35 mN/m and NaCl in water; B – chlorophorm solution of dipalmytoyl phosphatidyl choline labeled in the tail with NBD; C – chlorophorm solution of DP-NBD-PE.

On Fig. 8 and Fig. 9 are shown the polarization absorption spectra of two samples of LB films from DP-NBD-PE with identical thickness. Both normal incidence of the beam and incidence at 45° angle is used (index 45) to the substrate. The beam polarization is either parallel to the direction of withdrawal of the substrate (index p) or perpendicular to it (index s). Results for the area integral under the curves are summarized in Table. 2.

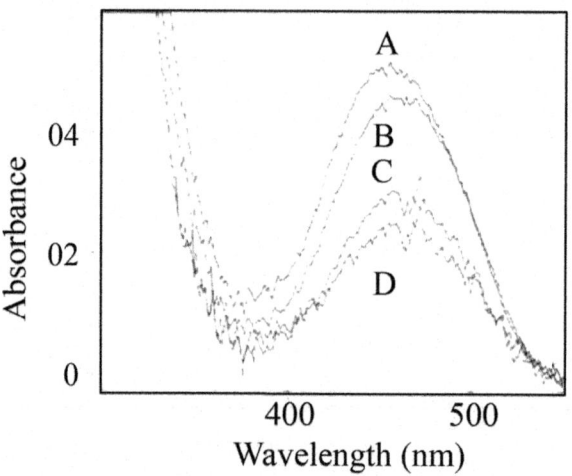

Figure. 8: Polarization spectroscopy of sample G5 - 11 LB layer structure of DP-NBD-PE deposited at Π=47 mN/m and fast withdrawal. A - p45; B - s; C - p; D - s45.

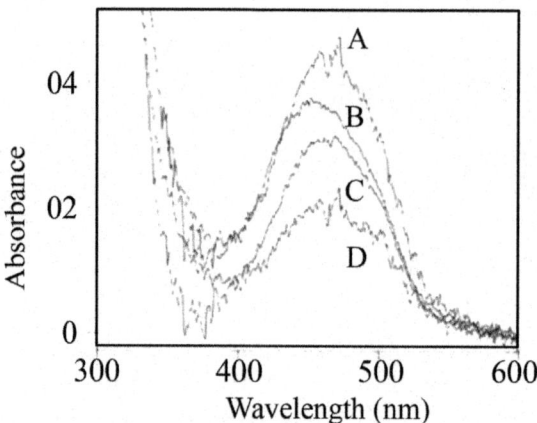

Figure. 9: Polarization spectroscopy of sample G9 - 11 LB layer structure of DP-NBD-PE deposited at Π = 35 mN/m with heat treatment and the presence of NaCl in the water. A - s45; B - p45; C - s; D - p.

Table 2: Integral areas of the 460 nm line of the absorption spectra at different polarization of the incident light for 2 different 11 layer LB films from DP-NBD-PE. G5 and G9 are the same films from Table 1. G9 was heat treated during the deposition.

LB film	p	s	p45	s45
G5	10,38	11,39	4,06	2,17
G9	1,93	2,27	2,55	3,35

From the polarization data it is seen that in both cases the optical dipoles of the molecules orient themselves with a slight advantage in the direction perpendicular to the direction of withdrawal (s component is larger than p component) From the data in table 2 order parameter for the molecules $<\cos^2\theta>$ can be calculated. The value for G5 film is 0.48. If the arrangement is perfect the value should be 1. We can see the weak ordering of molecules or more precisely of their chromophore heads. Fluorescence spectrum of LB multilayer structure of DP-NBD-PE deposited on a glass substrate in the solid phase was compared with the spectrum in chloroform solution in Fig. 10. Excitation was at the maximum absorption at 465 nm. About five-fold decrease in the fluorescence in the solid phase can be seen. When deposition is carried out in the liquid phase below surface pressure of 8 mN/m fluorescence is similar to that in a solution. In solid phase deposition the fluorescence quenching is almost complete. This is not due to reduced absorption, as absorption is increased by 10% in the solid phase deposited film. Interestingly, the addition of cobalt ions in the water subphase leads to almost complete recovery of fluorescence to its level in the solution (line not shown). This can be used in chemical sensors for heavy metal detection.

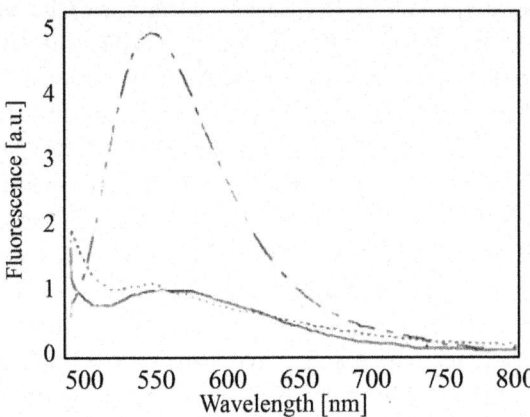

Figure. 10: Fluorescence spectroscopy of DP-NBD-PE of an LB film deposited in a solid phase (lower 2 curves) and in a solution. The spectra of an LB film deposited in a liquid phase is similar to the spectrum in solution.

An important question in view of sensor applications are the mechanisms of fluorescence selfquenching in NBD-labeled lipids. There is a similar study for octadecyl-rhodamine molecule (MacDonald, 1990), commonly used in Resonance Energy Transfer (RET) studies with NBD-molecules. The most obvious possibility for a mechanism to quench the fluorescence is the collision between an excited molecule and a quencher molecule. For this process is important the local concentration of fluorescently labeled molecules in the liquid phase. Calculations show that for collisions to occur with molecules with diffusion rate of 10 µm2/s and fluorescence lifetime of 4 ns then the distance between them must be less than 0.3 nm. And such small distance is impossible for molecules with 2 tails in a liquid phase, even assuming an increase in local concentration. In monolayers that undergo phase transition liquid - solid state diffusion rate decreases more than three orders of magnitude: from 50 µm2/s to 0,03 µm2/s (Peters and Beck, 1983). Fluorescence lifetime of dilauroil and dimiristoil NBD-PE in liposomes of egg lecithin is 6-8 ns (Arvinte et al., 1986). Thus, in the liquid phase diffusion during the excited state of the DP-NBD-PE is about 3 nm and in solid phase it is only about 0,02 nm. From the isotherm of DP-NBD-PE (Fig. 2) it can be seen that at room temperature the average area per molecule in the solid phase is 0,45 nm, and in liquid phase is 1,4 nm. Upon an assumption of cylindrical adjacent tightly packed molecules the distances between the centers of molecules in liquid phase is 1,33 nm and 0,78 nm in the solid phase. Obviously the collisional quenching mechanism of fluorescence is not applicable in solid phase. However, it is not clear why there was no significant fluorescence quenching in the liquid phase. The answer probably lies in the observation of Lin and Struve (1991) for extremely fast fluorescence lifetime of the NBD chromophore in water - 0,933 ns. Indeed, at the air-water interface the NBD-group in DP-NBD-PE molecule is positioned entirely in the water as shown by molecular conformational modeling (see Fig. 14). Another possible explanation is the dependence of the lifetime on the concentration of DP-NBD-PE (Brown et al., 1994). At all concentrations the lifetime is a two exponent function and the average ranged from 8,66 ns at 0,1% concentration; 5,39 ns at 10%; 1,32 ns at 40%; 0,97 ns at 50%. At 100% concentration (we work with films composed only from this molecule) the lifetime will probably be even smaller. The above mentioned distances between molecules in the liquid phase are several times larger than the distance which an excited molecule can diffuse for these small lifetimes. This explains why there is no fluorescence quenching in the liquid phase. Therefore, as a possible mechanism for fluorescence quenching remains energy transfer. For octadecyl-rhodamine molecule the distance at which energy transfer to monomer or dimer of the same molecule is 50% (Förster radius) is 5,5-5,8 nm for a transfer to monomer and 2,7 nm for a transfer to dimer. This calculation is done using a formula that takes into account the spectral overlap of the excitation and emission spectra for a given molecule. For NBDmolecules the Förster radius R0 is $2,55 \pm 0,15$ nm (Brown et at., 1994) . Research that shows anomalous long distance energy transfer (Draxler et al., 1989; Fromhertz and Reinbold, 1988) should also be taken into account. In LB films, they observed 20% efficiency of energy transfer over distances of 150 nm. Depending on mutual orientation of molecules, this distance may decrease to 30 nm. If the molecules, among which energy transfer takes place are different, then the lifetime of the donor molecule should increase with increasing its concentration. But in the case of octadecyl-rhodamine it decreases. Therefore, the basic mechanism of fluorescence quenching in octadecyl-rhodamine is emissionless energy transfer to the dimers of the same molecules. This lipid associated fluorescence label forms pre-bonded dimers. However this mechanism does not explain why fluorescence quenching in DP-NBD-PE molecules occurs only in the solid phase. As a most probable reason for the fluorescence quenching for DP-NBD-PE a model was developed that takes into account the likelihood of static quenching by forming emissionless traps consisting of pairs of statistical DP-NBD-PE molecules that are at critical distance Rc for trap formation (Brown et al., 1994; Shrive et al., 1995). Calculations give a value for R_c of 0,94 nm for this molecule. This value is greater than the intermolecular distances in solid phase layers of DP-NBD-PE and less than the distances in the liquid phase at room temperature. The addition of cobalt ions leads to increased intermolecular distances of up to 0,93 nm, which may explain the recovery of fluorescence intensity of the solid phase in presence of this ion which is known as good fluorescence quencher. Thus, this model explains fluorescence quenching in the solid phase, the lack of quenching in the liquid phase and the recovery of fluorescence in the solid phase with the addition of cobalt ions in the water. Given the anomalous long-distance energy transfer, observed in LB films (Draxler et al., 1989; Fromhertz and Reinbold, 1988) it is clear that critical is the formation of traps, which is precisely what we observed. In order to have a better understanding of the morphology of the LB films with submicrometer resolution we performed Atomic Force Microscopy (AFM) measurements. Fig. 11 shows AFM images with cross-sections along selected lines. On Fig. 11 A deposition was carried out at 7 mN/m, slightly above the transition from liquid to solid phase. The liquid – solid phase coexistence is clearly seen. Cross section through one of the solid domains reveals cylindrical structure with a height of 5,8 nm and a width of several tens of nanometers. Taking into account the height of the monolayer (3,1 nm), these structures can be interpreted as a bilayer in the solid phase. These structures are observed systematically in all our experiments. They occur above the main liquid-solid phase transition. Deposition in the liquid phase (Fig. 12) showed that there are no structures outside of the normal silicon wafer bumps of less than 0,5 nm in the monolayer. Cylindrical structures with bilayer height are observed also when the depositions are carried under the equilibrium spreading pressure of 19,6 mN/m (see Fig. 4). If the deposition at higher pressures is carried out these cylinders grow in height initially up to 13 nm for deposition at 33 mN/m (Fig. 11 B) and grow up to 35 nm, and in some cases to a hundred nanometers at 43 mN/m (Fig. 11 C). However, if the monolayer is allowed to relax under normal laboratory conditions for some time (50 days in this case), those cylinders again become with bilayer height (Fig. 11 D). These three-dimensional cylinders can not be obtained due to the deposition process and/or interactions with the substrate because their height depends on the deposition pressure. Layering in the vertical direction of the DP-NBD-PE molecule can be observed in these high structures, suggesting that the cylinders are made of DP-NBD-PE molecules rather than impurities. The question remains why it is energetically more favorable for part of the molecules to accumulate on one another and not to attach to the adjacent solid phase. Possible answer is that this can be due to kinetic effects if we are compressing the layer or depositing at higher speeds. Against this explanation is the lack of such structures in the liquid phase deposition under the same conditions (Fig. 12). Furthermore, our previous data from fluorescent microscopy at the air-water interface does not show the presence of kinetic effects at these speeds. Against this explanation is the fact that high structures relax to bilayer cylinders with time (Fig. 11 D). Data from Stark spectroscopy measurements show that DP-NBD-PE molecules tend to form centrosymmetrical non-polar structures. Thus for the second layer in the bilayer structures the molecules most probably flip over and we have a tail – tail contact. This is the first time that such 3D structures are observed when deposition is carried below the ESP. These structures are stable over time at least for several months. Their presence is very important for sensor applications because they ensure simultaneously high contact area and low film thickness. Thus high sensitivity at fast reaction times can be achieved (no slow diffusion needed). These self assembled 3D structures are part of our efforts in the hybrid assembly approach which combines the selfassembly technology with high performance robotic tools such as precise manipulators with submicron resolution and mechatronic handling (Kostadinov, 2010; Dantchev and Kostadinov, 2006).

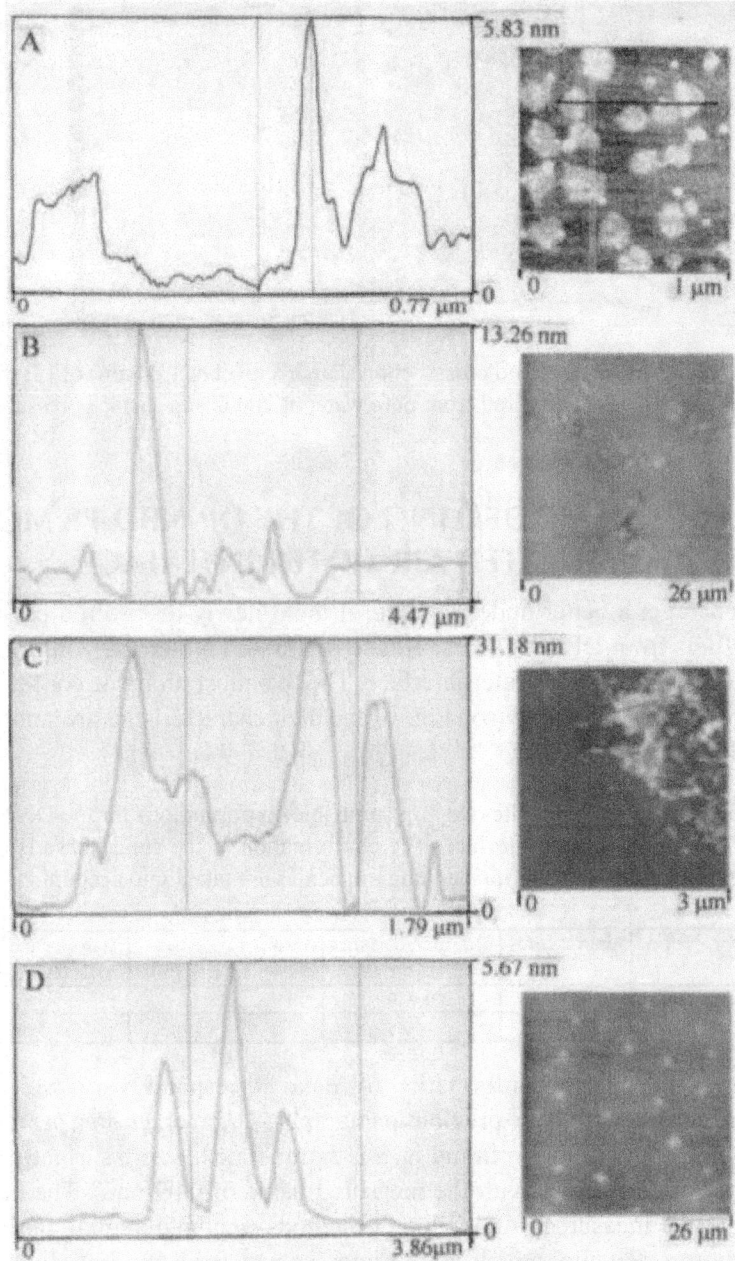

Figure. 11: AFM pictures and crosssections of LB monolayers from DP-NBD-PE deposited from pure water at 200 C and surface pressures of: (A) 7 mN/m; (B) 33 mN/m; (C) 43 mN/m; (D) 33 mN/m 50 days after the deposition.

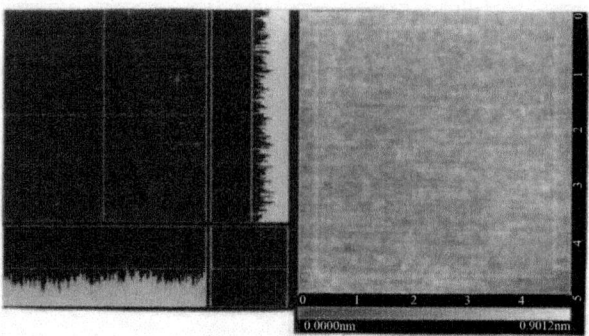

Figure. 12: AFM pictures and crosssection (left at a level of 1,04 nm) of LB monolayers from DPNBD-PE deposited from pure water at 200 C and surface pressures of 3,7 mN/m. Scan size is 5 μm.

MOLECULAR MODELLING OF THE DP-NBD-PE MOLECULE AT SIMULATED AIR-WATER INTERFACE

In order to get a better understanding of the 3 newly discovered phenomena in LB films from DP-NBD-PE we have performed molecular conformational analysis at simulated airwater interface. The two most probable conformations from this analysis are shown on Fig. 14 and their characteristics are summarized in Table. 3.

Table 3: Characteristics of the two most probable conformations for the DP-NBD-PE molecule at the air-water interface. The conformation in the solid phase is obtained only when interactions with surrounding molecules are taken into account.

Conformation	Height [nm]	phi - pho distance Δ [nm]	Area per molecule – single molecule [nm²]	Area per molecule – in a monolayer [nm²]
A – liquid phase	2,31	0,938	1,18	0,81
B – solid phase	3,25	1,396	-	0,69

The solid phase conformation A data correspond very well to the experimental data from the previous paragraphs. The average area per molecule is 0.66 nm2 if the isotherm from Fig. 2 is extrapolated to zero surface pressure, which almost coincides with the measured value of 0,69 nm2. The height of the molecule measured with different methods (including small angle X-ray diffraction) is 3,1 nm, which taking into account for some interdigitation of tails in the LB multilayer structure, matches the obtained here value of 3,25 nm. Also the predictions from the measurements of molecular orientation from polarized FTIR data are fulfilled. Particularly impressive is that the benzene ring is indeed perpendicular to the substrate. The liquid phase conformation in Fig. 13 A corresponds to the liquid film of DP-NBD-PE provided that the tails are even flatter and not in all-trans conformation. Indeed, the area per molecule in the liquid phase at zero pressure is about 2 nm, which is more than the 0,81 nm predicted by our model. Also AFM measurements showed a difference in height between the liquid and solid phase of about 1,6 nm. If the height of the molecule in the solid phase of 3,1 nm then for the height of the molecule in liquid phase remains 1,5 nm, which is less than the 2,31 nm in our model. So the assumption of greater tilting of the tails, leading to a lower height of the molecule and larger area is fully justified. Scanning surface potential microscopy measurements show that there is a big difference in the surface potential of the monolayer in liquid and in solid phase, most likely due to the different orientation of the strong dipole in the NBD-group. Indeed, the orientation of this group for the conformations on Fig. 14 show an angle of almost 90 ° between them, which may explain these results. In the liquid phase conformation A just over half of the chromophores are above the air-water interface, while the NBD group in the solid phase conformation B is deeply immersed in the water. The difference in dielectric constants of the environment in both cases leads to different fluorescence quenching by the water and may explain some phenomena observed by the fluorescence microscopy.

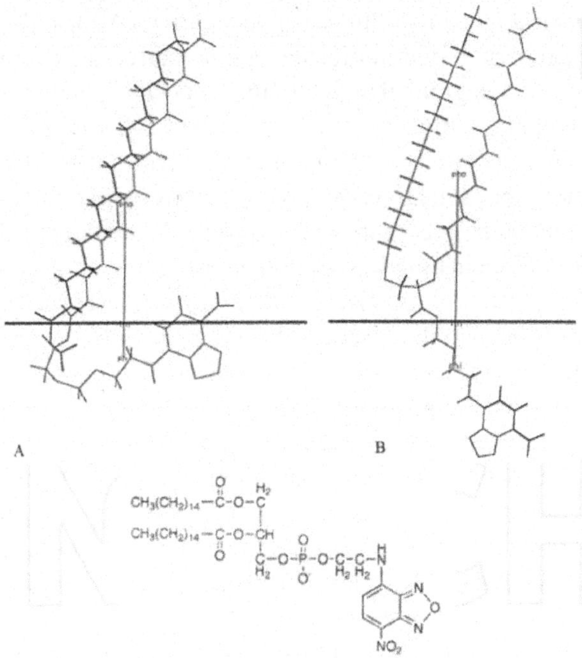

Figure. 13: The most probable conformations of DP-NBD-PE obtained from molecular conformational analysis. (A) The conformation in the liquid phase; (B) the confor-

mation in the solid phase. Also shown are the hydrophilic (phi) and hydrophobic (pho) centers. The line that connects them is the phi-pho distance Δ in Table 3. The chemical formula of the compound is shown below.

The hydrophilic-hydrophobic balance Φ for DP-NBD-PE was calculated to be 0,55. The distance Δ between the hydrophilic and hydrophobic center in the liquid phase conformation was 0,938 nm, while in the solid phase conformation it is 1,396 nm. According to the classification of Brasseur, 1990 (vol. 1, p. 210) both conformations fall in the zone with $\Phi > 0,2$ and $\Delta > 0,43$, which is characteristic of molecules that can self assemble in organized structures. Another prediction of Brasseur is that due to the large difference in Δ of the two conformations at the molecular level they can not mix and have to form two phases. Exactly this is what we observe in our experiments and there is coexistence between liquid and solid phase and well seen phase separated domains.

CONCLUSIONS

We were the first to start investigating systematically films at air-water interface and on solid support from fluorescently NBD-labeled phospholipids. Previous research has shown that this is the most promising fluorophore label for sensor applications. In our investigations we use the most advanced method for preparing supramolecular architectures from organic molecules – the method of Langmuir-Blodgett film deposition and research. This is a true nanotechnology process. Over the years we have discovered 3 new phenomena in these molecules which make them a promising candidate for chemical and biochemical sensor applications when fast response times, high sensitivity and selectivity are required. We were the first to observe fluorescence self quenching in insoluble monolayers at the air-water interface. Self quenching not only drastically decreases fluorescence intensity but also leads to a decrease in fluorescence kinetics times by an easily measurable change of over 30 %. Thus we have 2 independent channels to discriminate the effect in a sensor application. This phenomenon was understood in terms of molecular conformational change which leads to more dense molecular packing in the solid phase and radiationless energy transfer between the closely spaced molecular heads. So any change in the molecular environment which leads to this conformational change can be easily detected. The second new phenomenon describes the influence of heavy metals on fluorescent intensity in this type of molecules. Usually large paramagnetic metal atoms are strong fluorescence quenchers. But when they are dissolved in the water subphase during the deposition process the opposite effect was observed – the fluorescence intensity was increased. This was explained by the fact that these large atoms effectively increase intermolecular distances in the head of the molecules where they attach and thus decrease the fluorescence self quenching described above. This effect can be used for heavy metal detection. The third new phenomenon describes the possibility to deposit monolayers at some special conditions in which there is not only coexistence of solid and liquid phase but higher, bilayer or tens of nanometer high cylinders are deposited. This structure was very stable at least within several months period. It allows a much greater contact surface between the fluorescence molecules and the substances to be detected. Thus, high sensitivity sensors can be obtained without increasing their thickness. When the thickness is small so are the diffusion lengths which limit the sensor reaction time. Thus, very fast sensors with high sensitivity can be obtained. The possibility to mix selectively reacting proteins in this flexible phospholipid matrix can provide an unmatched selectivity. These properties are very important for environmental monitoring.

ACKNOWLEDGMENTS

This work is supported by the following contracts with the Bulgarian National Science Fund: VU-F1/2005, DO 0171/2008, DO 02-280/2008, DO 02-107/2008 and DO 02-167/2008.

REFERENCES

1. Arvinte, T., A. Cudd, and K. Hildenbrand. (1986). Biochim. Biophys. Acta, Vol. 860, p. 215

2. Barger, W.R., Means, J.C., (1985) Clues to the structure of marine organic material from the study of physical properties of surface films. In: Sigleo, A.C., Hattori, A. (Eds.), Marine and Estuarine Chemistry. Lewis Publishers, Chelsea, pp. 47–67

3. Brasseur, R. ed. (1990) Molecular Description of Biological Membranes by Computer Aided Conformational Analysis, CRC Press, Boca Raton Brennan, J.D. and UJ. Krull. (1992). Chemtech,Vol. 22, p. 227

4. Brennan, J.D., R.S. Brown, A. Delia Manna, K.M.R. Kallury, P.A. Piunno, and U.J. Krull. (1993). Sens. Act. B, Vol. 11, p. 109

5. Brennan, J.D., K.M.R. Kallury, and U.J. Krull. (1994). Thin Solid Films, Vol. 244, p. 898 Brennan, J.D., K.K. Flora, G.N. Bendiak, G.A. Baker, M.A. Kane, S. Pandey and F.V. Bright. (2000). Phys. Chem. B, Vol. 104, p. 10100

6. Brown, R.S., J.D. Brennan, and U.J. Krull. (1994). J. Chem. Phys., Vol. 100, p. 6019

7.   Chattopadhyay, A. and E. London. (1987). Biochemistry, Vol. 26, p. 39

8.   Czikklely, V., H.D. Forsterling and H. Kuhn. (1970 a). Chem. Phys. Lett., Vol. 6, p. 11

9.   Czikklely, V., H.D. Forsterling and H. Kuhn. (1970 b). Chem. Phys. Lett., Vol. 6, p. 207

10. Dantchev D. and K. Kostadinov (2006). On forces and interactions at small distances in micro and nano assembly process In: 4M 2006 Second International Conference on Multi-Material Micro Manufacture, Edited by: W. Menz, St. Dimov, B. Fillon. pp.241-245, Oxford: Elsevier

11. Draxler, S., M.E. Lippitsch and F.R. Aussenegg. (1989). Chem. Phys. Lett., Vol. 159, p. 231

12. Druffel, E.R.M., Bauer, J.E., (2000). Radiocarbon distributions in Southern Ocean dissolved and particulate organic matter, Geophys. Res. Lett., Vol. 27, pp. 1495–1498

13. Frew, N.M., Nelson, R.K. (1992). Scaling of marine microlayer film surface pressure-area isotherms using chemical attributes. J. Geophys. Res., 97, pp. 5291–5300

14. Fromhertz, P. and G. Reinbold. (1988). Thin Solid Films, Vol. 160, p. 347

15. Hunter, K.A., Liss, P.S., (1981). Organic sea surface films. In: Duursma, E.K., Dawson, R. (Eds.), Marine Organic Chemistry. Elsevier, New York, pp. 259–298.

16. Ivanov, G.R. (1992). First observation of fluorescence self-quenching in Langmuir films. Chem. Phys. Lett., Vol. 193, p. 323

17. Kostadinov, K. (2010) Robot Technology in Hybrid Assembly Approach for Precise Manufacturing of Microproducts, In: 12th Mechatronics Forum Biennial International Conference, Book 1, pp. 293-300, Swiss Federal Institute of Technology, Switzerland: ETH Zurich.

18. Lancet, D. and I. Pecht. (1977). Biochemistry, Vol. 16, p. 5150

19. Lin, S. and W.S. Struve. (1991). Photochem. Photobiol., Vol. 54, p. 361

20. Macdonald, R.I. (1990). J. Biolog. Chem., Vol. 265, p. 13533

21. Monti, J.A., S.T. Christian, and WA. Shaw, (1978). J. Lipid Research, Vol.19 p. 222

22. Morris, S.J., D. Bradley, and R. Blumenthal. (1985). Biochim. Biophis. Acta, Vol. 818, p. 365

23. Mukherjee, S., A. Chattopadhyay, A. Samanta and T. Soujanya. (1994). Phys. Chem., Vol. 98, p. 2809

24. Nikolelis, D.P, J.D. Brennan, R.S. Brown, and U.J. Krull. (1992). Anal.

Chim. Acta., Vol. 257, p. 49

25. Oida, T., T. Sako, and A. Kususmi. (1993). Biophys. J., Vol. 64, p. 676

26. Paprica, P. A., N. C. Baird and N. O. Petersen. (1993). J. Photochem. Photobiol. A: Chem., Vol. 70, p. 51

27. Peters, R. and K. Beck. (1983). Proc. Natl. Acad. Sci. US., Vol. 80, p. 7183

28. Pogorzelski S.J. and Kogut, A. D., (2003) Structural and thermodynamic signatures of marine microlayer surfactant films, J. Sea Research., Vol. 49 pp. 347–356

29. Pogorzelski, S.J., (2001). Structural and thermodynamic characteristics of natural marine films derived from force-area studies. Colloids Surfaces A: Physicochem. Eng. Aspects., Vol. 189, 163–176

30. Pogorzelski, S.J., Kogut, A.D., (2001). Static and dynamic properties of surfactant films on natural waters. Oceanologia, Vol. 43, 223–246

31. Pogorzelski, S.J., Stortini, A.M., Loglio, G. (1994). Natural surface film studies in shallow coastal waters of the Baltic and Mediterranean Seas. Cont. Shelf Res., Vol. 14, pp. 1621–1643

32. Rajarathnam, K., J. Hochman, M. Schindler, and S. Ferguson-Miller. (1989). Biochemistry, Vol. 28, p. 3168

33. Shrive, J.D.A., J.A. Brennan, R.S. Brown, and U.J. Krull. (1995). Appl. Spectroscopy, Vol. 49, p. 304

34. Suzuki, H. and H. Hiratsuka, (1988). In: Nonlinear Optical Properties of Organic Materials, SPIE Vol. 971, p. 97

35. Thompson, N. X., H. M. McConnell and T. P. Burghardt, (1984). Biophys. J., Vol. 46, p. 739

36. Van Vleet, E.S., Williams, P.M. (1983). Surface potential and film pressure measurements in seawater systems. Limnol. Oceanogr., Vol. 28, pp. 401–414

Chapter 7

RELEVANT ISSUES AND CURRENT DIMENSIONS IN GLOBAL ENVIRONMENTAL CHANGE

Julius I. Agboola[1,2]

[1]United Nations University, Institute of Advanced Studies, Operating Unit in Ishikawa/Kanazawa, 2-1-1 Hirosaka, Kanazawa, Ishikawa,Japan

[2]Department of Fisheries, Faculty of Science/ Centre for Environment and Science Education (CESE), Lagos State University, Ojo, Lagos Nigeria

INTRODUCTION

Global environmental change (GEC) includes both systemic changes that operate globally through the major systems of the geosphere-biosphere, and cumulative changes that represent the global accumulation of localized changes. The importance and awareness of GEC has greatly increased since the second UN Conference on Environment and Development (UNCED) in 1992. During the last two decades, GEC research programs around the world have advanced our understanding of the Earth's ever-changing physical, chemical, and biological systems and the growing human influences on these systems. On the basis of current knowledge attention is now focused on the critical unanswered scientific questions that must be resolved to fully understand and usefully predict future's GEC. It is hoped that measurable significant progress would be made in the forthcoming Earth Summit 2012, formerly known as United Nations Conference on Sustainable Development (UNCSD) or Rio+20 scheduled for Rio de Janeiro in June 2012. Generally, the earth's climate system varies naturally across a range of temporal scales, including seasonal cycles, inter-annual patterns such as the El Niño/La Niña-Southern Oscillation- ENSO, inter-decadal cycles such as the North Atlantic and Pacific Decadal oscillations, and multimillenial-scale changes such as glacial to inter-glacial transitions (Harley et al., 2006). This natural variability is reflected in the evolutionary adaptations of species and large-scale patterns of biogeography. In all, human activities play an important part in virtually all natural systems and are forces for change in the environment at local, regional, and even global scales. Human drivers of GEC include consumption of energy and natural resources, technological and economic choices, culture, and institutions. The effects of these drivers are seen in population growth and movement, changes in consumption, deor reforestation, land-use change, and toleration or regulation of pollution, and other issues highlighted in section two of this chapter. For instance, the Intergovernmental Panel on Climate Change (IPCC) reports that, if global average temperatures exceed 2°C there will be irreversible impacts on water, ecosystems, food, coastal zones and human health. We have a 50% chance of avoiding a 2°C warming if we stabilize greenhouse gases at 450 ppm CO_2 eq (parts per million carbon dioxide equivalents). Recent evidence suggests even more rapid change, which will greatly, and in some case irreversibly, affect not just people, but also species and ecosystems. The schematic framework on the current state of GEC, representing anthropogenic drivers, impacts of and responses to climate change, and their linkages is presented in Figure 1. In the light of the above overview on GEC, the rest of this chapter is organized as follows: First, in section 2, a review of relevant issues in GEC is given. Section three elucidates on pathways/indicators of GEC.

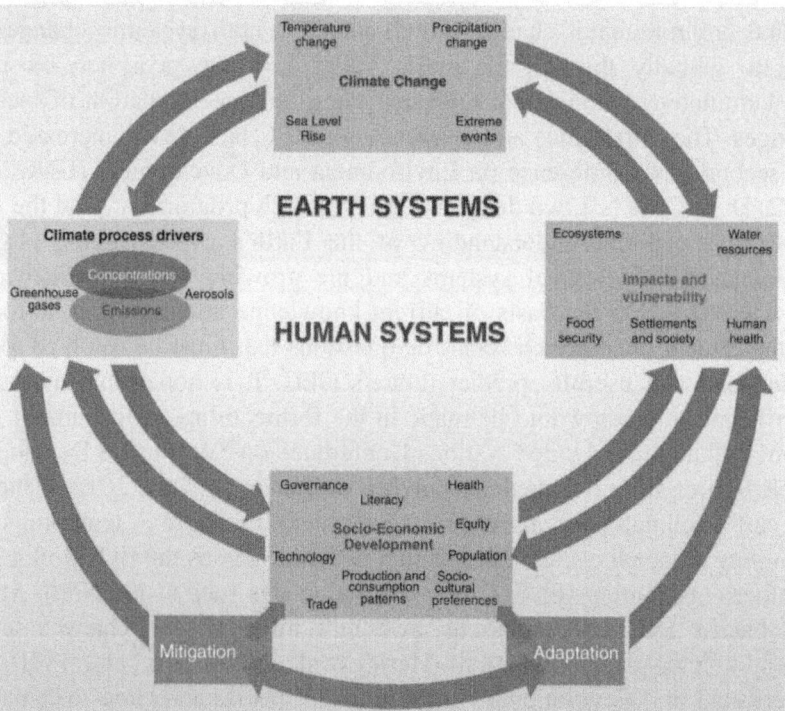

Figure. 1: Schematic framework of anthropogenic climate change drivers, impacts and response. Source: IPCC, 2007.

The fourth section centers on the interrelatedness of natural systems and human as driver of most GEC. Some of the consequences of GEC on natural and human systems are reviewed in section five. Section six dealt on global responses, ranging from positive policy drive and actions through innovations and a change where possible in institutional and environmental governance frameworks while weakening some implementation barriers ravaging existing institutions. Lastly, section seven concludes and foregrounds on the need for more interdisciplinary and integrative perspective on global environmental change issues. It hopes to proffer some definitions and answers to overarching questions in GEC.

RELEVANT ISSUES IN GLOBAL ENVIRONMENTAL CHANGE

In the past, several issues have been addressed on GEC. Very striking and concise in my opinion is Vitousek (1994). He concluded that three of the well-documented global changes are: increasing concentrations of carbon dioxide in the atmosphere; alterations in the biogeochemistry of the global nitrogen cycle; and ongoing land use/land cover change, and are perhaps the bedrock of other relevant issues in GEC mentioned in sub-sections 2.0 below. Human activity-now primarily fossil fuel combustion- has increased carbon dioxide concentrations from 280 ppm to 355 ppm since 1800; the increase is unique, at least in the past 160 000 yr, and several lines of evidence demonstrate unequivocally that it is human-caused (Vitousek, 1994; Ebi, 2011). The global nitrogen cycle has been altered by human activity to such an extent that more nitrogen is fixed annually by humanity (primarily for nitrogen fertilizer, also by legume crops and as a byproduct of fossil fuel combustion) than by all natural pathways combined. This added nitrogen alters the chemistry of the atmosphere and of aquatic ecosystems, contributes to eutrophication of the biosphere, and has substantial regional effects on biological diversity in the most affected areas. Finally, human land use/land cover change has transformed one-third to one-half of Earth's ice-free surface. This in and of itself probably represents the most important component of global change now and will for some decades to come; it has profound effects on biological diversity on land and on ecosystems downwind and downstream of affected areas. These three and other equally certain components of global environmental change are the primary causes of anticipated changes in climate, and of ongoing losses of biological diversity (Vitousek, 1994, Sala et al, 2000).

Climate Change

Climate change is one of several large-scale environmental changes to which human activities make a significant contribution, and that, in turn, affects human health and well-being. The Intergovernmental Panel on Climate Change (IPCC) concluded: "warming of the climate system is unequivocal, as is now evident from observations of increases in global average air and ocean temperatures, widespread melting of snow and ice, and rising global average sea level (IPCC 2007)." Over the past decade, the fact that emissions of greenhouse gases due to human activities are affecting the world's climate has become clear (Ebi 2011). In addition: "most of the observed increase in globally averaged temperatures since the mid-twentieth century is very likely due to the observed increase in anthropogenic greenhouse gas concentrations." These changes have begun to affect morbidity and mortality worldwide, with projections suggesting that overall health burdens will increase with increasing climate change. Although all countries are projected to experience increased health risks, those at greatest risk include the urban poor, older adults, children, traditional societies, subsistence farmers, and coastal populations, particularly in low-income countries (Ebi 2011). The task of understanding climate change and predicting future change would be complex enough if only natural forcing mechanisms were involved. It is significantly more daunting because of the introduction of anthropogenic forcing and even more so considering the limitations in available records.

Land Use

Landscapes are changing worldwide, as natural land covers like forests, grasslands, and deserts are being converted to human-dominated ecosystems, including cities, agriculture, and forestry. Between 2000-2010, approximately 13 million hectares of land (an area the size of Greece) were converted each year to other land cover types (FAO 2010). Developed regions like the US and Europe experienced significant losses of forest and grassland cover over the past few centuries during phases of economic growth and expansion. More recently, developing nations have experienced similar losses over the past 60 years, with significant forest losses in biologically diverse regions like Southeast Asia, South America, and Western Africa. Land use changes affect the biosphere in several ways. They often reduce native habitat, making it increasingly difficult for species to survive. Some land use changes, such as deforestation and agriculture, remove native vegetation and diminish carbon uptake by photosynthesis as well as hasten soil decomposition, leading to additional greenhouse gas release. Almost 20% of the global CO_2 released to the atmosphere (1.5–2 billion tons of carbon) is thought to come from deforestation. Landscapes are changing worldwide, as natural land covers like forests, grasslands, and deserts are being converted to human-dominated ecosystems, including cities, agriculture, and forestry. Between 2000-2010, approximately 13 million hectares of land (an area the size

Biodiversity

Biodiversity is the diversity of life on Earth and includes the richness (number), evenness (equity of relative abundance), and composition (types) of species, alleles, functional groups, or ecosystems. The period since the emergence of humans has displayed an ongoing biodiversity reduction and an accompanying loss of genetic diversity. Named the Holocene extinction, the reduction is caused primarily by human impacts, particularly habitat destruction. Currently, global biodiversity is changing at an unprecedented rate and scale in response to human-induced perturbation of the Earth System. Fossil records indicate that the background extinction rate (that is Pre-Industrial value) for most species is 0.1-1 extinctions per million species per year (Braimoh et al., 2010). Over the past years however, the species extinction rate has increased to more than 100 extinctions per million species per year (MA, 2005). There is a strong linkage between biodiversity loss and human-driven ecosystem processes from local to regional scales. Conversely, biodiversity impacts human health in a number of ways, both positively and negatively (Sala et al. 2009), and there is considerable evidence that contemporary biodiversity declines will lead to subsequent declines in ecosystem functioning and ecosystem stability (Naeem et al. 2009). Generally, observed changes in climate have already adversely affected biodiversity at the species and ecosystem level, and further changes in biodiversity are inevitable with further changes in climate (CBD 2009). While human actions have significantly contributed to the loss of biodiversity, in some cases, human actions have promoted biodiversity. Conservation strategies, such as creating parks to protect biodiversity hotspots, have been effective but insufficient (Bruner et al. 2001). For example, although biodiversity is often greater inside than outside parks, species extinctions continue. Specifically, biodiversity and ecosystem services are greater in restored than in degraded ecosystems but lower in restored than in intact remnant ecosystems (Benayas et al. 2009). Despite the positive effects of conservation and restoration efforts, biodiversity declines have not slowed (Butchart et al. 2010). Thus, further investigation is needed to determine new conservation and restoration strategies. Global agreements such as the Convention on Biological Diversity, give "sovereign national rights over biological resources" (not property). The agreements commit countries to "conserve biodiversity", "develop resources for sustainability" and "share the benefits" resulting from their use. Sovereignty principles can rely upon what is better known as Access and Benefit Sharing Agreements (ABAs). The Convention on Biodiversity implies informed consent between the source country and the collector, to establish which resource will be used and for what, and to settle on a fair agreement on benefit sharing. Theoretical and empirical studies have identified a vast number of natural processes that can potentially maintain biodiversity. Biodiversity can be maintained by moderately intense disturbances that reduce dominance by species that would otherwise competitively exclude subordinate species. For example, selective grazing by bison can promote plant diversity in grasslands (Collins et al. 1998). Now, it may be possible to predict future changes in biodiversity, ecosystem functioning, and ecosystem stability by considering how global ecosystem changes are currently influencing stabilizing species interactions. In this direction, the United Nations is currently developing an Intergovernmental Science-Policy Platform on Biodiversity and Ecosystem Services (IPBES) to monitor biodiversity and ecosystem services worldwide (Marris 2010). The IPBES will be modelled after the Intergovernmental Panel on Climate Change (IPCC), and there is great potential for ecologists to borrow strategies that have been successfully employed by climatologists. Another global effort in this direction (conservation of biodiversity) recently culminated in the United Nations designating 2011-2020 as the United Nations Decade on Biodiversity during the Aichi COP 10 meeting in 2010- the International Year of Biodiversity.

Pathways and Indicators of Global Environmental Change

Changes are occurring throughout the Earth System and are evident in the oceans, on the land and in the atmosphere. These changes are increasingly driven by human activities. There is the need to understand how these ecosystems react to global change so as to understand the consequences for their functioning and to manage ecosystems resources sustainably. Sala et al., (2000) recognize five major drivers of biodiversity loss, namely land use, climate, nitrogen deposition, biotic exchange and atmospheric carbon dioxide, and opined that the importance of these drivers varies from one ecosystem to the other. Here, I present and elucidate the pressures on three major pathways of global environmental change.

Terrestrial

Land-use change (especially deforestation) and climate change generally have the greatest impact for terrestrial ecosystems, whereas biotic exchange is more important for freshwater ecosystems. As earlier mentioned, human activity–now primarily fossil fuel combustion– has increased carbon dioxide concentrations from 280 ppm to 355 ppm since 1800; the increase is unique, at least in the past 160 000 yr, and several lines of evidence demonstrate unequivocally that it is human-caused (Vitousek, 1994). This increase is likely to have climatic consequences–and certainly it has direct effects on biota in all Earth's terrestrial ecosystems. Land use and land cover change has aroused increasing attention of scientists worldwide since 1990. Recognizing the importance of this change to other global environmental change and sustainable development issues, the International GeosphereBiosphere Programme (IGBP) and the Human Dimensions of Global Environmental Change Programme (IHDP) initiated a joint core project Land Use and Land Cover Change (LUCC) and published a Science/Research Plan for the project. This precipitated into a number of IGBP core projects, of which one is Global Change and Terrestrial Ecosystems (GCTE). Despite these achievements, agricultural activities have continued to be significant emitters of global greenhouse gases (GHGs) and as such agricultural activity is a major driver of anthropogenic climate change. Emissions from agricultural sources were 14% of global GHG emissions in 2000 with developing countries accounting for three quarters of agriculture emissions in the case of rice (WRI, 2006; Stern, 2007). Also, changes in vegetation structure influence the magnitude and spatial pattern of the carbon sink and, in combination with changing climate, also freshwater availability (runoff). The potential for terrestrial ecosystems to absorb significant amounts of CO_2, thus slowing the buildup of CO_2 in the atmosphere and reducing the rate of climate change, is a key issue in the debate on CO_2 emission controls. As more land is converted to agriculture, there is less area in natural ecosystems that can act as carbon sinks, thereby reducing the potential sink strength of the terrestrial biosphere.

Aquatic

The ocean is a vital component of the metabolism of the Earth and plays a key role in global change. In 1987 the World Commission on Environment and Development (Brundtland Commission) warned in the final report, Our Common Future, that water was being polluted and water supplies were overused in many parts of the world. The ocean is the source of most of the world's precipitation (rainfall and snowfall), but people's freshwater needs are met almost entirely by precipitation on land (see Figure 2), with a small though increasing amount by desalination. Due to changes in the state of the ocean, precipitation patterns are altering, affecting human well-being. Asides from this, ocean changes are also affecting marine living resources and other socio-economic benefits on which many communities depend, and anthropogenically induced global climate change has profound implications for marine ecosystems and the economic and social systems that depend upon them (Harley et al., 2006). Human pressures at global to basin scales are substantially modifying the global water cycle, with some major adverse impacts on its interconnected aquatic ecosystems- freshwater and marine - and therefore on the well-being of people who depend on the services that they provide. Like other ecosystems, marine and coastal areas are already adversely impacted by many stresses, which will be exacerbated by climate change (e.g., sea level rise) (see figure 3). At the same time, coastal ecosystems ranging from Polar Regions to Small Island developing States are essential to our capacity to respond to projected climate change impacts. In recent years, a major observed pressure on the aquatic ecosystem is Ocean acidification caused by decrease in the pH and increase in acidity of the Earth's oceans as a result of uptake of anthropogenic carbon dioxide (CO_2) from the atmosphere. Past, present and future predicted average surface ocean pH is shown in Table 1. Dissolving CO_2 in seawater increases the hydrogen ion (H^+) concentration in the ocean, and thus decreases ocean pH. Caldeira and Wickett (2003) placed the rate and magnitude of modern ocean acidification changes in the context of probable historical changes during the last 300 million years. Furthermore, in terms of resources, aquatic ecosystems continue to be heavily degraded, putting many ecosystem services at risk, including the sustainability of food supplies and biodiversity. Global marine and freshwater fisheries show large scale declines, caused mostly by persistent overfishing. Freshwater stocks also suffer from habitat degradation and altered thermal regimes related to climate change and water impoundment. A continuing challenge for the management of water resources and aquatic ecosystems is to balance environmental and developmental needs. It requires a sustained combination of technology, legal and institutional frameworks, and, where feasible, market-based approaches.

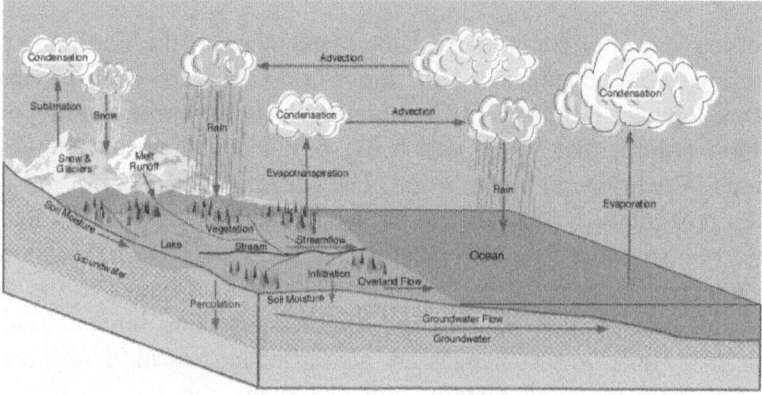

Figure. 2: Hydrologic cycle (Adapted from Pidwirny, 2006).

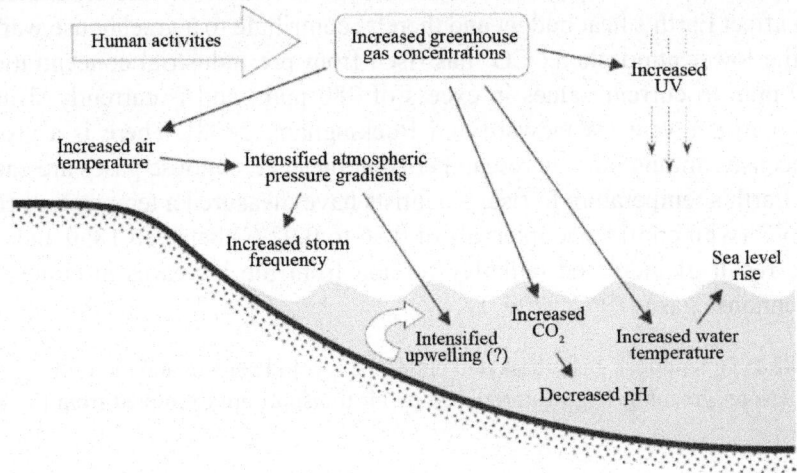

Figure. 3: Important abiotic changes associated with climate change. Human activities such as fossil fuel burning and deforestation lead to higher concentration of greenhouse gases in the atmosphere, which in turn leads to a suit of physical and chemical changes in coastal ocean. (Adapted from Harley et al., 2006).

Table 1: Average surface ocean pH (Adapted from Orr et al 2005)

Time	pH	pH change	Source	H+ concentration change relative to pre-industrial
Pre-industrial (18th century)	8.179	0.000	*Analyzed field* Key et al (2004)	0%
Recent past (1990s)	8.104	-0.075	*Field* Key et al (2004)	+18.9%
Present levels	~8.069	-0.11	*Field* Hall-Spencer et al (2008)	+28.8%
2050 (2×CO2 = 560 ppm)	7.949	-0.230	*Model* Orr et al (2005)	+69.8%
2100 (IS92a) (IPCC 2001)	7.824	-0.355	*Model* Orr et al (2005)	+126.5%

Atmospheric

In the last two centuries, human have released ever greater quantities of carbon dioxide (CO_2), methane (CH_4), nitrous oxide (N_2O) and other greenhouse gases, into the Earth's atmosphere (Woodward and Buckingham, 2008). Scientists have found that the four most important variable greenhouse gases, whose atmospheric concentrations can be influenced by human activities, are carbon dioxide (CO_2), methane (CH_4), nitrous oxide (N_2O), and chlorofluorocarbons (CFCs). Greenhouse gases basketed under the Kyoto Protocol and their main generators are shown in Table 2. Historically, CO_2 has been the most important, but those other atmospheric trace gases are also radiatively active, in that they can affect Earth's heat budget and thereby contribute to a greenhouse warming of the lower atmosphere. CO_2 has risen from pre-industrial concentration of 280 ppm to current values in excess of 380 ppm, and is currently rising by 1.9 ppm per year (Woodward and Buckingham, 2008). There is a growing concensus among climate researchers that these greenhouse gases are causing the Earth's temperature to rise. Scientists have measured a temperature rise of 0.76°C (with confidence intervals of 0.56 to 0.92°C) between 1850 and 2005, as a result of increased radiative forcing from the increases in atmospheric greenhouse gases (IPCC 2007).

Table 2: Greenhouse gases basketed under the Kyoto Protocol and their main generators (Note: greenhouse gases produced by air transport are exempted from the Protocol)

Greenhouse Gas	Main Sources
Carbon dioxide (CO_2)	Fossil fuel combustion (e.g. road transport, energy industries, other industries, residential, commercial and public sector); forest clearing
Methane (CH_4)	Agriculture, landfill, gas leakage, coal mines
Nitrous oxide (N_2O)	Agriculture, industrial processes, road transport, others
Perfluorocarbons (PFCs)	Industry (e.g. aluminium production, semi-conductor industry)
Hydrofluorocarbons (HFCs)	Refrigeration gases, industry (as perfluorocarbons)
Sulphur hexafluoride ($SF/_6$)	Electrical transmissions and distribution systems, circuit breakers, magnesium production

Current developments in atmospheric chemistry are revealing the close links between chemistry, radiation, dynamics, and climate. Examples include the powerful role played by aerosol formation in both the boundary layer and the upper troposphere, chemical initiation of subvisible cirrus in the region of the tropopause, the control exerted by water vapor and temperature on the sharply nonlinear partitioning of halogen and hydrogen radicals in the lower stratosphere, and the importance of stratosphere-troposphere exchange on the composition and meteorology of the upper troposphere and lower stratosphere.

HUMAN DIMENSIONS OF GLOBAL ENVIRONMENTAL CHANGE

According to Jager (2002), research on the human dimensions of global environmental change is concerned with the human causes of change, the consequences of such changes for individuals and societal groups, and the ways in which humans respond to the changes. The human causes include emissions of pollutants into the atmosphere, especially carbon dioxide, chlorofluorocarbons and acidifying substances, as well as land-use and land-cover changes. It is now established wisdom that humans are the prime drivers of change on Earth, and it is this recognition that underpins the discussion of the Anthropocene. Social, economic, and cultural systems are changing in a world that is more populated, urban, and interconnected than ever. Such large-scale changes increase the resilience of some groups while increasing the vulnerability of others. With the impetus of global change research, study of large-scale ecosystems has become a rapidly maturing field of science and has shown major successes over the past decade. Improved fundamental understanding of marine and terrestrial ecosystems and hydrology has already led to practical applications in weather and climate modeling, air quality, and better management and natural hazards responses for water, forest, fisheries, and rangeland resources. The development of spatially resolved global-scale ecosystem models has occurred only during the past five years. Computing capability and remote sensing technology have further driven change in the nature of the field. The capability has emerged not only to model at global scales but also to exploit data at these scales. Such capability is increasingly important for developing our economy, protecting our environment, safeguarding our health, and negotiating international agreements to ensure the sustainable development of the global community of nations. As earlier mentioned, if global average temperatures exceed 2°C there will be irreversible impacts on water, ecosystems, food, coastal zones and human health (IPCC 2007). We have a 50% chance of avoiding a 2°C warming if we stabilize greenhouse gases at 450 ppm CO2 eq (parts per million carbon dioxide equivalents). This means we must start radically reducing emissions now and stay on a low emissions pathway to avoid increasing the amount of CO2 in the atmosphere. The good management of ecosystems such as wetlands and forests remains an effective mitigation option given the high sequestration potential of natural systems. The permanence of carbon sinks is also tied to the maintenance or enhancement of the resilience of ecosystems (CBD 2009).

CONSEQUENCES FOR NATURAL AND HUMAN SYSTEMS AND RESPONSES

There are consequences for natural and human systems in GEC events. Studies have focused, for example, on global environmental change impacts on agriculture and human health and on particular locations, such as the coastal zone. Rising sea levels; increased temperatures; increased risk of droughts, floods and fires; stronger storms and increased storm damage; changing landscapes; forced environmental migrations and food insecurity are but a few of the issues linked to a changing climate. However, while a large proportion of climate change impacts will be negative, some will be positive too. For certain societies these will include, among others, increased agricultural growing periods and lower winter mortalities (warmer winters), although it is generally accepted that the negatives will significantly outweigh the positives (Nelson et al, 2009). The Stern Review suggests that all countries will be affected by climate change, but it is the poorest countries that will suffer earliest and most. Unabated climate change may risks raising average temperatures by over 5°C from pre-industrial levels. Such changes would transform the physical geography of our planet, as well as the human geography- how and where we live our lives (Stern Review, 2007). Some examples of these consequences are elucidated.

Food Security

Food security is the ability of people to have access to sufficient, nutritious food. Global environmental change (GEC), including land degradation, loss of biodiversity, changes in hydrology, and changes in climate patterns resulting from enhanced anthropogenic emission of greenhouse gas emissions, will have serious consequences for food security, particularly for more vulnerable groups (Ericksen et al. 2009). Growing demands for food in turn affect the global environment because the food system is a source of greenhouse gas emissions and nutrient loading, and it dominates the human use of land and water. The speed, scale and consequences of human-induced environmental change are beyond previous human experience, and thus science has a renewed responsibility to support policy formation with regard to food systems (Carpenter et al., 2009; Steffen et al., 2003). Most research linking global environmental change and food security focuses solely on agriculture: either the impact of climate change on agricultural production, or the impact of agriculture on the environment, e.g. on land use, greenhouse gas emissions, pollution and/ or biodiversity (Ericksen et al. 2009). Although, we currently grow enough food to feed the global human population, a population rising to 9 billion by 2050, combined with climate changes, will strain the capacity of some regions to feed people, thereby raising the risks of food insecurity (Godfray et al. 2010). Thus, the effects of global environmental change (GEC) are increasingly making the practical achievement of food security more difficult in some of the world's poorest communities. It is important to note that while technical fixes are important, they will not alone solve the food security challenges. Adapting to the additional threats to food security arising from major environmental changes requires an integrated food system approach, not just a focus on agricultural practices. In this line, Ericksen et al (2009) further highlighted on six key issues that has emerged for future research: (i) adapting food systems to global environmental change requires more than just technological solutions to increase agricultural yields; (ii) tradeoffs across multiple scales among food system outcomes are a pervasive feature of globalized food systems; (iii) within food systems, there are some key underexplored areas that are both sensitive to environmental change but also crucial to understanding its implications for food security and adaptation strategies; (iv) scenarios specifically designed to investigate the wider issues that underpin food security and the environmental consequences of different adaptation options are lacking; (v) price variability and volatility often threaten food security; and (vi) more attention needs to be paid to the governance of food systems.

Human Security

The concept of human security came to prominence through the 1994 Human Development Report, which defined human security as a "concern with human life and dignity" (UNDP 1994, 22), and which adopted a comprehensive approach by identifying economic, food, health, environmental, personal, community, and political components to human security. In the 21st century three key issues facing humankind are environmental degradation, impoverishment, and the insecurity that can result from either of these two. A review of environment and security work indicates that there is an ongoing need for conceptual and theoretical discussions on the nature of the relationship between environment and security. It is also important to build upon early empirical work that focused on environment and conflict and to provide additional empirical studies on environmental change and its relationship to a broader conception of security. At the same time expanded research networks and improved communication among researchers, policy makers, and NGOs are required in order to develop integrated research projects on environmental change and human security.

Human Health

It is well established that human health is linked to environmental conditions, and that changes in the natural environment may have subtle, or dramatic, effects on health. Timely knowledge of these effects may support our public health infrastructure in devising and implementing strategies to compensate or respond to these effects. Ecosystems, human health and economy are all sensitive to changes in climate- including both the magnitude and rate of climate change. Climate change is likely to affect human health and well-being through a variety of mechanisms. For example, it can adversely affect the availability of freshwater, food production, and the distribution and seasonal transmission of vector-borne infectious diseases such as malaria, dengue fever and schistosomiasis. The additional stress of climate change will interact in different ways across regions. It can be expected to reduce the ability of some environmental systems to provide, on a sustained basis, key goods and services needed for successful economic and social development, including adequate food, clean air and water, energy, safe shelter and low levels of diseases (IPCC 2001). In this chapter, current state of knowledge of the associations between weather/climate factors and health outcome(s) for the population(s) concerned, either directly or through multiple pathways is as outlined in Figure 4. The figure shows not only the pathways by which health can be affected by climate change, but also shows the concurrent direct-acting and modifying (conditioning) influences of environmental, social and health system factors.

Natural Disturbances

Changing climate has the potential to increase risks from sea level rise, extreme storm events, and drought. About 25% of the world's population lives with 100 km of the coast. In 2007, the IPCC projected an 18-59 cm sea level rise by 2100, but many scientists argue that this range is too low, and that sea level rise could be as great as 1-6 m (Kopp et al 2009; Jevrejeva et al. 2010). People living in low lying regions, such as Bangladesh and Pacific island nations (e.g., Tuvalu and the Maldives) are already experiencing the effects of salt water incursion in their agricultural fields and fresh water supplies. Arctic Inuit communities are battling the loss of coastal villages as a result of increased storm surges from sea level rise.

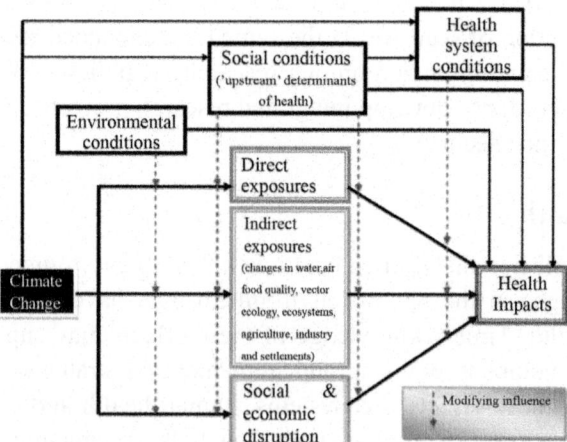

Figure. 4: Schematic diagram of pathways by which climate change affects health, and concurrent direct-acting and modifying (conditioning) influences of environmental, social and health-system factors (Modified from Confalonieri et al 2007).

Extreme precipitation events may become more common in mid-to-high latitude regions, consistent with the prediction that warmer air temperatures, as a result of climate warming, will increase the moisture content of the atmosphere (IPCC 2007). Besides catastrophic flooding and associated property damage, extreme storms are a concern for infrastructure, with cities now faced with the prospect of significant costs associated with roads, dams, and levees being washed out by floods. Regions such as the Sahel are particularly vulnerable to increases in rainfall variability and extremes; the timing of rainfall is as important as the amount of rainfall for agriculture in the Sahel. Extreme rainfall events in semi-arid regions are also likely to lead to increased soil erosion; analysis of interannual variability in rainfall and atmospheric dust content in the Sahel indicates that episodic rainfall events play an important role in generating the erodible material that is necessary for the development of dust storms (Brooks and Legrand, 2001). Atmospheric dust is a major cause of respiratory problems in regions such as the Sahel (Shinn, 2001), and a shift towards higher rainfall variability and intensity might therefore have a negative impact on health.

GLOBAL RESPONSE

"The scientific evidence is now overwhelming: climate change presents very serious global risks, and it demands an urgent global response" (Stern Review, 2007). Human perceptions of the natural environment, and the way we use the environment, are socially constructed. The human responses are mitigation and adaptation. However, environmental problems must be addressed from a broader perspective that includes issues of impoverishment and issues of (in) equity, and recognizing the fact that "Space Matters". In the context of global environmental change, it is important to consider the various spatial levels at which both environment and security concerns can be addressed. It is believed that these needs can be best met through an international research program that focuses both on guiding future research and assisting in policy development (at all levels). Given the severity of global environmental change impacts, various adaptation and mitigation measures are being used in the fight against climate change. The effective targeting of these measures across different sectors such as water, agriculture, tourism, infrastructure development and others requires the use of both practical and innovative strategies. Adaptation measures fall within a broad range, among others, from expanded water harvesting, storage and conservation techniques to the diversification of tourism activities. An example of an adaptation strategy to prevent damage from climate change is shore protection (e.g., dikes, bulkheads, beach nourishment), which can prevent sea level rise from inundating low-lying coastal property, eroding beaches, or worsen flooding. If the costs or environmental impacts of shore protection are high compared with the property being protected, an alternative adaptation strategy would be a planned retreat, in which structures are relocated inland as shores retreat. Also, with current climate fluctuations, a good example of adaptation and coping strategies include farmers planting different crops for different seasons, and wildlife migrating to more suitable habitats as the seasons change. Mitigation measures include "climate friendly" technological innovations; alternative fuels; sustainable land management practices to name but a few (Nelson et al, 2009). Also, the resilience of biodiversity to climate change can be enhanced by reducing non-climatic stresses in combination with conservation, restoration and sustainable management strategies. Conservation and management strategies that maintain and restore biodiversity can be expected to reduce some of the negative impacts from climate change; however, there are rates and magnitude of climate change for which natural adaptation will become increasingly difficult (CBD 2009).

Furthermore, from the highlights of Stern Review, emerging schemes that allow people to trade reductions in CO_2 have demonstrated that there are many opportunities to cut emissions for less than \$25 a tonne. In other words, reducing emissions will make us better off. According to one measure, the benefits over time of actions to shift the world onto a low-carbon path could be in the order of \$2.5 trillion each year. The shift to a low-carbon economy will also bring huge opportunities. Markets for low-carbon technologies will be worth at least \$500bn, and perhaps much more, by 2050 if the world acts on the scale required. Tackling climate change is the pro-growth strategy; ignoring it will ultimately undermine economic growth (Stern Review, 2007). Lastly, international policy responses to the global challenges described above have been many and varied. Most of them are based on several UN Conventions such as the UN Conference on Environment and Development, Framework Convention on Climate Change (UNFCC), Convention on Biological Diversity (CBD) or the large number of agreements tied to the UN Convention on the Law of the Sea. At the same time, institutional and organizational weaknesses in some countries and the complex interaction among myriad authorities responsible for ecosystems (including marine and coastal) and environmental management make the implementation of policies a difficult task (Agboola and Braimoh,2009). Changes in institutional and environmental governance frameworks are sometimes required to create the enabling conditions for effective management of ecosystems, while in other cases existing institutions could meet these needs but face significant implementation barriers (MA, 2005). A lot of commitments to sustainable development have been made at past UN conferences, including Agenda 21 (1992), the Rio Declaration on Environment and Development (1992) and the Johannesburg Plan of

Implementation (2002). In all these, there is the need to examine how far we have come in achieving these commitments before we can channel a better way to moving forward.

CONCLUSION

This chapter dealt on some relevant issues and current dimensions in GEC. It presents certainties as explained by some GEC indicators and predictions, and future uncertainties that may require strategic interventions. In all of these, it foregrounds the need for a broader perspective in tackling the myriads of global environmental challenges. Although, research on the human dimensions of global change concerns human activities that alter the Earth's environment, the driving forces of those activities, the consequences of environmental change for societies and economies, and human responses to the experience or expectation of global change; such research is essential both to understand global change and to inform public policy. This review suggests the need for a more interdisciplinary and integrative perspective on environmental change issues. A more integrated understanding of the complex interactions of human societies and the Earth system is essential if we are to identify vulnerable systems and pursue options that take advantage of opportunities and enhance resilience.

REFERENCES

1. Agboola, J.I., Braimoh, A.K., (2009) Strategic partnership for sustainable management of aquatic resources. Water Resour. Manage., 23: 2761-2775

2. Benayas, J. M. R., Newton, A. C., Bullock, J. M., (2009) Enhancement of biodiversity and ecosystem services by ecological restoration: A meta-analysis. Science 325, 1121–1124

3. Braimoh, A. K., Subramanian, S. M., Elliott, W. S., Gasparatos, A., (2010) Climate and HumanRelated Drivers of Biodiversity Decline in Southeast Asia. UNU-IAS Policy Report. Brooks, N., Legrand, M., (2000) Dust variability over northern Africa and rainfall in the Sahel, in Linking climate change to land surface change, S. McLaren and D. Kniveton (eds.), Dordrecht, Kluwer Academic Publishers, 1-25.

4. Bruner, A. G., Gullison, R.E., Rice, R. E., da Fonseca, G. A. B., (2001) Effectiveness of parks in protecting tropical biodiversity. Science 291, 125–128.

5. Butchart, S.H.M., Walpole, M., Collen, B., Strien, A.V., Scharlemann, J.P.W., Almond, R.E.A., Baillie, J.E.M., Bomhard, B., Brown, C., Bruno,

J., Carpenter, K.E., Carr, G.M., Chanson, J., Chenery, A.M., Csirke, J., Davidson, N.C., Dentener, F., Foster, M., Galli, A., Galloway, J. N., Genovesi, P., Gregory, R. D., Hockings, M., Kapos, V., Lamarque, J. F., Leverington, F., Loh, J., McGeoch, M.A., McRae, L., Minasyan, A., Morcillo, M. H., Oldfield, T.E.E., Pauly, D., Quader, S., Revenga, C., Sauer, J.R., Skolnik, B., Spear, D., Stanwell-Smith, D., Stuart, S.N., Symes, A., Tierney, M., Tyrrell, T.D., Vie, J.C., Watson, R., (2010) Global biodiversity: indicators of recent declines. Science, 328, 1164–1168.

6. Caldeira, K., Wickett, M.E. (2003). "Anthropogenic carbon and ocean pH". Nature 425 (6956): 365-365. Carpenter, S.R., Mooney, H.A., Agard, J., Capistrano, D., DeFries, R.S., Diaz, S., Dietz, T., Duraiappah, A.K., Oteng-Yeboah, A., Pereira, H.M., Perring, C., Reid, W.V., Sarukhan, J., Scholes, R.J., Whyte, A., (2009) Science for managing ecosystem services: beyond the millennium ecosystem assessment. PNAS 106, 1305–1312.

7. Collins, S. L., Knapp, A. K., Briggs, J. M., Blair, J. M., Steinauer, E. M., (1998) Modulation of diversity by grazing and mowing in native tallgrass prairie. Science 280, 745–747.

8. Confalonieri, U., Menne, B., Akhtar, R., Ebi, K.L., Hauengue, M., Kovats, R.S., Revich, B., Woodward, A., (2007) Human health. Climate Change 2007: Impacts, Adaptation and Vulnerability. Contribution of Working Group II to the Fourth Assessment Report of the Intergovernmental Panel on Climate Change, M.L. Parry, O.F. Canziani, J.P. Palutikof, P.J. van der Linden and C.E. Hanson, Eds., Cambridge University Press, Cambridge, UK, 391-431.

9. Convention on Biological Diversity- CBD, (2009) Connecting Biodiversity and Climate Change Mitigation and Adaptation: Report of the Second Ad Hoc Technical Expert Group on Biodiversity and Climate Change. Montreal, Technical Series No. 41, 126 pages Ebi, K.L., (2011) Climate Change and Health. Encyclopedia of Environmental Health, pp 680-689

10. Ericksen, P. J., Ingram, J. S. I., Liverman, D.M., (2009) Food security and global environmental change: emerging Challenges. Environmental Science & Policy, 12: 373-377

11. FAO, (2010) Global Forest Resources Assessment. FAO, Rome, Italy. Godfray, H.C.J., Beddington, J.R., Crute, I.R., Haddad, L., Lawrence, D., Muir, J.F., Pretty, J., Robinson, S., Thomas, S.M., Toulmin, C. (2010) Food security: the challenge of feeding 9 billion people Science 327, 812-818.

12. Hall-Spencer, J. M., Rodolfo-Metalpa, R., Martin, S., Ransome, E., Fine, M., Turner, S. M., Rowley, S., Tedesco, D., Buia, M. C., (2008) Volcanic carbon dioxide vents reveal ecosystem effects of ocean acidification. Nature 454: 96-99.

13. Harley, C. D. G., Hughes, A. R., Hultgren, K. M., Miner, B. G., Sorte, C. J. B., Thornber, C. S., Rodriguez, L. F., Tomanek, L., Williams, S. L., (2006) The impacts of climate change in coastal marine systems. Ecology Letters 9, 228–241

14. IPCC, (2001) Climate Change 2001: Impacts, Adaptation and Vulnerability. Contribution of Working Group II to the Third Assessment Report of the Intergovernmental Panel on Climate Change. Cambridge, United Kingdom, and New York, United States, Cambridge University Press.

15. IPCC, (2001) IPCC Special Report on Emissions Scenarios IPCC, (2007) Core Writing Team; Pachauri, R.K; and Reisinger, A., ed., Climate Change 2007: Synthesis Report, Contribution of Working Groups I, II and III to the Fourth Assessment Report of the Intergovernmental Panel on Climate Change, IPCC, ISBN 92-9169-122-4.

16. Jager, J., (2002) Global Environmental Change: Human Dimensions. International Encyclopedia of the Social & Behavioral Sciences. pp 6227-6232. DOI:10.1016/B0- 08-043076-7/04137-1

17. Jevrejeva, S., Moore, J. C., Grinsted, A., (2010) How will sea level respond to changes in natural and anthropogenic forcings by 2100? Geophysical Research Letters 37, L07703, 5 pp.

18. Key, R.M., Kozyr, A., Sabine, C.L., Lee, K., Wanninkhof, R., Bullister, J., Feely, R.A., Millero, F., Mordy, C., Peng, T. H. (2004) "A global ocean carbon climatology: Results from GLODAP". Global Biogeochemical Cycles 18 (4): GB4031.

19. Kopp, R.E., Simons, F.J., Mitrovica, J.X., Maloof, A.C., and Oppenheimer, M., (2009) Probabilistic assessment of sea level during the last interglacial stage. Nature 462, 863-868

20. MA, (2005) Millennium Ecosystem Assessment (MA). Ecosystems and Human Well-being: Synthesis. Washington, DC: Island Press.

21. Marris, E., (2010) UN body will assess ecosystems and biodiversity. Nature 465, 859. Naeem, S., Bunker, D. E., Hector, A., Loreau, M. & Perrings, C., (2009) Introduction: The ecological and social implications of changing biodiversity. An overview of a decade of biodiversity and ecosystem functioning research. In Biodiversity, Ecosystem Functioning, and Human Wellbeing: An Ecological and Economic Perspective, S.

Naeem, D. E. Bunker, A. Hector, M. Loreau & C. Perrings Eds, Oxford University Press, Oxford, United Kingdom, pp. 3-13.

22. Nelson, G.C., Rosegrant, M.W., Koo, J., Robertson, R., Sulser, T., Zhu, T., Ringler, C., Msangi, S., Palazzo, A., Batka, M., Magalhaes, M., Valmonte-Santos, R., Ewing, M., Lee, D., (2009) Climate Change: Impact on Agriculture and Costs of Adaptation. International Food Policy Research Institute (IFPR).

23. Orr, J. C., Fabry, V. J., Aumont, O., Bopp, L., Doney, S. C., Feely, R. A., Gnanadesikan, A., Gruber, N., Ishida, A., Joos, F., Key, R. M., Lindsay, K., Maier-Reimer, E., Matear, R., Monfray, P., Mouchet, A., Najjar, R.G., Plattner, G., Rodgers, K. B., Sabine, C. L., Sarmiento, J. L., Schlitzer, R., Slater, R. D., Totterdell, I. J., Weirig, M., Yamanaka, Y., Yool, A., (2005) "Anthropogenic ocean acidification over the twenty-first century and its impact on calcifying organisms". Nature 437 (7059): 681–686.

24. Pidwirny, M., (2006) "The Hydrologic Cycle". Fundamentals of Physical Geography, 2nd Edition. Date Viewed: 20/09/2011. http://www. physicalgeography.net/fundamentals/8b.html

25. Sala, O.E., Chapin III, F.S., Armesto, J.J., Berlow, E., Bloomfield, J., Dirzo, R., HuberSannwald, E., Huennecke, L.F., Jackson, R.B., Kinzig, A., Leemans, R., Lodge, D.M., Mooney, H.A., Oesterheld, M., Poff, N.L., Sykes, M.T., Walker, B., Walker, M., Wall, D.H., (2000) Global biodiversity scenarios for the year 2100. Science, 287, 1770–1774.

26. Sala, O. E., Meyerson, L. A., Parmesan, C., (2009) Biodiversity change and human health: from ecosystem services to spread of disease. Island Press. pp. 3–5. ISBN 9781597264976. Retrieved 28 June 2011.

27. Shinn, Eugene A., (2001) African dust causes widespread environmental distress, Environmental Geology

28. Steffen, W., Sanderson, A., Tyson, P.D., Jager, J., Matson, P.A., Moore, III, B., Oldfield, F., Richardson, K., Schellnhuber, H.J., Turner, II, B.L., Wasson, R.J. (Eds.), 2003. Global Change and the Earth System: A Planet Under Pressure. SpringerVerlag, Berlin/New York.

29. Stern, N., (2007) The Economics of Climate Change. The Stern Review. Cambridge University Press, Cambridge. UNFCC, (2003) United Nations Framework Convention on Climate Change. Bonn UNFCCC 2003

30. UNUDP, (1994) Human Development Report 1994. New York Oxford, Oxford University Press

31. Vitousek, P. M., (1994) Beyond Global Warming: Ecology and Global Change. Ecology 75:1861–1876. [doi: 10.2307/1941591]

32. Woodward, J. and Buckingham, S., (2008) 'Global climate change'. In Buckingham, S. and Turner, M. (eds.) Understanding environmental issues. London; Sage, pp. 175-206

33. WRI., (2006) Climate Analysis Indicators Tool (CAIT) on-line database version 3.0. World Resources Institute, Washington DC. Available at <www.cait.wri.org> [Accessed 15 March 2009]

Chapter 8

ACID STRESS SURVIVAL MECHANISMS OF THE CARIOGENIC BACTERIUM STREPTOCOCCUS MUTANS

Yoshihisa Yamashita and Yukie Shibata

Section of Preventive and Public Health Dentistry, Kyushu University Faculty of Dental Science Japan

INTRODUCTION

Streptococcus mutans, the major etiological agent in human dental caries, is capable of forming a biofilm, or dental plaque, on the tooth surface (Loesche, 1986; Tanzer et al., 2001). S. mutans generates large amounts of acid within dental plaque from fermentable dietary carbohydrates. During meals, the ingestion of carbohydrates causes the pH of the dental plaque to fall below 4.0. Acid accumulation can eventually destroy the crystalline structure of teeth that is the hardest tissue in the human body, leading to the formation of a carious lesion (Quivey et al., 2001). The ability of S. mutans to survive in such a severe environment represents one of the most important virulence factors of this microorganism. The mechanisms of acid tolerance that are most common among Gram-positive bacteria have been proposed to be: i) proton pumps; ii) protection and/or repair of macromolecules; iii) cell-membrane changes; iv) production of alkali; v) regulators; vi) cell density and biofilms; and vii) alteration of metabolic pathways (Fig. 1) (Cotter & Hill, 2003). Many researchers have sought to explain the mechanisms of acid tolerance in S. mutans, and various genes contributing to aciduricity in S. mutans have been identified. In this chapter, we review those genes that have been reported to be involved in S. mutans aciduricity, including those participating in two-component systems and others, especially targeting the dgk homolog.

TWO-COMPONENT SYSTEM

Two-component systems (TCSs), prokaryote-specific signal transduction systems, are widespread in prokaryotes and play extensive roles in adaptation to environmental changes. The TCS operon (tcs) consists of hk, which encodes a sensory histidine kinase (HK), and rr, which encodes its cognate response regulator (RR). The HK undergoes autophosphorylation on a histidine residue in response to a specific environmental signal and relays this phosphate group to an aspartic acid residue on the cognate RR. The phosphorylated RR then binds target DNA elements with greater affinity, inducing or repressing the transcription of target genes (Hoch, 2000; Rampersaud et al., 1994). In this way, bacteria are able to adapt to the changes in external environment and to modulate gene expression. TCSs may be responsible for the acid tolerance of S. mutans.

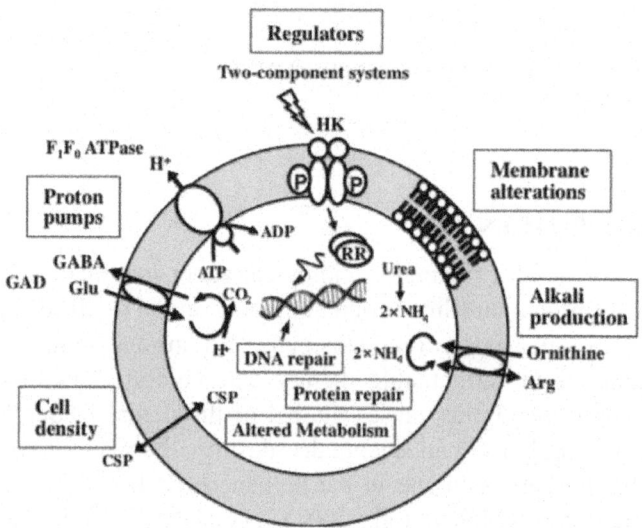

Figure. 1: Acid tolerance mechanisms proposed for gram-positive bacteria. The figure is taken from Cotter & Hill (2003) with some modification.

Analysis of the complete genome sequence of S. mutans UA159 suggested the presence of 13 hk-rr homologs and one orphan rr homolog (Table 1, smtcs02-15) (Ajdic et al., 2002). The roles of some specific tcs genes in acid tolerance have been evaluated. Li et al. showed that disruption of smhk13 or smrr13 resulted in a diminished log-phase acid-tolerance response in S. mutans BM71 (Li et al., 2001) and that only smhk02 of smtcs02 was involved in the acidtolerance response of S. mutans NG8 (Li et al., 2002). Qi et al. (2004) and Ahn et al. (2006) reported that the smhk08 mutant exhibited a significant growth defect, whereas the growth of both the smrr08 mutant and smhk-rr08 double mutant was similar to that of wild-type UA159 when grown at pH 6.4. Ahn et al. (2006) also showed that all smhk08, smrr08, or smhkrr08 mutants presented growth defects when grown at pH 5.5. Then, Lévesque et al. (2007) systematically inactivated each of the 13 hk, but not rr, genes in S.

mutans UA159 and evaluated the roles of the hk genes in acid tolerance. They showed that smhk09 and smhk14 were involved in S. mutans acid tolerance. Furthermore, Biswas et al. (2007) found an additional tcs (Table 1, smtcs01) in the genome of S. mutans UA159 and examined the involvement of 14 hk genes in acid tolerance. They showed that only smhk08 was involved in aciduricity. However, these studies focused only on the role of HKs and so did not provide a comprehensive overview of the role of TCSs in acid tolerance.

Table 1: The tcs genes identified in the S. mutans UA159 genome

tcs genes	hk gene, rr gene	GenBank Locus Tag[a]	Gene order
smtcs01	smhk01	SMU.45	hk-rr
	smrr01	SMU.46	
smtcs02	smhk02	SMU.486	hk-rr
	smrr02	SMU.487	
smtcs03	smhk03	SMU.577c	hk-rr
	smrr03	SMU.576c	
smtcs04	smhk04	SMU.660	rr-hk
	smrr04	SMU.659	
smtcs05	smhk05	SMU.928	rr-hk
	smrr05	SMU.927	
smtcs06	smhk06	SMU.1009	rr-hk
	smrr06	SMU.1008	
smtcs07	smhk07	SMU.1037c	rr-hk
	smrr07	SMU.1038c	
smtcs08	smhk08	SMU.1128c	rr-hk
	smrr08	SMU.1129c	
smtcs09	smhk09	SMU.1145c	rr-hk
	smrr09	SMU1146c	
smtcs10	smhk10	SMU.1516c	rr-hk
	smrr10	SMU.1517c	
smtcs11	smhk11	SMU.1548c	hk-rr
	smrr11	SMU.1547c	
smtcs12	smhk12	SMU.1814c	rr-hk
	smrr12	SMU.1815c	
smtcs13	smhk13	SMU.1916c	rr-hk
	smrr13	SMU.1917c	
smtcs14	smhk14	SMU.1965c	hk-rr
	smrr14	SMU.1964c	
smtcs15	smrr15	SMU.1924c	rr

tcs, two-component system; hk, histidine kinase; rr, response regulator.
[a] GenBank locus tag was associated with the S. mutans genome at the Oral Pathogen Sequence Database site (http://www.stdgen.lanl.gov/oragen).

Therefore, we systematically constructed rr deletion mutants and hk-rr double mutants of S. mutans UA159 and examined the effect on acid tolerance (Kawada-Matsuo et al., 2009). Thirteen rr mutants and twelve hk-rr double mutants were obtained, the exceptions being smrr10, smtcs10, smrr12, and smtcs12. The derivation of null mutations of these genes was unsuccessful, probably due to a loss of viability of these mutants. To examine the effects of these rr mutations on the acid tolerance of S. mutans, wild-type UA159 and the 25 mutants were grown in brain–heart infusion (BHI) broth adjusted to pH 7.2 or pH 5.5. Growth curves were generated, and the mid-log-phase doubling time was determined. All rr and hk-rr mutants grew similarly to wild-type UA159 at pH 7.2. However, as shown in Table 2, deletion of four rr genes (smrr03, smrr05, smrr08, and smrr13) caused significantly decreased growth rates compared with that of wild-type UA159 when grown at pH 5.5. The growth rates of the hk-rr double mutants were similar to those of the corresponding rr mutants, and the differences in doubling time between them were not significant.

Table 2: Doubling times of rr or hk-rr deletion mutants at pH 5.5

	Doublimg time (min)[a] in:	
UA159[b]	123.8 ± 9.5	
tcs genes	hk+rr−	hk−rr−
smtcs01	116.8 ± 10.0	117.7 ± 10.3
smtcs02	131.5 ± 8.8	132.6 ± 5.9
smtcs03	146.4 ± 7.5*	148.5 ± 8.3*
smtcs04	127.4 ± 7.6	122.5 ± 3.7
smtcs05	160.1 ± 11.6**	155.7 ± 7.8*
smtcs06	132.8 ± 3.2	125.1 ± 10.2
smtcs07	131.0 ± 3.2	122.3 ± 7.2
smtcs08	146.3 ± 11.9*	145.2 ± 7.7*
smtcs09	126.3 ± 5.2	129.9 ± 5.0
smtcs11	126.4 ± 7.1	131.7 ± 11.0
smtcs13	141.5 ± 8.8*	140.9 ± 2.6*
smtcs14	132.5 ± 3.7	131.7 ± 12.1
smtcs15[c]	127.6 ± 7.1	−

[a]Doubling time (Td) was calculated based on the formulas $\ln Z - \ln Z0 = k (t - t0)$, where k is the growth rate, and $g = 0.693/k$, where g is the doubling time. Values are the mean ± standard deviation obtained from three independent experiments.
[b]Wild-type strain
[c]Orphan rr
* Significant increase from Td of wild-type UA159 by Tukey's HSD, $p < 0.05$
** Significant increase from Td of wild-type UA159 by Tukey's HSD, $p < 0.01$

The finding that smrr08 and smrr13 were involved in the acid tolerance of S. mutans is consistent with previous findings (Ahn et al., 2006; Li et al., 2001). On the other hand, smrr03 and smrr05 were, for the first time, demonstrated to be involved in S. mutans UA159 acid tolerance. However, Lévesque et al.

(2007) and Biswas et al. (2007) showed that inactivation of their cognate hk genes did not affect acid tolerance. To confirm whether only the rr gene is involved in acid tolerance in the smtcs03 and smtcs05 mutants, the hk genes of smtcs03 and smtcs05 were individually inactivated, and the acid tolerance ability of these mutants compared. As shown in Table 3, the smhk03 mutant exhibited a decreased growth rate compared with the wild type when grown at pH 5.5. This was not consistent with previous results. In contrast, the smhk05 mutant grew similarly to wild-type UA159 at pH 5.5, as shown in previous studies, whereas the smrr05 mutant and smhk-rr05 double mutant exhibited reduced growth rates.

Generally, a signal sensed by a HK is thought to be transmitted to the cognate RR via transfer of phosphoryl groups, and deletion of either the hk or rr should generate a similar phenotype. However, we found that only the rr of smtcs05 was involved in S. mutans acid tolerance. Furthermore, as mentioned above, Li et al. (2002) reported that the hk, but not the rr, of smtcs02 was involved in the acid tolerance of S. mutans NG8. Qi et al. (2004) and Ahn et al. (2006) also reported that the smhk08 mutant showed a significant growth defect, whereas the growth of both the smrr08 mutant and smhk-rr08 double mutant was similar to that of wild-type UA159. These results suggested involvement of several TCSs in S. mutans aciduricity via cross-talk between different TCS components. After all, among 15 TCSs, seven (Smtcs02, 03, 05, 08, 09, 13, and 14) appear to be involved in S. mutans aciduricity. Nevertheless, inactivation of no single TCS caused a complete loss of acid tolerance. Therefore, other TCSs that are definitively related to S. mutans acid tolerance may exist, but they cannot be identified by homology searching.

Table 3: Doubling times of hk, rr, and hk-rr deletion mutants of smtcs03 and smtcs05 at pH 7.2 and pH 5.5.

tcs genes	Strain	hk/rr	Doubling time (min) in:	
			BHI pH 7.2	BHI pH 5.5
smtcs03	SMHK03	-/+	53.6 ± 4.1	153.4 ± 7.9
	SMRR03	+/-	50.0 ± 2.7	146.4 ± 7.5
	SMTCS03	-/-	57.1 ± 8.6	148.5 ± 8.3
smtcs05	SMHK05	-/+	51.0 ± 3.4	128.8 ± 12.9
	SMRR05	+/-	49.7 ± 2.5	160.1 ± 11.6
	SMTCS05	-/-	51.5 ± 4.6	155.7 ± 7.8

*p < 0.05 (Tukey's HSD)

GENES OTHER THAN TCS INVOLVED IN S. MUTANS ACID TOLERANCE

Many studies have implicated genes other than those involved in TCS in S. mutans aciduricity. Table 4 summarizes the characteristics of 14 such genes, of which the functions of the products of only nine have been experimentally verified. These are aguA, encoding an agmatine deiminase that is involved in alkali production (Griswold et al., 2004); dltC, involved in the synthesis of D-alanyl-lipoteichoic acid, which is associated with alteration of membrane composition (Boyd et al., 2000); gluA, encoding a glucose-l-phosphate uridylyltransferase involved in the synthesis of UDP-D-glucose (Yamashita et al., 1998); lgl, encoding a lactoylglutathione lyase involved in the detoxification of methylglyoxal (Korithoski et al., 2007); luxS, encoding a Sribosylhomocysteine lyase involved in autoinducer AI2 synthesis (Wen and Burne, 2004); and uvrA, encoding an excinuclease ABC subunit A that is involved in DNA repair (Hanna et al., 2001). Mutation of three genes (ffh, ftsY, yidC2 genes) that are involved in the signal recognition particle pathway, significantly reduced H+/ATPase specific activity compared with that of the wild type (Hasona et al., 2005). However, how these functions contribute to S. mutans aciduricity remains unclear. Furthermore, the functions of the five remaining gene products were predicted based on DNA sequence homology, and so estimation of their role in S. mutans aciduricity is much more difficult To examine the extent to which a gene contributes to S. mutans acid tolerance, 14 mutants in which one of the genes listed in Table 4 was inactivated and the dgk mutant were constructed from S. mutans UA159, and their growth at pH 7.4 and at pH 5.5 was compared with that of the wild type (Fig. 2) (Shibata et al., 2011). Inactivation of aguA, brpA, glrA, htrA, lgl, luxS, ropA, or uvrA did not significantly affect the acid tolerance of S. mutans as compared with wild-type UA159 when grown at pH 5.5. Three types of method of assessing the acid tolerance of S. mutans are available. Simple acid tolerance is evaluated by a procedure in which over-night cultures or log-phase cells grown at neutral pH are subcultured in media or on plates at neutral and acidic pH. Another method is acid killing, in which log-phase cells grown at neutral pH are incubated at a lethally acidic pH, and then viability is determined by plate counts. The final method is determination of the acid tolerance response in which viability is estimated by plating after incubation at neutral or acidic pH of log-phase cells grown at neutral pH followed by incubation at killing pH. We used the first method, and so any discrepancy between this and other studies may derive from differences in the method used. However, comparing all of the mutants in terms of the most basic characteristics of acid tolerance is a worthwhile endeavor. Of course, a comparison using the other criteria is important, and will be performed as part of our next effort. Notably, the dgk and the gluA mutants grew extremely slowly at pH 5.5, although the clpP, dltC, ffh, ftsY, and yidC2 mutants also displayed significant reductions in growth rate at pH 5.5 compared with the wild-type UA159. However, only the brpA, dgk, dltC, htrA, lgl, luxS, ropA, and uvrA mutants showed growth rates comparable to the wild-type strain at pH 7.4. These findings suggest that the reduction in growth rates of the clpP, ffh, ftsY, yidC2, and gluA mutants at acidic pH values might be derived from reduced viability and not specifically related to acid tolerance. Therefore, dltC and dgk are likely specifically involved in acid tolerance. Of these two genes, the striking reduction in growth rate of the dgk mutant at acidic pH indicates that dgk is of great interest for elucidating the acid-tolerance mechanism of S. mutans.

Table 4: Genes reported to be involved in S. mutans acid tolerance

Gene	GenBank Locus Tag	Function	Evidence	Determination of the response against low pH	Reference
aguA	SMU.264	Agmatine deiminase	Testified experimentally	Not determined	Griswold et al., 2004
brpA	SMU.410	Transcriptional regulator	Putative	Acid tolerance & Acid killing	Wen et al., 2006
clpP	SMU.1672c	Clp protease (Serine protease)	Putative	Acid tolerance	Lemos & Burne, 2002
dltC	SMU.1689c	D-alanyl carrier protein	Testified experimentally	Acid tolerance & Acid killing & Acid tolerance response	Boyd et al., 2000
ffh	SMU.1060c	Signal recognition particle	Testified experimentally (partially)	Acid tolerance	Kremer et al., 2001
ftsY	SMU.744	Signal recognition particle receptor	Testified experimentally (partially)	Acid tolerance	Hasona et al., 2005
glrA	SMU.1035	ABC transporter ATP-binding-protein	Putative	Acid tolerance	Cvitkovitch et al., 2000
gluA	SMU.322c	Glucose-l-phosphate uridylyltransferase	Testified experimentally	Acid tolerance	Yamashita et al., 1998
htrA	SMU.2164	Serine protease	Putative	Acid tolerance (agar plate)	Biswas & Biswas, 2005
lgl	SMU.1603	Lactoylglutathione lyase	Testified experimentally	Acid tolerance & Acid tolerance response	Korithoski et al., 2007
luxS	SMU.474c	S-ribosylhomo-cysteine lyase	Testified experimentally	Acid tolerance & Acid killing & Acid tolerance response	Wen & Burne, 2004
ropA	SMU.91	Peptidyl-prolyl isomerase, trigger factor	Putative	Acid killing	Wen et al., 2005
uvrA	SMU.1851c	UV repair excinuclease	Testified experimentally	Acid tolerance & Acid tolerance response	Hanna et al., 2001
yidC2	SMU.1727	Oxa1(or A)-like protein	Testified experimentally (partially)	Acid tolerance	Dong et al., 2008

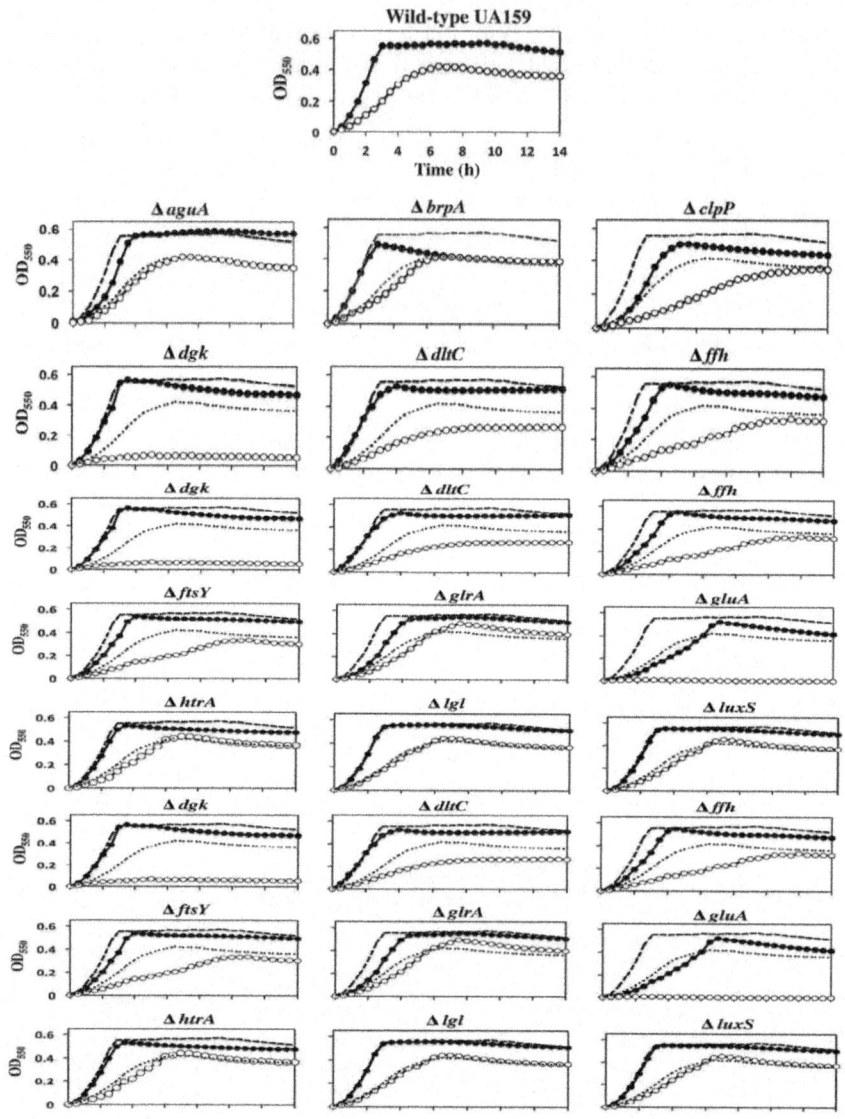

Figure. 2: Growth curves of S. mutans UA159 and mutant strains grown in BHI medium at pH 7.4 ( ) or pH 5.5 ( ). Growth was defined as the increase in OD550, and was calculated by subtraction of OD550 at the initiation of growth from that at the times indicated. Data represent the means of three independent experiments. Graphs of mutant strains also represent the growth curves of wild-type UA159 at pH 7.4 ( ) and pH 5.5 ( ) as controls.

DIACYLGLYCEROL KINASE

Diacylglycerol kinase (Dgk) catalyses the ATP-dependent phosphorylation of sn-1,2- diacylglycerols, resulting in production of phosphatidic acid. In eukaryotic cells, diacylglycerol and phosphatidic acid are immediate second cellular messengers responding to extracellular signals, suggesting that Dgk is a key enzyme in cellular signal transduction (Moolenaar et al., 1986; Murayama & Ui, 1987; Nishizuka, 1984; Topham & Prescott, 1999). Among bacterial Dgk, only that of Escherichia coli has been well-characterized, and it is a small integral membrane protein with a molecular mass of 13.2 kDa. This enzyme functions in the recycling of the diacylglycerol produced during turnover of membrane phospholipids (Hasin & Kennedy, 1982; Rotering & Raetz, 1983) and plays an important physiological role in responding to environmental stress as well as its role in eukaryotic cells (Raetz & Newman, 1979; Walsh et al., 1986). On the other hand, the S. mutans Dgk homolog has a molecular mass of 15.3 kDa and comprises 137 amino acids. It is interesting that insertion of the transposon Tn916 into the codon for the tenth amino acid from the C terminus of the Dgk homolog resulted in defective growth of the mutant (GS5Tn1) at acidic pH values (Yamashita et al., 1993). In addition to attenuation of aciduricity, this mutant possessed reduced resistance to high osmolarity and temperature (Yamashita et al., 1993). The C terminus of the Dgk homolog may thus play an important role in signal transduction during environment stress. To evaluate how the C terminus of Dgk contributes to S. mutans acid tolerance, we sequentially truncated amino acids from the C terminus of Dgk and finally constructed 11 mutants termed UADGK0–10, expressing Dgk0–10 (Fig. 3) (Shibata et al., 2009). The mutants showed no significant difference in growth rate at neutral pH (doubling times: 53.8 to 61.6 min; Table 5). Most, with the exception of UADGK0 to UADGK2, showed a reduction in growth rate at pH 5.5 compared with the wild type (Table 5 and Fig. 4). UADGK3, in which three amino acid residues had been deleted from the C-terminus of Dgk, showed a slight reduction in growth rate. Subsequent deletion of amino acids from the C-terminus resulted in further reductions in growth rate at acidic pH. Indeed UADGK4, UADGK5, and UADGK6 had significantly increased doubling times ($p < 0.05$, $p < 0.001$, and $p < 0.0001$, respectively) compared with UADGK0. UADGK7, in which seven amino acid residues had been deleted, showed extremely limited growth in the first 9 hours. Further truncation of the C-terminus of Dgk (UADGK8 to UADGK10) resulted in no growth at pH 5.5. These results suggest that the C-terminal of the Dgk homolog is indispensable for its function in aciduricity of S. mutans. We further constructed two additional UA159 dgk mutants, UADGK11 and UADGK12 (Fig. 3) to evaluate the function of truncated Dgk. There were only negligible differences in the growth rates of these two mutants at pH 5.5, 5.8, or 6.3, compared with that of UADGK10.

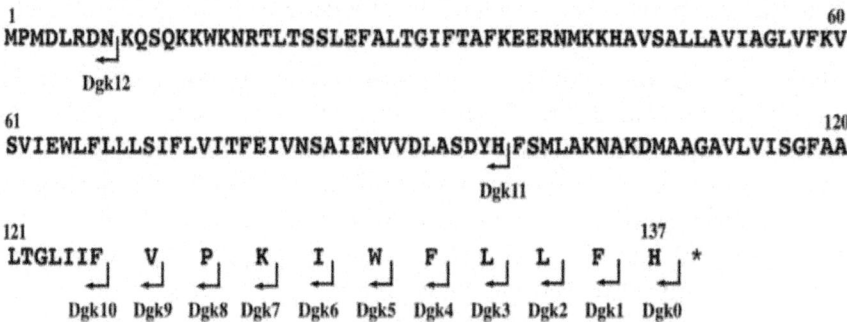

```
1                                                              60
MPMDLRDN⌐KQSQKKWKNRTLTSSLEFALTGIFTAFKEERNMKKHAVSALLAVIAGLVFKV
        └┘
        Dgk12

61                                                            120
SVIEWLFLLLSIFLVITFEIVNSAIENVVDLASDYH⌐FSMLAKNAKDMAAGAVLVISGFAA
                                    └┘
                                   Dgk11

121                                                       137
LTGLIIF⌐  V⌐  P⌐  K⌐  I⌐  W⌐  F⌐  L⌐  L⌐  F⌐  H⌐  *
       └┘  └┘  └┘  └┘  └┘  └┘  └┘  └┘  └┘  └┘  └┘
Dgk10 Dgk9 Dgk8 Dgk7 Dgk6 Dgk5 Dgk4 Dgk3 Dgk2 Dgk1 Dgk0
```

Figure. 3: Representation of the truncated Dgk proteins used. The deduced amino acid sequence of the dgk gene from S. mutans UA159 is presented. The terminal amino acid of the truncated Dgk expressed in each UA159 dgk mutant and each E. coli RZ transformant is indicated along the sequences by a curved arrow, which indicates that the sequence is deleted from the right up to this site. All the truncated Dgk proteins names were changed from those of previous paper (Shibata et al., 2009) to help readers understand.

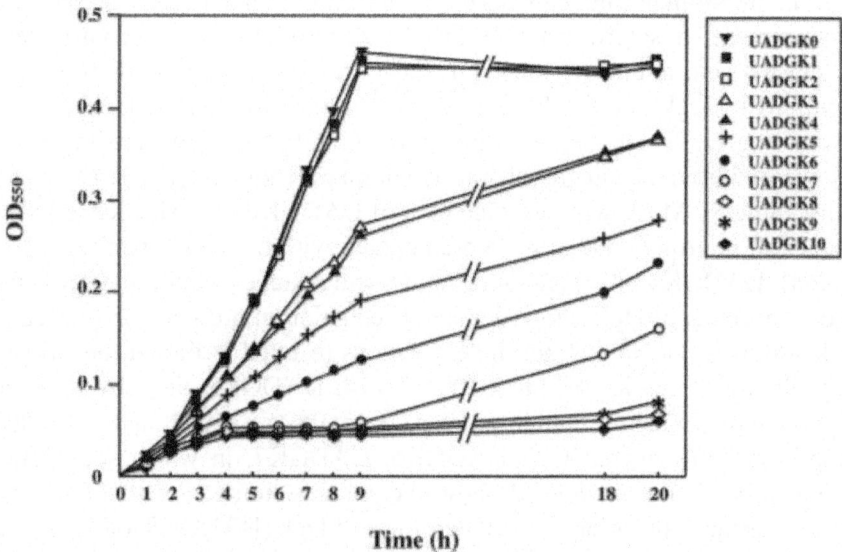

Figure. 4: Growth curves of S. mutans UA159 dgk mutants grown in BHI medium at pH 5.5. Growth was defined as the increase in OD550, and was calculated by subtraction of OD550 at the initiation of growth from that at the times indicated. Data repre-

sent the means of three independent experiments. All the mutants names were changed from those of previous paper (Shibata et al., 2009) to help readers understand.

Table 5: Effect of low pH on growth of S. mutans UA159 dgk mutants

Strain	Doubling time (min) in:	
	BHI pH 7.45	BHI pH 5.5
UADGK0	57.8 ± 4.4	132.3 ± 12.9
UADGK1	58.2 ± 6.4	131.2 ± 11.7
UADGK2	61.2 ± 5.8	133.6 ± 11.7
UADGK3	61.6 ± 1.8	177.6 ± 19.0
UADGK4	55.5 ± 3.4	202.5 ± 24.4*
UADGK5	57.5 ± 2.5	248.9 ± 32.2**
UADGK6	55.3 ± 5.4	271.3 ± 34.6 ***
UADGK7	55.1 ± 6.0	>1000 [a]
UADGK8	53.8 ± 3.7	>1000 [a]
UADGK9	58.4 ± 6.7	>1000 [a]
UADGK10	57.0 ± 6.1	>1000 [a]

Differences in the doubling time between UADGK0 and UADGK1-10 were analyzed by Bonferroni multiple comparison test (*, $p < 0.05$; **, $p < 0.001$; ***, $p < 0.0001$).
[a]Statistical analyses were not carried out because of too slow growth rates.
All the mutants names were changed from those of previous paper (Shibata et al., 2009) to help readers understand.

We next constructed recombinant Dgk proteins corresponding to S. mutans strains UADGK10, UADGK7, UADGK5, and UADGK0 utilizing E. coli strains RZDGK10, RZDGK7, RZDGK5, and RZDGK0, respectively. The kinase activity in cell lysates of E. coli transformants was examined by an octyl glucoside mixed-micelle assay (Preiss et al., 1986), using undecaprenol as a substrate because of it has a higher substrate specificity for the S. mutans Dgk homolog compared with diacylglycerol (Lis & Kuramitsu, 2003). As shown in Fig. 5A, whereas the full-size S. mutans Dgk protein expressed in RZDGK0 catalyzed a high level of phosphorylation of undecaprenol, the Dgk missing five amino acid residues from the C terminus expressed in RZDGK5 exhibited markedly reduced kinase activity. Furthermore, RZDGK7 (seven amino acids missing from the C terminus) exhibited much weaker kinase activity than did RZDGK5. The deletion of 10 C-terminal amino acid residues of Dgk in RZDGK10 resulted in a total lack of kinase activity. These differences were confirmed by quantitative analysis (Fig. 5B). These data indicate that the C-terminus of the S. mutans Dgk homolog plays an important role in kinase activity and may harbor residues required for catalysis. Alternatively, incorrect folding of the protein due to the missing C-terminal residues may cause loss of kinase activity. Therefore, its catalysis of undecaprenol phosphorylation is closely related to S. mutans acid tolerance.

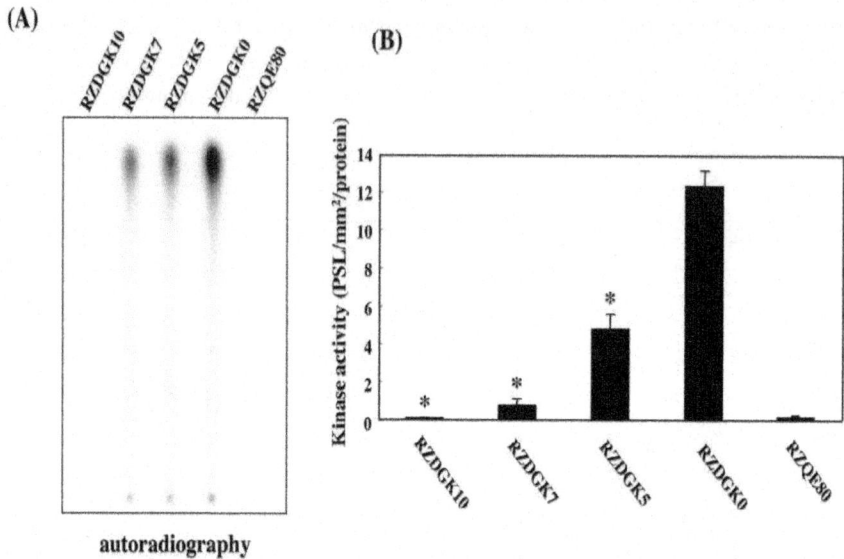

autoradiography

Figure. 5: Effect of deletion of the C-terminal tail of Dgk on undecaprenol kinase activity. (A) Comparison of undecaprenol kinase activity of the full-size Dgk and various C-terminally truncated forms of Dgk. The undecaprenol kinase activity in the lysates from E. coli RZ cells was determined using an octyl glucoside mixed-micelle assay. (B) Quantification analysis of the kinase assay. Quantification was carried out by normalization of radioactive bands in the kinase assay using the protein level. Vertical bars represent standard deviation. Differences in kinase activity between RZDGK0 and RZDGK5, RZDGK7, or RZDGK10 were analyzed by Student's t test (*, $p < 0.0001$). All the E.coli strains names were changed from those of previous paper (Shibata et al., 2009) to help readers understand.

Moreover, the importance of the C-terminal end of Dgk in S. mutans acid tolerance was examined in a specific pathogen-free animal model (Table 6). The dgk mutant strain clearly displayed a significant reduction in smooth-surface caries compared with the wild type ($p < 0.005$). In contrast, no significant difference in plaque extent was observed between the wild-type and dgk mutant strains. These results suggest that aciduricity regulated by the dgk gene product might play a critical role in S. mutans virulence.

Table 6: Influence of dgk deletion on smooth-surface plaque extent, initial and advanced dentinal fissure lesions, smooth-surface caries, and colonization properties

Treatment	Plaque extent (Δ)	Initial dentinal fissures ($\Delta\Delta$)	Advanced dentinal fissures ($\Delta\Delta$)	Smooth-surface caries ($\Delta\Delta\Delta$)	Total bacteria CFU (10^7)	Total streptococci CFU (10^7)	Total S. mutans CFU (10^7)
Water control	2.8±0.63[a][b]	9.5±1.72[a]	6.6±2.80[a][b]	0.5±0.97[a]	4.4±2.36[a]	2.7±2.26[a]	ND[a][b]
UA159 (wt)	1.1±0.32[a]	11.5±1.27[a]	10.8±1.62[a]	9.5±6.55[a][c]	7.9±3.21[a]	7.4±2.79[a][c]	4.0±2.11[a][c]
UADGK10 (dgk)	1.5±0.53[b]	10.6±0.84	9.3±1.2[b]	2.5±2.68[c]	6.4±2.40	4.4±2.53[c]	2.3±1.20[b][c]

ND, Not determined. Δ, 4 units at risk; $\Delta\Delta$, 12 fissures at risk; $\Delta\Delta\Delta$, 20 units at risk.
[a]Significant difference between water control and UA159 (wt), $p < 0.05$.
[b]Significant difference between UADGK1 (dgk) and water control, $p < 0.05$.
[c]Significant difference between UA159 (wt) and UADGK1 (dgk), $p < 0.05$.
The mutant name was changed from those of previous paper (Shibata et al., 2009) to help readers understand.

Table 6. Influence of dgk deletion on smooth-surface plaque extent, initial and advanced dentinal fissure lesions, smooth-surface caries, and colonization properties.

A POTENTIAL TARGET FOR ANTI-CARIES CHEMOTHERAPY

Development of an effective anti-caries agent is the ultimate goal of our work. Considering the characteristics of known mutants, the Dgk homolog seems to be the most promising target for anti-caries agents. Dgks have been extensively studied in mammals, and several inhibitory compounds, e.g., R59022 and R59949 (Fig. 6), have been reported. In contrast, inhibitors of prokaryotic Dgk have not yet been elucidated.

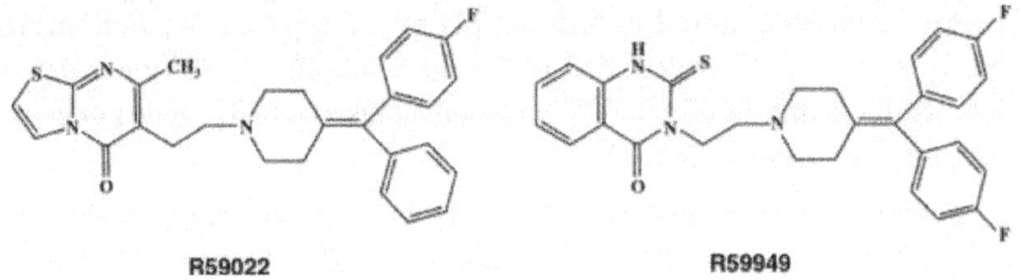

R59022   R59949

Figure. 6: Structures of R59022 and R59949.

When first attempting to discover inhibitors of prokaryotic Dgk, we tested the effects of R59022 and R59949 on the growth of S. mutans (Shibata et al., 2011). Although neither R59022 nor R59949 influenced growth at pH 7.4, R59949, but not R59022, showed a significant inhibitory effect at acidic pH (Fig. 7). Inhibition by R59949 increased by 13, 29, 58, 68, and 78% at pH 5.4,

5.3, 5.2, 5.1, and 5.0, respectively (Fig. 7A). These findings were particularly interesting because R59022 and R59949 were used at concentrations of 100 µM and 25 µM, respectively, due to the limited solubility of R59949.

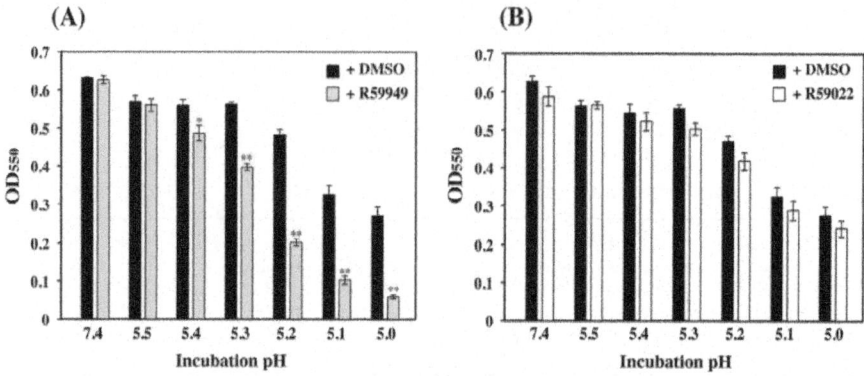

Figure. 7: Effect of R59949 (A) and R59022 (B) on the growth of S. mutans. Data represent the mean ± standard deviation. Differences in growth rate between cells cultured in the presence and absence of Dgk inhibitor were analyzed using Student's t test. *, $p < 0.05$; **, $p < 0.0001$.

Furthermore, we examined the inhibitory effects of R59022 and R59949 on the kinase activity of S. mutans Dgk. Neither R59022 nor R59949 inhibited kinase activity at pH 7.4; this is in agreement with their lack of effect on S. mutans growth at neutral pH. As mentioned above, R59949 significantly inhibited the growth of S. mutans at acidic pH values (below 5.4). When evaluating the effect of R59949 on enzyme activity at acidic pH, it is important to know the intracellular pH of S. mutans cells; the intracellular pH of S. mutans cells was 6.4 when cultured in broth at pH 5.2. Therefore, we determined the inhibitory effect of R59949 and R59022 on S. mutans Dgk kinase activity at pH 6.4. R59949, but not R59022, inhibited kinase activity with undecaprenol as a substrate by around 20% (Fig. 8).

S. mutans Dgk is inherently different from mammalian Dgk in terms of its molecular size, molecular structure, and substrate specificity. However, it is interesting that R59949 inhibits the enzymatic activity of S. mutans Dgk even with undecaprenol as the substrate. Additionally, the difference in inhibitory activity between R59949 and R59022 means that a comparison of their molecular structure may lead to discovery of further potent Dgk inhibitors specific for prokaryotic enzymes, that is, new anti-caries agents.

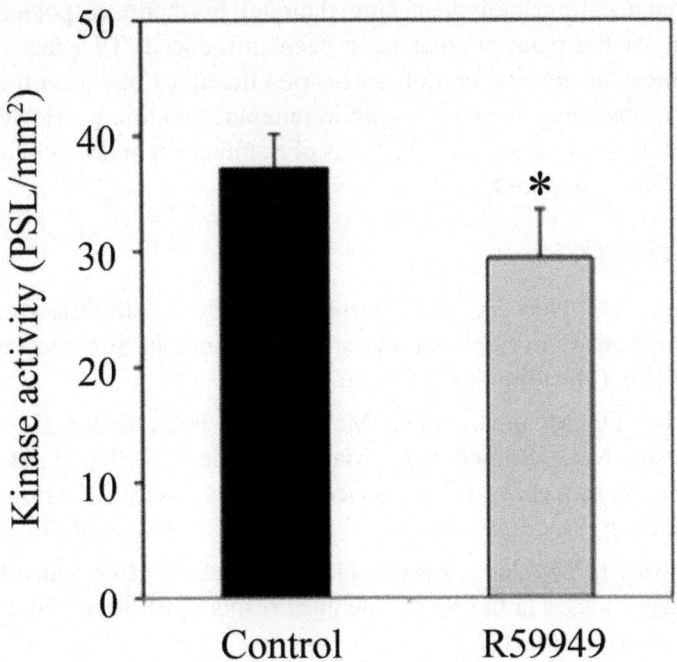

Figure. 8: Effect of R59949 on kinase activity with undecaprenol as a substrate. Data represent the mean ± standard deviation. Differences in kinase activity between cells cultured in the presence and absence of R59949 were analyzed by Student's t test. *, p < 0.05.

CONCLUSION

The reduction of environmental pH in dental plaque by the cariogenic microorganisms is important step in the development of dental caries. The cariogenic microorganisms should survive in a relentless environment produced by themselves in order to exhibit their maximum virulence. In this chapter, we described the acid tolerance characteristics of the cariogenic microorganism, S. mutans. TCSs seem to be the most suitable system for adaptation to environmental conditions. However, no TCS seems to be definitively responsible for S. mutans acid tolerance. At present, identification of TCS depends on gene homology searching, which may not identify all genes encoding TCS that contribute to S. mutans acid tolerance. We focus on dgk because it is the most promising contributor to S. mutans acid tolerance when assessed using a simple acid tolerance assay. Although the precise mechanism by which the gene product is involved in acid tolerance has not yet been elucidated, dgk is the only gene whose product has been definitively implicated in cariogenicity in an animal model. Furthermore, potential specific inhibitors of the gene product have been introduced. This fact may aid in development of next-generation anti-caries therapies based on the ability of this microorganism to adapt to environmental conditions. However, much detail of the acid tolerance mechanisms of S. mutans remains unknown, and so further study is required.

REFERENCES

1. Ahn, S.J., Wen, Z.T., & Burne, R.A. (2006). Multilevel control of competence development and stress tolerance in Streptococcus mutans UA159. Infect Immun, 74, 3, 1631-1642

2. Ajdic, D., McShan, W.M., McLaughlin, R.E., Savic, G., Chang, J., Carson, M.B., Primeaux, C., Tian, R., Kenton, S., Jia, H., et al. (2002). Genome sequence of Streptococcus mutans UA159, a cariogenic dental pathogen. Proc Natl Acad Sci U S A, 99, 22, 14434-14439

3. Biswas, I., Drake, L., Erkina, D., & Biswas, S. (2007). Involvement of sensor kinases in the stress tolerance response of Streptococcus mutans. J Bacteriol, 190, 1, 68-77

4. Biswas, S. & Biswas, I. (2005). Role of HtrA in surface protein expression and biofilm formation by Streptococcus mutans. Infect Immun, 73, 10, 6923-6934

5. Boyd, D.A., Cvitkovitch, D.G., Bleiweis, A.S., Kiriukhin, M.Y., Debabov, D.V., Neuhaus, F.C., & Hamilton, I.R. (2000). Defects in D-alanyl-lipoteichoic acid synthesis in Streptococcus mutans results in acid sensitivity. J Bacteriol, 182, 21, 6055-6065

6. Cvitkovitch, D.G., Gutierrez, J.A., Behari, J., Youngman, P.J., Wetz, J.E., Crowley, P.J., Hillman, J.D., Brady, L.J., & Bleiweis, A.S. (2000). Tn917-lac mutagenesis of Streptococcus mutans to identify environmentally regulated genes. FEMS Microbiol Lett, 182, 1, 149-154

7. Cotter, P.D. & Hill, C. (2003). Surviving the acid test: responses of gram-positive bacteria to low pH. Microbiol Mol Biol Rev, 67, 3, 429-453

8. Dong, Y., Palmer, S.R., Hasona, A., Nagamori, S., Kaback, H.R., Dalbey, R.E. & Brady, L.J. (2008). Functional overlap but lack of complete cross-complementation of Streptococcus mutans and Escherichia coli YidC orthologs. J Bacteriol, 190, 7, 2458-2469

9. Griswold, A.R., Chen, Y.Y., & Burne, R.A. (2004). Analysis of an agmatine deiminase gene cluster in Streptococcus mutans UA159. J Bacteriol, 186, 6, 1902-1904

10. Hanna, M.N., Ferguson, R.J., Li, Y.H., & Cvitkovitch, D.G. (2001). uvrA is an acid-inducible gene involved in the adaptive response to low pH in Streptococcus mutans. J Bacteriol, 183, 20, 5964-5973

11. Hasin, M. & Kennedy, E.P. (1982). Role of phosphatidylethanolamine in the biosynthesis of pyrophosphoethanolamine residues in the lipopolysaccharide of Escherichia coli. J Biol Chem, 257, 21, 12475-12477

12. Hasona, A., Crowley, P.J., Levesque, C.M., Mair, R.W., Cvitkovitch, D.G., Bleiweis, A.S., & Brady, L.J. (2005). Streptococcal viability and diminished stress tolerance in mutants lacking the signal recognition particle pathway or YidC2. Proc Natl Acad Sci U S A, 102, 48, 17466-17471

13. Hoch, J.A. (2000). Two-component and phosphorelay signal transduction. Curr Opin Microbiol, 3, 2, 165-170

14. Kawada-Matsuo, M., Shibata, Y., & Yamashita, Y. (2009). Role of two component signaling response regulators in acid tolerance of Streptococcus mutans. Oral Microbiol Immunol, 24, 2, 173-176

15. Korithoski, B., Levesque, C.M., & Cvitkovitch, D.G. (2007). Involvement of the detoxifying enzyme lactoylglutathione lyase in Streptococcus mutans aciduricity. J Bacteriol, 189,21, 7586-7592

16. Kremer, B.H., van der Kraan, M., Crowley, P.J., Hamilton, I.R., Brady, L.J. & Bleiweis, A.S. (2001). Characterization of the sat operon in Streptococcus mutans: evidence for a role of Ffh in acid tolerance. J Bacteriol, 183, 8, 2543-2552

17. Lemos, J.A. & Burne, R.A. (2002). Regulation and Physiological Significance of ClpC and ClpP in Streptococcus mutans. J Bacteriol, 184, 22, 6357-6366

18. Lévesque, C.M., Mair, R.W., Perry, J.A., Lau, P.C., Li, Y.H., & Cvitkovitch, D.G. (2007). Systemic inactivation and phenotypic characterization of two-component systems in expression of Streptococcus mutans virulence properties. Lett Appl Microbiol, 45, 4, 398-404

19. Li, Y.H., Hanna, M.N., Svensater, G., Ellen, R.P., & Cvitkovitch, D.G. (2001). Cell density modulates acid adaptation in Streptococcus mutans: implications for survival in biofilms. J Bacteriol, 183, 23, 6875-6884

20. Li, Y.H., Lau, P.C., Tang, N., Svensater, G., Ellen, R.P., & Cvitkovitch, D.G. (2002). Novel two-component regulatory system involved in biofilm formation and acid resistance in Streptococcus mutans. J Bacteriol, 184, 22, 6333-6342

21. Lis, M. & Kuramitsu, H.K. (2003). The stress-responsive dgk gene from Streptococcus mutans encodes a putative undecaprenol kinase activity. Infect Immun, 71, 4, 1938-1943

22. Loesche, W.J. (1986). Role of Streptococcus mutans in human dental decay. Microbiol Rev, 50, 4, 353-380

23. Moolenaar, W.H., Kruijer, W., Tilly, B.C., Verlaan, I., Bierman, A.J., & de Laat, S.W. (1986). Growth factor-like action of phosphatidic acid. Nature, 323, 6084, 171-173

24. Murayama, T. & Ui, M. (1987). Phosphatidic acid may stimulate membrane receptors mediating adenylate cyclase inhibition and phospholipid breakdown in 3T3 fibroblasts. J Biol Chem, 262, 12, 5522-5529

25. Nishizuka, Y. (1984). The role of protein kinase C in cell surface signal transduction and tumour promotion. Nature, 308, 5961, 693-698

26. Preiss, J., Loomis, C.R., Bishop, W.R., Stein, R., Niedel, J.E., & Bell, R.M. (1986). Quantitative measurement of sn-1,2-diacylglycerols present in platelets, hepatocytes, and rasandsis-transformed normal rat kidney cells. J Biol Chem, 261, 19, 8597-8600

27. Qi, F., Merritt, J., Lux, R., & Shi, W. (2004). Inactivation of the ciaH Gene in Streptococcus mutans diminishes mutacin production and competence development, alters sucrose-dependent biofilm formation, and reduces stress tolerance. Infect Immun, 72, 8, 4895-4899

28. Quivey, R.G., Kuhnert, W.L., & Hahn, K. (2001). Genetics of acid adaptation in oral streptococci. Crit Rev Oral Biol Med 12, 4, 301-314

29. Raetz, C.R. & Newman, K.F. (1979). Diglyceride kinase mutants of Escherichia coli: inner membrane association of 1,2-diglyceride and its relation to synthesis of membranederived oligosaccharides. J Bacteriol, 137, 2, 860-868

30. Rampersaud, A., Harlocker, S.L., & Inouye, M. (1994). The OmpR protein of Escherichia coli binds to sites in the ompF promoter region in a hierarchical manner determined by its degree of phosphorylation. J Biol Chem 269, 17, 12559-12566

31. Rotering, H. & Raetz, C.R. (1983). Appearance of monoglyceride and triglyceride in the cell envelope of Escherichia coli mutants defective in diglyceride kinase. J Biol Chem, 258, 13, 8068-8073

32. Shibata, Y., van der Ploeg, J.R., Kozuki, T., Shirai, Y., Saito, N., Kawada-Matsuo, M., Takeshita, T., & Yamashita, Y. (2009). Kinase activity of the dgk gene product is involved in the virulence of Streptococcus mutans. Microbiology, 155, 557-565

33. Shibata, Y., Kawada-Matsuo, M., Shirai, Y., Saito, N., Li, D., & Yamashita, Y. (2011). Streptococcus mutans diacylglycerol kinase homologue: a potential target for anticaries chemotherapy. J Med Microbiol, 60, 625-630

34. Tanzer, J.M., Livingston, J., & Thompson, A.M. (2001). The microbiology of primary dental caries in humans. J Dent Educ 65, 10, 1028-1037

35. Topham, M.K. & Prescott, S.M. (1999). Mammalian diacylglycerol kinases, a family of lipid kinases with signaling functions. J Biol Chem, 274, 17, 11447-11450

36. Walsh, J.P., Loomis, C.R., & Bell, R.M. (1986). Regulation of diacylglycerol kinase biosynthesis in Escherichia coli. A trans-acting dgkR mutation increases transcription of the structural gene. J Biol Chem, 261, 24, 11021-11027

37. Wen, Z.T. & Burne, R.A. (2004). LuxS-mediated signaling in Streptococcus mutans is involved in regulation of acid and oxidative stress tolerance and biofilm formation. J Bacteriol, 186, 9, 2682-2691

38. Wen, Z.T., Suntharaligham, P., Cvitkovitch, D.G., & Burne, R.A. (2005). Trigger factor in Streptococcus mutans is involved in stress tolerance, competence development, and biofilm formation. Infect Immun, 73, 1, 219-225

39. Wen, Z. T., Baker, H. V. & Burne, R. A. (2006). Influence of BrpA on critical virulence attributes of Streptococcus mutans. J Bacteriol, 188, 8, 2983-2992

40. Yamashita, Y., Takehara, T., & Kuramitsu, H.K. (1993). Molecular characterization of a Streptococcus mutans mutant altered in environmental stress responses. J Bacteriol, 175, 6220-6228

41. Yamashita, Y., Tsukioka, Y., Nakano, Y., Tomihisa, K., Oho, T., & Koga., T. (1998). Biological function of UDP-glucose synthesis in Streptococcus mutans. Microbiology, 144, 1235- 1245

Chapter 9

CHARACTERIZATIONS OF ENVIRONMENTAL COMPOSITES

Ali Hammood [1] and Zainab Radeef[1]

[1]Department of Materials Engineering-University of Kufa, Iraq

INTRODUCTION

Recently, environmental preservation issues have been critical between the chemical pollution matters and the development technology requirements. However, the renewable and friendly materials come to use.

Numerous researches have richly studies the natural fiber reinforcement polymer composites. This fact, based on both fibers and matrixes are derived from renewable resources. Therefore, the formed composites have more compatibility with the environmental preservation issues. [1] Isabel investigated of the most natural fibers are used such as palm, cotton, silk, coconut, wool and wood fibers. A significant development in the lignocelluloses fiber in thermoplastics realized the distinct researches presented by [2-5] Composite-reinforcing fibers can be categorized by chemical composition, structural morphology, and commercial function. Natural fibers, such as kenaf, ramie, jute, flax, sisal, sun hemp and coir are derived from plants that used almost exclusively in PMCs. Aramid fibers[6] are crystalline polymer fibersare mostly used to reinforce PMCs. The compounds percentage of composite have the essential role for verify the designed values according to applications, therefore the mechanical properties of PMCs predicated by Mohamed (2007).

The primary function of a reinforcing fiber is to increase the strength and stiffness of a matrix material. The fibers reinforced composite have the essential role in this investigation for its significant property advantages as high stiffness, lightweight, easily recycled material, availability, low manufacturing cost, the environment effect and lifetime rupture behavior. Various types of natural fibers are available to combine with other mineral fiber for construct composite material. Essentially, the fiber can be classified as vegetable, animal, and man-made fibers. The main disadvantages of natural fibers are their high level of moisture absorption, poor and interfacial adhesion, relatively low heat resistance. [7-8] investigated high speed impact events using (PKV, PRM) composites. This research was indicated significant improvements in the penetration resistance. This fact comes from the improvement of target geometry structure. Numerous researches have been carried out on the ballistic impact on high strength fabric structures [9-11].In the airport and marine applications, the dynamic loads effect and the chemo interactions were attracted the researchers and many methods are employed for computing the surface topography parameters, thereby numerous estimations were covered the erosion-corrosion behavior of PMCs [12-13]. The most impinging parametersfocusedon environment effects and impinging angle [14-17]. There were many instruments and electronic microscopesdeveloped with the time for measuring the roughness parameters and drawing the surface topography [19]. The Erosion and corrosion of composite must be determent for the accelerator objects, whereby this values will be indicator for measuring the life time rupture of composite [18].

IMPORTANT

The automotive and aerospace industries have both shifted for using natural fiber reinforced composites as a factor to reduce the weight and getting significant properties of composite components. As matter of fact, the impinging liquids of the naval and aerospace applications have a direct effect on surface topography. Therefore, advanced studies focused on corrosion and erosion behavior. The impinging angle, velocity versus time, composite morphology represented the essentialparameters of this field of study. In this investigation, surface roughness versus time was the indicator for erosion and corrosion effects.

EXPERIMENT PROCEDURE AND SAMPLES PREPARATION

In order to develop new composite material with high impact resistance and high erosion resistance, characterization study for two sets of composites materials have been computed. The specific composites materials in this research are: polyester resin-matrix and Kevlar reinforced fiber (PKV) with V_f (42%), polyester resin–matrix and ramie reinforced fiber (PRM) with V_f (42%). Experimental program was carried out to study the erosion and corrosion behavior by computing the surface roughness parameters of (PKV,PRM) before and after impinging operation. Hence, Polymer Matrix Composites (PMCs) were examined by impingement using water jets when the aqueous solution was 3.5 wt% NaCl. Erosion and corrosion tests were impinging at 90 angles at velocity 30 m/s and the impinging period was 12 hours.

The composite subject study consists from five Kevlar layers and five ramie layers were impregnated with unsaturated polyester. The layers aligned alternatively according to the expectation performance. The synthetic ramie fiber was weaved as shown in Figure 1(a)& (b).

TENSILE TESTING OF COMPOSITES

Tensile specimen (145 x 15 x 7) mm3 are caught according to ASTM D 3039 /D 3039M-95M standard (ASTME D 3039, 2003). Tensile test has been conducted and the data acquired digitally. Tensile stress, tensile strain and Young modulus of PKV with Vf (42%) and PRM with Vf (42%) were calculated and the test was performed by instron machine 10 KN, series 2716 and 2736 under stable speed rate 2 mm/min. The test has verified all the specific test conditions to determine the tensile properties of specimen according to the ISO 527. Tensile test has been performed to estimate the yield stress, young modulus and tensile extension at yield point. Additionally, Poisson's ratio has been calculated by evaluate transverse strain and longitudinal strain of composite. Hence, the transverse strain calculated by using strain gage that supplied from Tokyo Sokki Kenkyuio co, Ltd. The used gage type was BFLA-2-8 with gage resistance 120±0.3 Ω. Then, the strain is connected to DAQ bridge system for reading the transverse strain.

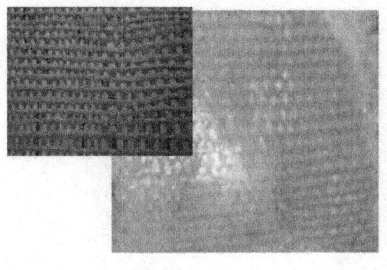
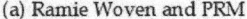

(a) Ramie Woven and PRM   (b)Kevlar Woven

Figure 1: Ramie and Kevlar Woven

The composite behavior under test was an important subject for this investigation and the specimen geometry was constructed from five layers Ramie fiber and five layers of Kevlar as clarified in theFigure 2.

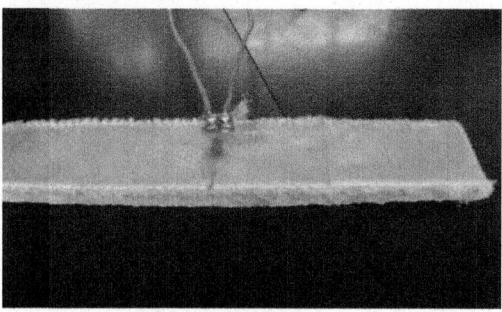

Figure 2: Tensile Test Specimen with Strain Gage

BALLISTIC LIMIT AND NUMBER OF LAYERS

The ballistic limit is commonly defined as a 50% probability of penetrating a target at a given impact velocity. The energy absorption is related to the impact velocity, interpreted by the effect of the striking velocity on the amount of kinetic energy that is absorbed by the composite material. Hence, the energy absorbed by the fabric is equal to the residual energy amount subtracted from total impact energy.

In this test, the target impinged by using gas gun machine supported by a high speed camera for record the impact event at 30,000 frames per second with image size 512 x 64 pixels per image [7]. Figure 3 (a) shown the gas gun machine. Thus,The velocity before the target and the residual velocity after the target were estimated. All the targets were impinged by Semi-conical bullets as shown in the Figure 3(b).

(a) Ballistic Panel

(b) Semi-conical bullets

Figure 3: Gus gun device of University Putra Malaysia [7,8]

EROSION TEST AND CORROSION

Accurate estimation is carried out for calculating the erosion and corrosion percentage of (PKV, PRM) samples. The tests were conducted using a jet at velocity 30 m/s and impinging angles 90°. Essentially, the exposure period was the major factor of erosion estimation. Hence, the impinging periods were from 3 to 6 and from 6 to 12 hours under room temperature. Samples roughness data were recorded before and after the tests by Image processing software for scanning probe microscope

RESULT AND DISCUSSIONS

Tensile Test

The tensile test of this composite was conducted for specifying the mechanical properties of the composite. Generally, the test results recorded high tensile strength. The brittle manner was the first stage of composite under test due to the low elongation ability of matrix. The second stage was the ductile behavior that embodied of high elongation ability of Kevlar layer. From stress – strain curve topography, the specimen have extension at maximum load up to 2.49 mm was recorded. There are continuously reductive of the curve as a result to the elongation of Kevlar layers. The extension at maximum tensile strain observed at 43.25 mm that was evident in the ductile behavior at front face of the specimen. Practically, through the extension test, the back face of the specimen that contented from ramie layers reinforced polyester was separated gradually as result to the brittle behavior of ramie – polyester matrix. Automatic Young's modulus in this composite is 4930.5 MPa. This fact plotted in stress – strain curve in the Figure 4. The composite properties depend at compounds types and volume fraction. Therefore, the volume fraction of fiber or matrix could be calculated as it is clarified in the following equation.

$$V_m = \frac{v_m}{vc}$$

(1)

Where

V_m = volume fraction of matrix

m_n = volume of matrix

c_n = total volume of the composite

Volume fraction of fiber could be evaluated from

$$V_m = 1 - V_f$$

(2)

Where

V_f = volume fraction of fiber

The volume of ramie fiber could be calculated from the following equation

$$V_f = V_K + V_R$$

(3)

Where

V_k = volume fraction of Kevlar

V_R = volume fraction of ramie Fiber density can be determined experimentally by weighting the Kevlar and ramie fibers and calculated the volume fraction for the fibers and resin. Therefore, from the following expression could be calculated the density.

$$\delta_f = \frac{\delta_c - V_m \delta_m}{V_f}$$

(4)

Where

f_d = Fiber density

c_d = Composite density

m_d = Matrix density

The Poisson coefficient represents the contraction in the transverse direction and could be calculated by using the follow expression.

$$V_{lt} = v_f V_f + v_m V_m$$

(5)

$$V_{lt} = 0.34$$

(6)

V_{lt} = Poisson ratio

Where the Kevlar Poisson ratio is equivalent to 0.34 [6], ramie Poisson ratio is equal to 0.3 [20] and 0.4 for unsaturated polyester resin [21].

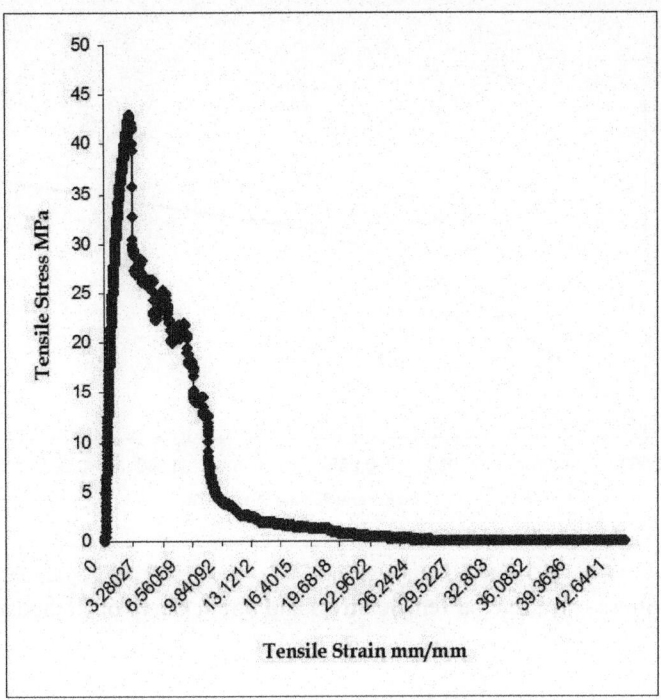

Figure 4:. Stress – Strain Curve of PKV +PRM Composite

Modulus of elasticity (E$_{11}$) can be calculated in (7).

$$E_{11} = \frac{\sigma_L}{\varepsilon}$$

(7)

σ_L=longitudinal tensile stress

ε =strain

(E$_{22}$) is calculated by using gage strain that recorded the shrinking displacement and more than 550 data point recorded load and displacement. The transverse Young's Modulus is the initial slope of σ tra. ε_2 curve.

The Young's modulus can be calculated by using the retrieved data from Figure 5as clarified in the following equation.

$$E_{22} = \frac{\sigma_{tra}}{\varepsilon_2}$$

(8)

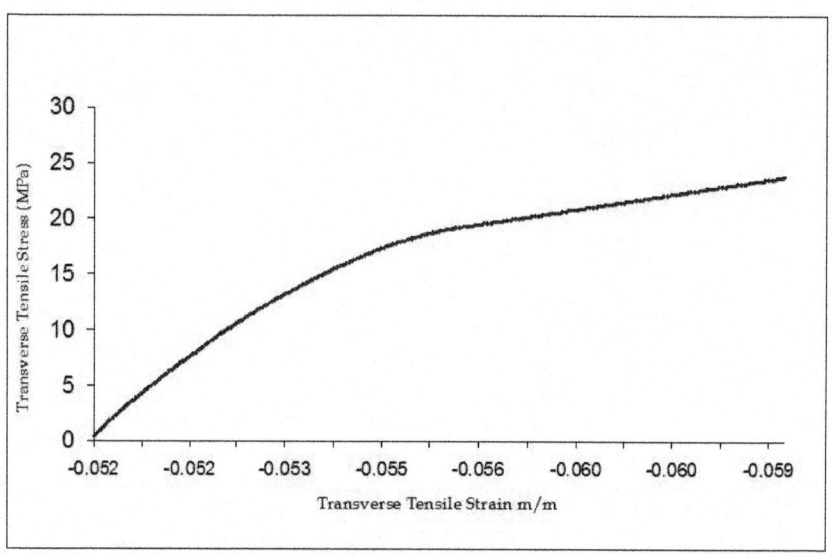

Figure 5: Stress – Transverse Tensile Stress and Strain Curve of Composite

Where

$E_{22} = $ Transverse Young's Modulus

$\sigma_{tra} = $ Transverse tensile stress

$\varepsilon_2 = $ Transverse strain

The maximum shear stress can be obtained from the following equation.

$$\tau_{max} = \frac{\sigma_y}{2}$$

(9)

Where

$y_s = $ Stress in the yield point max

t = Maximum shear stress Shear modulus

(G_{12}) can be calculated from the load data that plotted in the shear stress curve

$$G_{12} = \frac{\tau}{\gamma}$$

(10)

Where

t = shear stress

g = shear strain

In addition, the Poisson ratio can be estimated from the following equation and the result seems equivalent to the Poisson ratio that was calculated according to the compounds volumes fraction. Table 1 presents the longitudinal and transverse averages of data.

$$\gamma_{12} = \frac{\varepsilon_{lat}}{\varepsilon_{long}}$$

(11)

Where

e $_{lat}$ =Transverse strain

e $_{long}$ =Longitudinal strain

In addition, the Poisson ratio can be estimated from the following equation and the result seems equivalent to the Poisson ratio that was calculated according to the compounds volumes fraction. Table 1 presents the longitudinal and transverse averages of data.

$$\gamma_{12} = \frac{\varepsilon_{lat}}{\varepsilon_{long}}$$

(11)

Where

ε_{lat} =Transverse strain

ε_{long} =Longitudinal strain

Table 1: Tensile Test Data

The average of longitudinal data for Kevlar – Ramie –Polyester composite				
E_{11} (GPa)	$(\sigma_1)_{ult}$ (MPa)	$(\varepsilon_1)_{ult}$	ζ_{max} (MPa)	Poisson's Ratio
3.9489±0.5	66.75±5.4	0.125±0.04	33.37±3	0.37

The average of transverse data for Kevlar – Ramie –Polyester composite				
E_{22} (MPa)	$(\sigma_2)_{ult}$ (MPa)	$(\varepsilon_2)_{ult}$	γ_{12}	G_{12} (MPa)
244.74±2.5	22.77±1.3	0.0735±0.03	0.257±0.05	132.65±15

High Speed Impact Results

Understanding the impact response of composites has be come an area of great academic and practical interest. The major advantages of composite materials are their high strength and stiffness, light weight, corrosion resistance, crack, fatigue resistance and flexibility. Ramie – Kevlar reinforced polyester resin present high resistance. The high level from resistance could be realized by increasing of Kevlar layers.

$$[E_{abs} = \frac{1}{2}m\ (V_{imp}^2 - V_{res}^2)]$$

(12)

Where

E_{abs}= Energy absorption

m= projectile mass

V_{imp} = Impact velocity

Vres= Residual velocity

Ballistic limit can be identified as the limit between the penetration and the fully arrested. Thus, the composite with five layers (PKV) and five layers (PRM) couldn't meet the specific requirements of ballistic resistance. In fact, the absorption of energy will be increased with increase the number of layers [7, 8].

In the event when no perforation occurs, the energy absorbed by the target will be equal to the initial impact energy. The high speed impact data for PKV & PRM are shown in the table 2. The following equation has been verified to the ballistic limit or fully arrested action.

$$[E_{abs} = \frac{1}{2}m\ V_{imp}^2]$$

(12)

Where

E_{abs}= absorption of energy

V_{imp}= impact velocity

In this prospective must be mention the most high speed impact parameters that represented rich fields of studies are, target geometry, projectile type, target thickness, composite compounds.

Table 2: High Speed Impact of PKV&PRM

Humanity:53%	Bullet type: Semi-conical
Specimen type: TSP	Camera temperature:40°C
Target area:15 × 10 mm	Temperature: 32 °C
Material: PKV& PRM	Camera resolution: 512 × 48

Layer No.	Gas gun Pressure Psi	Initiation Velocity m/s	Residual velocity m/s	Absorption of Energy J
5K-5R	300	273.9	125.9	147.926
	250	275.255	150.17	133.035
	250	255.07	120.9	126.109

Erosion Result

The surface roughness of engineering applications has interacted with the environment. Therefore, the studies pay attention for estimating the surfaces roughness of materials with respect to essential parameters to limit the wear mechanism of materials. Roughness value can either be calculated on a profile or on a surface Rz, Rq, and Sa is the arithmetic average of the 3D roughness. Hence, the impinging test conduct through specific periods is illustrated in the Table (3).

Table 3: Roughness values

Kevlar roughness by (nm)				Ramie roughness by (nm)			
0 hours	3 hours	6 hours	12 hours	0 hours	3 hours	6 hours	12 hours
8.92	11	9.36	9.94	12.9	11	6.71	10.8
16.7	8.25	8.06	10.4	7.37	9.82	6.85	10.4
15.6	8.44	7.94	11	10	8.32	7.78	10.3
7.53	9.26	6.35	8.78	9.12	11	7.82	8.43
6.06	8.69	6.24	11.6	10.6	10.9	8.25	8.8
6.44	12.9	9.45	10.4	11.3	8.31	9.23	9.18
6.28	12.9	8.81	14	10.6	11.3	9.97	9.57

The roughness parameters rates, such as (Amplitude parameters, Hybrid Parameters and Functional Parameters) were estimated by all morphology images with image size (100.00nm X 80.00nm)Figure 6 illustrate the morphology images before PKV tests.

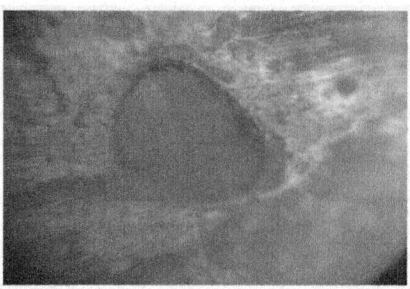 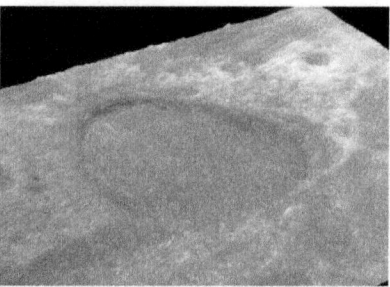

Figure 6: MorphologyImages after Tests of PKV.

The erosion wear loss was determined using probe microscope. Typically, scanning probe microscope image process software provided reliable means to evaluate the erosion volume lose. This technique characterizes and quantifies the surface roughness parameters, the surface profile and topographical features in three-dimension using high precision range observed at 100 nm. All measurements were made with an effective magnification of X 12.5. Excellent data were recorded for average size and average height for surface grains. Hence, the average volume of grains can be computed by the grain size analyzer. Randomly, erosion areas were elected for scanning morphologic image. The erosion volume loss was derived from analyzing the erosion surface at three dimensions. Therefore, the erosion volume loss (V_{loss}) can be expressed as the average volume of surface at zero erosion time (V_0) subtracted from the average volume of surface after specific time (V_t). Eroded area was randomly measured at seven locations. Then, the average of erosion volume was calculated.

$$V_{loss} = V_0 - V_T$$

(14)

According to the V_{loss} formula, the averages of erosion volume loss were illustrated in Figure 7.

Ramie reinforced polyester matrix present higher value of volume erosion loss than Kevlar reinforced polyester. This fact derives from the poor of adhesionbetween Kevlar filaments and polyester resin.Table 4 illustrates the grains size versus time.

Table 4: Grains size versus time

volume loss	Kevlar			Ramie		
	avg. size (nm²)	avg.height (nm)	avg. vol-ume	avg. size (nm²)	avg.height (nm)	avg. volume
Zero Time	1.13	7.14	8.21	1.24	8.97	11.18
after 3 hours	1.11	10.334	10.44	0.82	9.36	7.846
after 6 hou	1.72	8.91	11.55	0.9	5.41	4.25.003

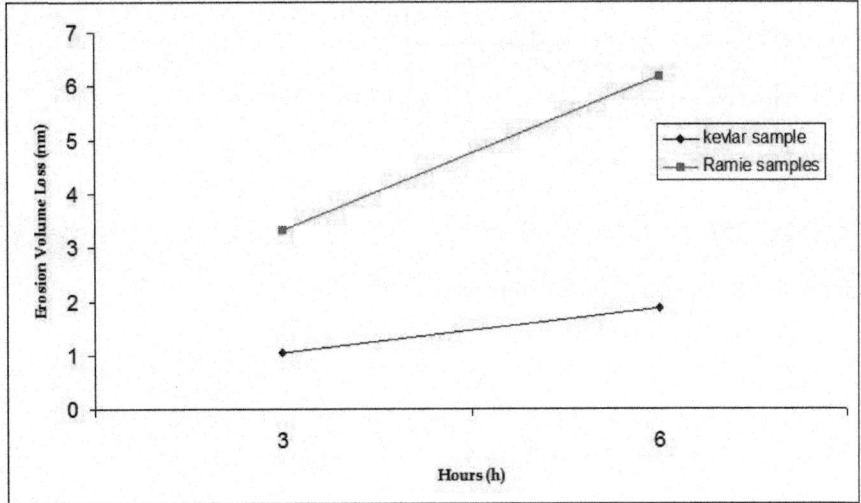

Figure 7: Erosion volume loss concomitantwith time

Transforming image to 3D topography with 200% zoom area for PKV sample at zero time (before test) illustrated accurate 3D surface profile with 243 peaks number and the maximum height 46 nm as shown in the Figure 8 and Figure 9 shown the PKV after 12 hours illustrated in 3D surface with 238 peaks number and maximum height 73.6 nm.

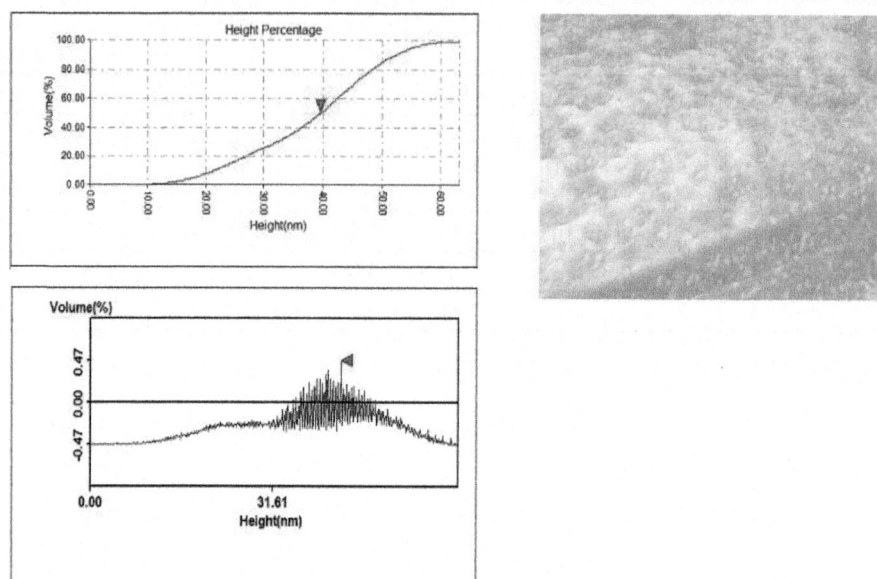

Figure 8: PKV at Zero Time (before test) in 3D Image

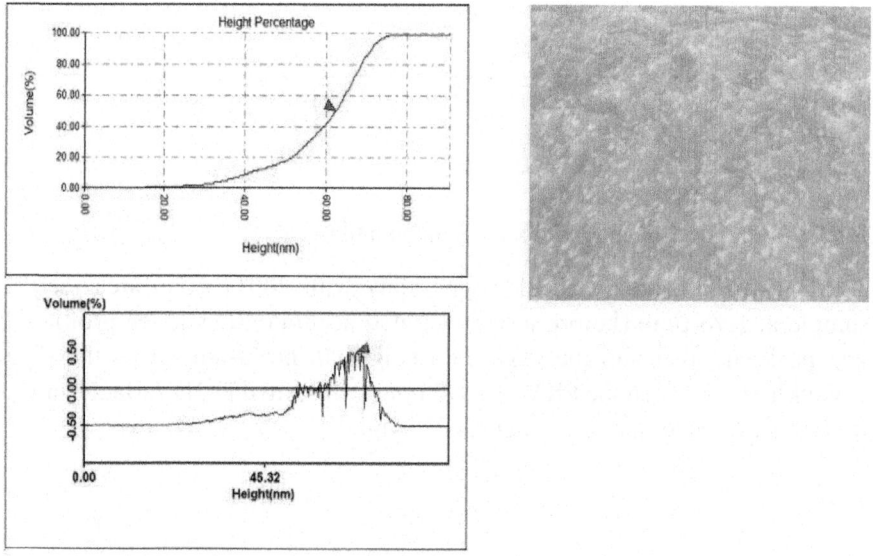

Figure 9: PKV after 12 Hours in 3D Image

Accurate roughness parameters were established for PKV and PRM versus the time before and after test as illustrates in the table 5. PKV & PRM morphologies images versus time were illustrated inFigure 10 & 11.

Table 5: The roughness parameters rates

Image size:100.00nm X 80.00nm (PKV) at zero time					
Area 4	Area 4	Area 3	Area 2	Area 1	Amplitude parameters:
16.7	15.6	8.92	7.53	6.06	Sa(Roughness Average) [nm]
20.2	18.8	12.3	10.7	9.1	Sq(Root Mean Square) [nm]
-0.386	-0.451	-1.58	-1.67	-2.32	Ssk(Surface Skewness)
2.44	2.41	6.75	8.92	12.7	Sku(Surface Kurtosis)
91.4	90.3	89.5	90.2	87.8	Sy(Peak-Peak) [nm
88.5	89.9	85.8	89	85.2	Sz(Ten Point Height) [nm]
					Hybrid Parameters:
-149	-146	-146	-161	-155	Ssc(Mean Summit Curvature) [1/nm]
8.02	8.69	8.12	10.7	10.1	Sdq(Root Mean Square Slope) [1/nm]
2.08E+03	2.44E+03	2.17E+03	3.62E+03	3.29E+03	Sdr(Surface Area Ratio)
					Functional Parameters:
14.1	2.45	0.779	1.05	0.624	Sbi(Surface Bearing Index)
1.32	1.34	0.949	1.18	0.983	Sci(Core Fluid Retention Index)
0.112	0.103	0.211	0.17	0.192	Svi(Valley Fluid Retention Index)
0.355	7.98	4.61	7.41	4.79	Spk(Reduced Summit Height) [nm]
55.1	44.4	22.4	20.1	16.4	Sk(Core Roughness Depth) [nm]
19.6	19.9	25.2	20.8	21.5	Svk(Reduced Valley Depth) [nm]
1.43	7.67	15.7	10.2	14.6	Sdc 0-5(0-5% height intervals of Bearing Curve) [nm]
0.803	4.32	1.31	3.08	1.72	Sdc 5-10(5-10% height intervals of Bearing Curve)[nm]

26.9	19.4	10.3	9.69	7.63	Sdc 10-50(10-50% height intervals of Bearing Curve) [nm]
35.7	36.1	26	19.9	16.6	Sdc 50-95(50-95% height intervals of Bearing Curve) [nm]
					Spatial Parameters:
5.56E+06	5.67E+06	6.02E+06	4.23E+06	4.28E+06	Sds(Density of Summits) [1/um2]
2	2.37	2.81	3	3	Fractal Dimension

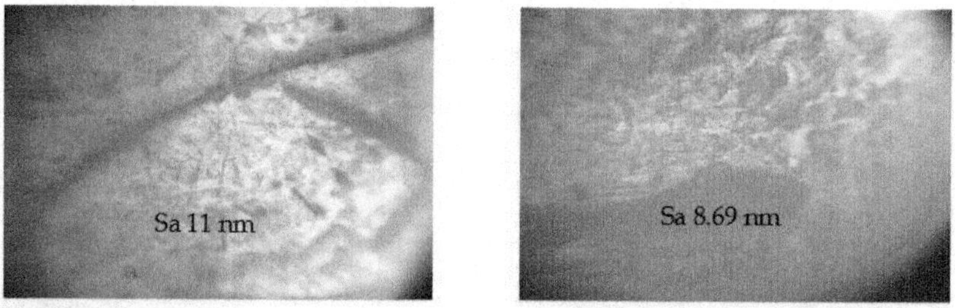

(a) After 3 hours

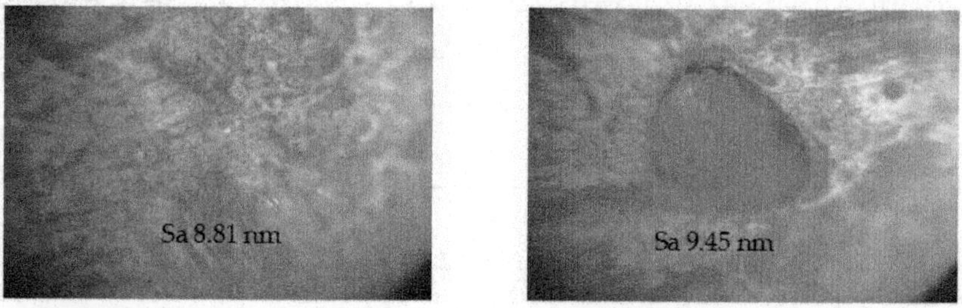

(b) After 6 hours

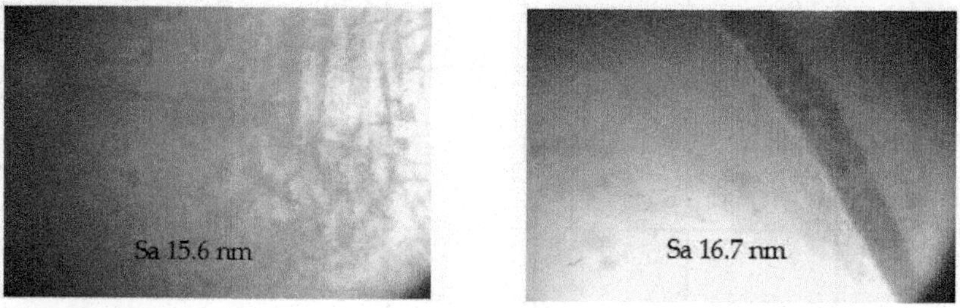

(c) Before test

Figure 10: Erosion and corrosion test for Kevlar – polyester composite morphology and Sa(Roughness Average)

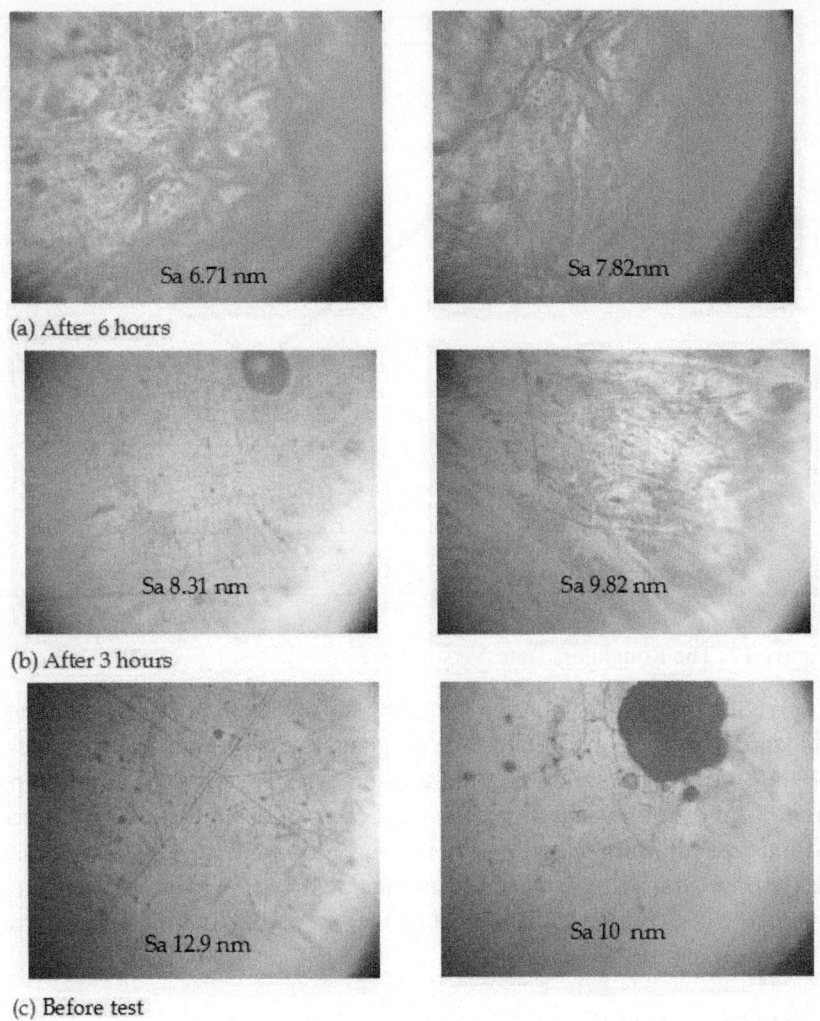

Sa 6.71 nm

Sa 7.82nm

(a) After 6 hours

Sa 8.31 nm

Sa 9.82 nm

(b) After 3 hours

Sa 12.9 nm

Sa 10 nm

(c) Before test

Figure 11: Ramie – polyester composite morphology and Sa(Roughness Average) after erosion-corrosion test

The roughness rate of the Kevlar–polyester and ramie-polyester concomitant time tented to be equivalent values. Really, the polyester reinforced faced the impinged water has the major role in the erosion and corrosion resistance. The Kevlar and ramie fibers have specific interfacial adhesion with the matrix. The ramie fibers presented high interfacial adhesion with resin as a result to plant fiber nature. On the other hand, the Kevlar presented a poor linkage with polyester resin.

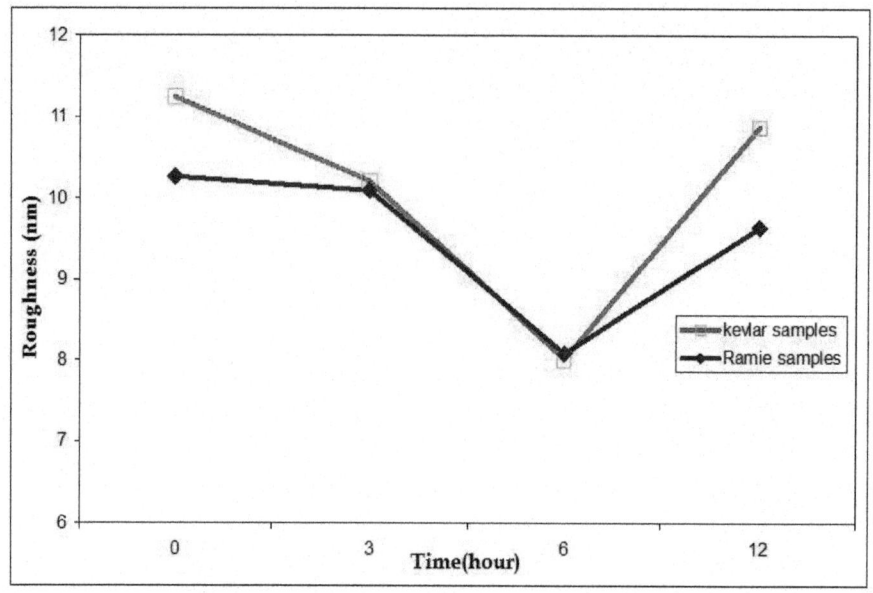

Figure 12: The Roughness Rates Versus Time

All the fiber layers were protected by the matrix mass. Figure 12 illustrated a declining in the polyester surface roughness versus time. The roughness will be the function for estimating the erosion rate. Therefore, within the period, from 3 until 6 hours the impinging effect assisted of soften the surface, but within period from 6 until 12 hours the erosion rate will be increase and the surface tent to be rough. Electrochemical test have been conducted and there are no significant corrosion data recorded, this finding derived from the compound nature of composite.

CONCLUSIONS

Generally, plain weave represented the most common fabric, due to the significant properties that embodied their tensile strength to weight ratio. The study of the stress – strain response, high speed impact evaluation, and erosion corrosion behavior leads to the following conclusion:

1.   The Kevlar –polyester composites behave ductile manner but the ramie-polyester composites behave brittle manner.

2.   The roughness rate of the Kevlar-polyester and ramie-polyester decreases through 3 -6 hours but there is an increase in the roughness rate through 6-12 hours due to the increasing in erosion rate of polyester matrix.

3. All roughness parameter computed accurately.
4. PKV and PRM with five layers PKV and five layers PRM present fully penetration.
5. No significant corrosion has been recorded.

ACKNOWLEDGEMENT

The authors would like to express their gratitude and sincere appreciation to the department of Mechanical and Manufacturing Engineering of the University Putra Malaysia and Material Engineering Department- College of Engineering- University of Kufafor scientificassistance and support.

REFERENCES

1. B. Isabel, Wingate, 1976Textile fabrics and their selection. Library of congress cataloging in publication data.

2. R. T. Woodhams, G. Thomas, D. K. Rodgers, 1984wood Fiber as Reinforcing Fillers forpolyolefins.polym.eng.Sci. 24

3. C. Klason, J(198. Kubat, KubatJ(1989Cellulose in polymer composites. In composite Systems from Natural and synthetic polymers, SalmenL, de Ruvo, ASefe-ris, J C Stark, E B Eds; Elsevier Science: Amsterdam

4. G. E. Myers, C. M. Clemons, J. J. Balatinecz, R. Woodhams, T(199, T(1992Effects of composite and polypropylene Melt Flow on polypropylene-Waste Newspaper Composites. proceedings on the Annual Technical Coneference; Society of plastics Industry 602

5. B. V. Kokta, R. G. Raj, C(198. Daneault, C(1989Use of Wood Flour as Filler in Polypropylene; Studies on Mechanical PropertiesPolymplast. Tecnol.Eng. 28, 247 EOF259 EOF

6. Yang H H1993Kevlar Aramid Faiber. west Sussex PO191UD,England.

7. Aidy, Z. R. Shaker, A. Kahalina, 2010Development of anti-ballistic board, fiber polymer-plastics technology and engineering 50

8. Zainab Shaker Radif, Aidy AliKhalinaAbdan(2010DEVELOPMENT OF A GREEN combat armour from rame-kevlar- polyester composite, Pertanika Journal of Science and Technology. 19339348

9. Lee B L,Song J W, Ward J E J1994Compos Mater.2812021226

10. W. Goldsmith, C. K. Dharan, H. Chang, 1995Quasi-static and ballistic perforation of carbon fiber laminates.Int J Solid StructVOL(32):89-103.

11. Almohandes AA, Abdel- Kader MS, Eleiche, AM1996Experimental investigation of the ballistic resistance of steel-fiberglass reinforced polyester laminated platesComposites:Part B 2744758

12. E. Bardal, T. G. Eggen, T. Rogne, T. Solem, 1995The erosion and corrosion properties of thermal spray and other coatingsin: Proceedings of the Int. Therm. Spray. Conf., Kobe, Japan.

13. Y. Puget, K. R. Tretheway, R. J. K. Wood, 1998The performance of cost effective coatings in aggressive saline environmentsNACE Corrosion 688

14. G. T. Burstein, K. Sasaki, (200, 2000Effect of impact angle on the slurry erosion-corrosion of 304 L stainless steel, Wear 240

15. Dawson J L, Shih C C, John C C, Eden D A1987Electrochemical testing of differential flow induced corrosion using jet impingement rigsNACE Corrosion, Paper 453

16. H. M. Clark, K. K. Wong, angle. Impact, 1995particle energy and mass loss in erosion by dilute slurries, Wear 186-187 454-464.

17. M. M. Stack, S. Zhou, R. C. Newman, of. Identifications, in. transitions, regimes. erosion-corrosion, aqueous. in, Wear. 1. environments, 1995

18. Sherrington. (1988),modern measurement techniques in surface metrology,wear 125 :271-288.

19. Y. Matsuno, H. Yamada, M. Harada, A. Kobayashi, 1975The microtopography of the grinding wheel surface with SEM,Ann. CIRPVOL(24):237242

20. Y. Nakamura, et al.2003Neurosci. Abst. 608.5

21. M. A. Girardi, M. G. Phill, 1993Microstructure and properties of polyester/urethane acrylate thermosetting blends, and their use as composite matrices,Journal of Materials Science,28DOI:BF00354718.

22. Mohamed Thariq.2007High velocity Impact analysis of glass epoxy-laminate plates. Thesis,university Putra, Malaysia, Malaysia.

Chapter 10

GEOLOGICAL CARBON DIOXIDE STORAGE IN MEXICO: A FIRST APPROXIMATION

Oscar Jiménez, Moisés Dávila, Vicente Arévalo, Erik Medina and Reyna Castro

Comisión Federal de Electricidad México

INTRODUCTION

Carbon dioxide (CO_2) is one of the industrial gases that contribute to the greenhouse gas (GHG) effect. During the last decades, the emissions of CO_2 due to human activity have increased significantly all over the world. There are different and important efforts to reduce or stabilize the concentrations of greenhouse gases in the atmosphere, such as improvements in the efficiency of power plants and the development of renewable energies. However, those approaches cannot deliver the level of emissions reduction needed, especially against a growing demand for energy that promotes economic growth and prosperity. Carbon capture and storage (CCS) approach encompasses the processes of capture and storage of CO_2 that would otherwise reside in the atmosphere for long periods of time. Among the different carbon capture and storage options currently in progress all over the world, the geological storage option is defined as the placement of CO_2 into an underground repository in such a way that it will remain permanently stored. Mexico is one of the countries which are signatories of different international treaties which call for stabilization of atmospheric gases emissions at a level that prevent anthropogenic interference with the world's regional climates. In Mexico CO_2 represents almost 70% of the total greenhouse gases em

issions where the primary sources of CO_2 are the burning of fossil fuels for power generation. CCS is a technological approach that holds great promise in reducing atmospheric CO_2 concentrations in Mexico. This is the first coordinated assessment of carbon storage potential across the country.

GEOGRAPHICAL LOCATION OF MEXICO

Mexico is a country located in the southern portion of North America, and is bordered to the north by the United States, to the southeast by Guatemala,

Belize and the Caribbean Sea, to the west and south by the Pacific Ocean, and to the east by the Gulf of Mexico (Figure 1). The country's total area is about 1 972 550 square kilometers.

PREVIOUS WORK

With the aim of searching for places where to store carbon dioxide, Mexico was subdivided into three exclusion zones and four inclusion zones [1](Figure 2). The exclusion zones are zones A, B and G. Zone A is composed by igneous rocks with high seismic and volcanic hazard, and is not recommended for storage. Zone B encompasses also igneous rocks with less seismic and volcanic hazards than zone A, but not yet recommended for CO_2 storage. The zone G is a marine zone of exclusion comprising the ocean floor, deep marine sediments and high seismic and tectonic hazardous processes in the Pacific Ocean.

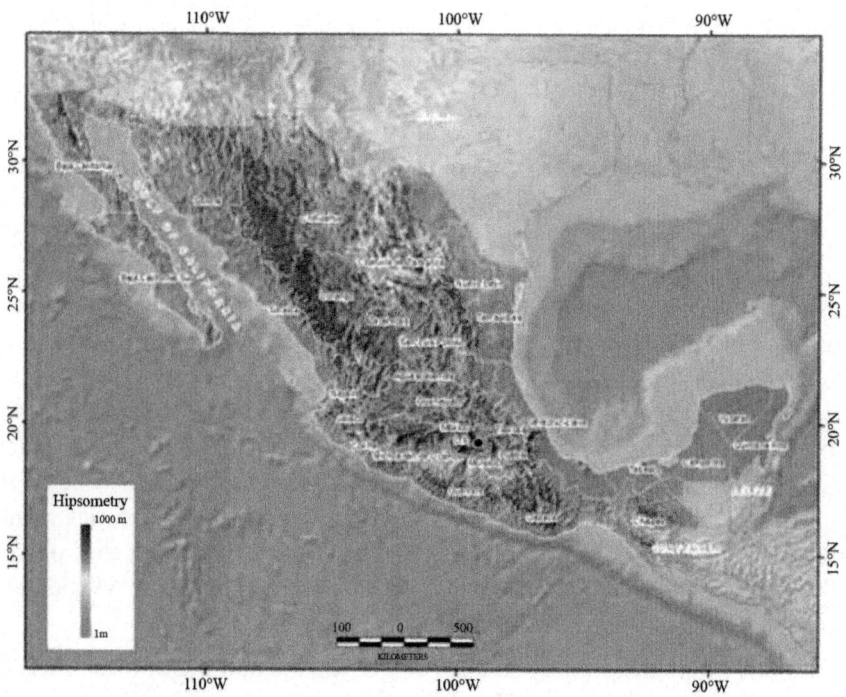

Figure. 1. Hypsographic map of Mexico displaying federal states divisions and countries' borderlines.

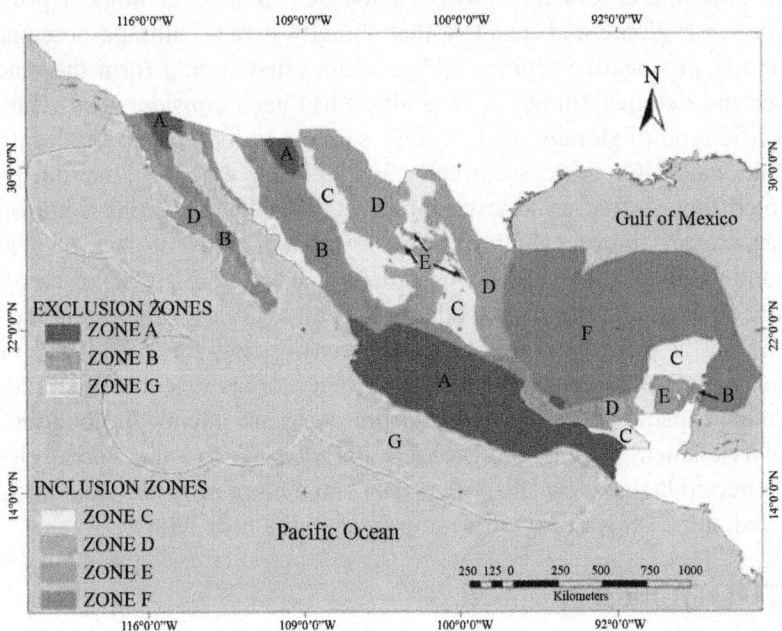

Figure. 2. Exclusion and inclusion zones for geologic CO_2 storage in Mexico. After [1].

The inclusion zones are zones C, D, E and F. Zone C represents terrigenous geological formations and mainly carbonate sedimentary rocks cropping out in the area. Zone D includes terrigenous as well as carbonate sedimentary rocks sequences. Zone E is composed of evaporitic deposits and associated sedimentary rocks. And zone F reflects sediments deposited in the marine continental shelf, slope and deep waters beneath the Gulf of Mexico. All of these zones were outlined taking into account surficial lithological features, large geological subsurface structures and recent volcanic and tectonic activity in a country scale assessment. The exclusion zones were not recommended for geologic carbon storage due to its high seismic, geothermic and active volcanic hazardous potential. On the contrary, the inclusion zones yielded the best CO_2 storage potential and were recommended for further detailed studies in order to find geological provinces with a good CCS capacity.

PURPOSE AND SCOPE

The purpose of this chapter is to present the analysis of different geological provinces to address the possibility of storing anthropogenic CO_2 in deep underground geologic formations, particularly in eastern continental Mexico.

Up to now, the assessment has been focused on five geological provinces in order to evaluate and quantify theoretically its CO_2 storage potential and to identify prospective regions and/or sectors that should form the object of further and detailed studies. The analysis has been considered in relation to a specific type of storage, that is, deep saline aquifers and to the location of the stationary CO_2 sources currently available for the whole nation. It must be noted though that an assessment of CO_2 storage potential is surrounded by large uncertainties, which increase in number with the lack of available data and detailed information. The proposed work in this chapter recognizes this uncertainty, and the envisaged output is an overview of possible scenarios rather than the quantification of specific areas or sites for CCS. The aim is to provide a high level summary of CO_2 geologic storage potential across Mexico where the capacity resource estimates presented are intended to be used as an initial assessment of potential geologic storage prior to a local area selection. It is expected that as new subsurface data and a more refined methodology are acquired, the CCS studies will be improved in the near future.

METHODOLOGY

The total CCS process is frequently analyzed from several viewpoints which include very wide technological, economic and environmental issues. Some of the issues are well constrained while others are poorly understood. In the particular case of CO_2 storage potential there are also various aspects involved, such as the separation and capture of CO_2 at the point of emission, the mass of CO_2

emitted by the point of emission, the infrastructure and transportation of CO_2, and the storage of CO_2 in deep underground geologic formations [2]. However, here we are only concerned with the types of CO_2 emission sources, the searching of suitable geologic reservoir rock sequences and their location, and the quantification of the theoretical capacity of storing a given volume or mass of CO_2 in selected sectors across Mexico. This pragmatic methodology was based on the public domain accessible data and present-day geological knowledge, and it does not incorporate geological constraints in the theoretical capacity estimations, nor does it incorporate risk factors, environmental hazards, solubility and mineral trapping of CO_2, or quantification of injectivity of the potential storage rock sequences.

The first phase included a survey of CO_2 points of emission, production information, source category, emissions factors, and annual CO_2 emissions that were obtained from the mexican Pollutant Release and Transfer Inventory (RETC by its Spanish acronym) and the Ministry of the Environment and Natural Resources (SEMARNAT, by its Spanish acronym) databases [3,4].

These databases consider the stationary sources. A compilation for the United Nations Framework Convention on Climate Change (UNFCCC) [4] includes the stationary and the non-stationary source emissions. The non-stationary source emissions such as those that come from the transportation sector, the change of land use and forestry, and some others like landfills were excluded from the analysis. The CO_2 stationary sources included power plants, oil and natural gas processing facilities, cement plants, agricultural processing facilities, iron and steel production facilities, and other industry processing facilities. The spatial location of the stationary CO_2 emission sources were calculated and compiled through different mapping tools that contain latitude and longitude information for various Mexican locations. The analysis of CO_2 stationary sources was done to provide reliable emission estimations, identify major CO_2 emission sources within each region, and to asses the applicability of the data in subsequently infrastructure analyses. The second phase consisted of the identification of geological storage provinces through the careful analysis and screening of available geological data. In this regard, there are different proposed methodologies that are similar [5, 6, 7, 8, 9, 27]. Only minor differences are evident depending upon the used weights that show the relative importance of the criteria. Therefore, our selection of candidate storage provinces was according to the basin level of the assessment scale [10] (Figure 3). This "basin scale" exploration assessment required a little more local data categories and a better level of detail than the "country scale".

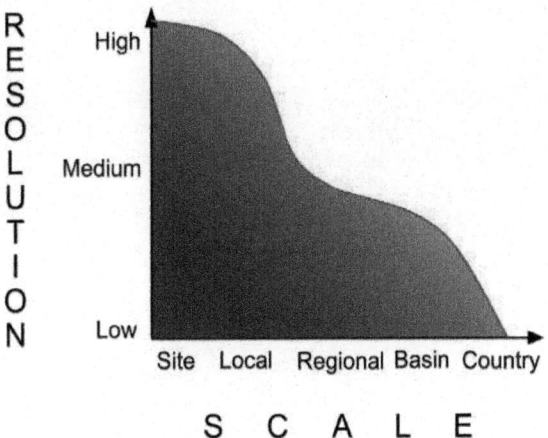

Figure. 3. Data and assessment scales for CCS geological screening studies. After [10].

In this "basin scale" assessment, both terms, basin and province, are considered synonyms. The term basin has different meanings depending upon geologic features of the region, such as geothermal regime, size, age, boundaries, type and thickness of sedimentary fill, geologic deformation, tectonic context, and many others parameters that can change with time [11, 12, 13, 14]. However, these variable geologic features are also possible to be applied to the meaning of the term province. The assessment was focused on the previously identified inclusion zone. Within the inclusion zone, twelve provinces were defined taking into consideration the types of geomorphological developments, stratigraphic successions, major structural deformation patterns, homogeneous tectonic history, and known subsurface geological boundaries between all of them (Figureure 4). Actually, their outlined boundaries are very similar with those of the petroleum basins previously named for those areas of Mexico [15, 16, 17, 18, 19]. From the twelve established provinces, at the moment, only five of them were considered to be studied in greater detail to estimate the geological resource for storing CO_2. These provinces are: Burgos, Tampico-Misantla, Veracruz, Sureste and Yucatan, all of them located in the continental and marine platform areas along Gulf of Mexico

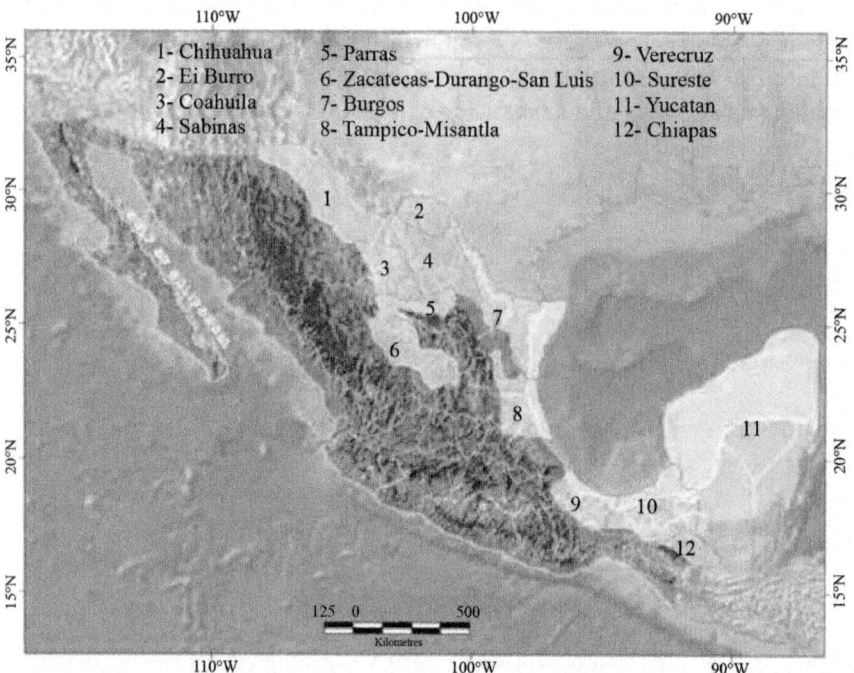

Figure. 4. Mexican geological provinces identified according to their underground potential for CO_2 storage.

The screening and selection of the provinces was based on the published geologic maps from a scale of 1:250,000 to 1:4,000,000 and reports about surface geology, stratigraphic and structural features, regional geologic cross-sections (50-200 km in length and 500m to 3 km in thickness), geophysical information and available public oil well data within each province. Three main groups of sedimentary formations for underground geologic carbon storage were observed. These groups of sedimentary formations are referred to as carbonate, evaporite and terrigenous sequences depending upon the main, respectively, carbonated, evaporitic and clastic content of the rock units. It is worth to mention that the stratigraphic uncertainty is high since the specific subsurface geologic information is quantitatively scarce and sometimes restricted and/or no detailed. Otherwise, the disposal of CO_2 in geological formations, generally, includes unmineable coal seams, oil and gas reservoirs, and deep saline reservoirs. In Mexico unmineable coal areas are not considered as a CCS option because they are located inside the exclusion zone, that is, they are affected seismo-tectonically and located close to the surface. On the contrary, the oil and gas reservoirs are the best option, particularly the EOR (Enhanced Oil Recovery) technique in the exhausted oil fields. But, at the moment, this prospect is ruled out due to the inaccessibility to the public domain of the oil databases and information. Only PEMEX (Petróleos Mexicanos) the oil governmental industry could carry out such studies. So, based on the fact that subsurface layers of porous rocks are generally saturated with brine and that they form deep saline aquifers characterized by high concentrations of dissolved salts and unsuitable for agriculture or human consumption, they were envisaged as the favorable option for CO_2 storage in Mexico. The storing CO_2 in saline formations is achievable since there are examples from such projects [20, 21]. The third phase dealt with the estimation of theoretical capacity within each identified geological province. At present, various calculation methods have been proposed to know the storage capacity of a rock formation [22, 10, 23, 20, 24, 25, 2]. They have been applied to different country projects within their respective areas and still there is uncertainty. The reasons for this uncertainty are diverse but they broadly comprise key aspects such as financial support, CCS technology research and development, and a real partnership between country organizations and academic teams [26, 28]. The concept of storage capacity was referred to a completely free phase of the CO_2, which means without taking into account the CO_2 reaction with the walls of the reservoirs or formations. It is considered only the volume of CO_2 that can be retained in the available porous space of the storage formation or reservoir at depths between 800 and 2500 meters. At such depths the CO_2 has some properties like a gas and some like a liquid due to the changes in temperature and pressure conditions [64]. These are known as the CO_2 supercritical conditions or the critical point of the CO_2. The huge advantage of storing CO_2 in the supercritical condition is that the required storage volume is much less if the CO_2 were at standard pressure conditions.

For the estimation of the theoretical capacity of storing CO_2, it was used an approach here called "parameterization". The parameterization refers to observations, deductions, and calculations derived from the physical parameters obtained from geological maps, regional stratigraphic and structural cross-sections, and well data from the public petroleum industry. Different geological variables were taken into account since the estimation was done with respect to general storage capacity resources and following the standards used in the petroleum industry, that is, stratigraphic and structural traps, as well as seal (cap) rocks that play a decisive role within any geological province. One first step in the parameterization approach was the determination of important geological features that would fulfill the storage requirements such as structural or stratigraphic trap, seal formation, stratigraphic discontinuities, geological faults, depth conditions, appropriate porosity and thickness of the target sedimentary sequence. The critical features were: reservoir depth (more than 800 m and less than 2500 m), thickness, porosity, lithological composition (predominantly carbonates and clastic deposits) and, for effects of the volume calculation, the relationship between "net thickness" versus "total thickness". All of this, with the goal of having an expression Figureure of the fraction of the geological formation susceptible to become a reservoir. The previous information had to be homogeneously similar within the area with a radius between 10 and 20 kilometers around each oil well considered and the nature of trap boundaries. When the information was assumed to be minimally sufficient and it was valued as an attractive target from the point of view of the depth, thickness, porosity, and permeability, then it was selected to quantify its potential capacity to become a CO_2 storing sector. Otherwise, the portion of the regional section including the wells was discarded. One second step of the approach was the direct application of an equation whose variables were fulfilled with the information above mentioned for deep saline aquifers. Therefore, the critical parameters obtained in the previous step were substituted in the formula proposed by Bachu et al in 2007 [10]:

$$VCO_2t = V\varphi(1-S_{wirr}) \equiv Ah\varphi(1-S_{wirr}) \qquad (1)$$

Where A is the trap area, h is the average thickness, VCO_2t is the theoretical volume available, φ is the effective porosity, V is the volume and Swirr is the irreducible water saturation. The solving of the equation yielded the theoretical storage capacity volume of the sector under consideration.

ESTIMATED CO_2 EMISSIONS FROM STATIONARY SOURCES

The most recent update on the mexican national inventory (SEMARNAT) was compiled in 2006 (UNFCCC)[4]. This document shows that the total annual GHG in Mexico are above 709 million metric tons (Mt) of CO_2 equivalent. The carbon dioxide represents 69.5% out of a total of 492 Mt of emissions from stationary and non-stationary sources. There were estimated 285 Mt of CO_2 emissions from stationary sources (Figureure 5).

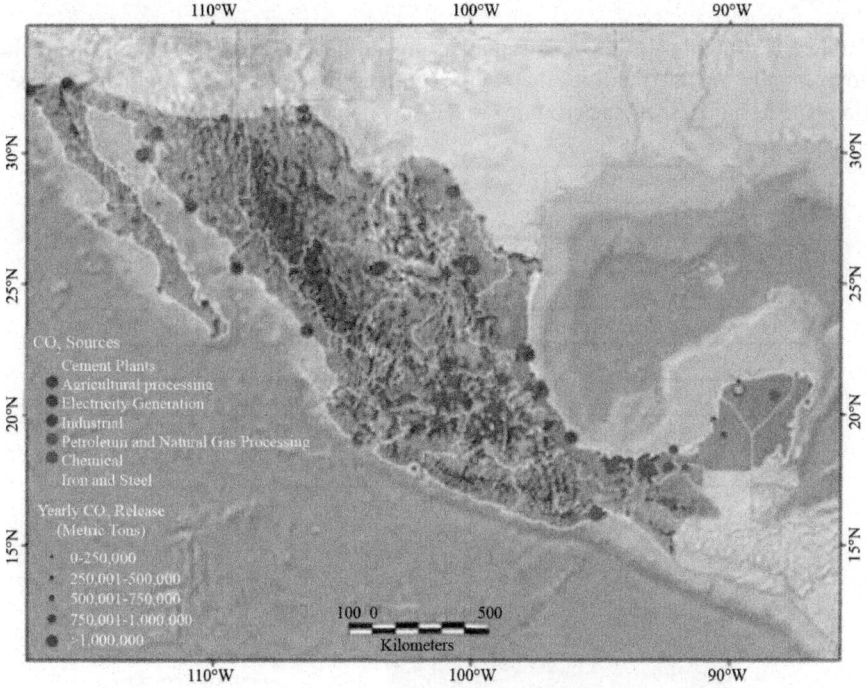

Figure. 5. Main CO_2 stationary source emissions in Mexico. Each colored dot represents a different type of stationary source by category. Dot size represents the relative magnitude of CO_2 emissions released per year.

In addition, RETC data shows approximately 216 Mt of CO_2 emitted from 1,860 stationary sources, according to the different industrial and economic activities in Mexico (Table 1). From the above data it is evident that the electricity supplier sector is the most important contributor to CO_2 emissions from stationary sources. It releases to the atmosphere 107 Mt of CO_2, roughly 50% of the total. It includes emissions from the Federal Commission for Electricity (CFE, by its Spanish acronym) which is the national public service agency, as well as from private small electricity suppliers companies. The oil & petrochemicals facilities add another 22% and, therefore, the whole energy sector is responsible for 72% (154 Mt) of CO_2 emissions in the country. The cement, metallurgical, iron & steel industries are also major contributors to the overall CO_2 country emissions, though they are smaller in comparison to the energy industry. In fact, the electricity production industry is the largest contributor, and it does from a small number of stationary sources (Figureure 6). The industrial and chemical sectors show a much larger number of identified sources, but the relative share of their CO_2 emissions, compared to those of the energy sector, is lower.

Table 1. Estimations of CO2 emissions from stationary sources by sectors. The point sources only include facilities that were reported via the Annual Certificate of Operation (COA, by its Spanish acronym) to RETC, managed by SEMARNAT [3]

SECTOR	CO₂ EMISSIONS (metric tons)	No. OF SOURCES
Electricity Generation	107 351 754	113
Oil & Petrochemical	47 556 986	273
Cement	26 016 726	60
Metallurgical, Iron & Steel	21 367 965	261
Industrial	8 764 815	709
Chemical	4 027 475	438
Agriculture Processing	735 319	6
TOTAL	215 821 040	1 860

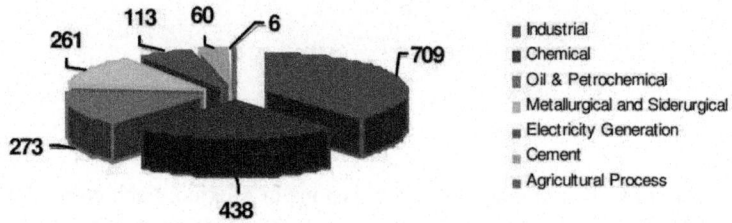

Figure. 6. Number of reported emissions from stationary sources by sector.

From the geographical point of view, the areas with higher CO_2 emissions are located in the northeastern portion of Mexico and in the ferderal states around the Gulf of Mexico. The state of Coahuila tops the list with more than 23 Mt of CO_2 released per year (Table 2). This is mostly due to the deployed coal-fired power plants and metallurgical, iron and steel facilities. The states of Nuevo León and Tamaulipas release approximately 25 Mt n of CO_2 that come from a scattered high number of source points. In the southeastern part, the states of Veracruz and Campeche together attain almost 40 Mt of CO_2. In this context, it is advisable to apply CCS technologies in such industries, since on the one hand, the fewer number of stationary sources with a high level of CO_2 emissions, the better the opportunity to deploy CO_2 capture, injection and storage facilities. On the other, the scenario leads to an economic feasibility projects particularly at the Gulf Costal region where power generation plants, oil & petrochemical, industrial and chemical facilities share the large CO_2 emissions.

Table 2. Estimated CO2 emissions by mexican state and number of point sources

STATE	CO₂ EMISSIONS (metric tons/year)	SOURCES
Coahuila	23 219 675	66
Campeche	21 946 705	25
Veracruz	17 962 809	80
Hidalgo	16 362 111	46
San Luis Potosí	13 580 498	42
Nuevo León	12 725 855	145
Tamaulipas	12 554 901	123
Sonora	9 596 070	46
Michoacán	9 568 763	35
México	9 286 971	284
Chihuahua	8 016 227	265
Guerrero	7 286 999	4
Colima	7 040 064	11
Guanajuato	5 751 629	62
Tabasco	5 676 613	67
Baja California	4 672 787	34
Yucatán	4 214 110	13
Oaxaca	4 108 894	9
Puebla	3 982 865	53
Querétaro	3 466 122	67
Jalisco	3 301 123	87
Sinaloa	3 079 872	11
Durango	2 961 072	18
Morelos	1 805 748	18
Baja California Sur	959 132	9
Aguascalientes	799 295	32
Distrito Federal	746 588	123
Chiapas	732 172	26
Tlaxcala	203 851	43
Quintana Roo	136 962	8
Zacatecas	74 555	7
Nayarit	2	1
TOTAL	215 821 040	1 860

GEOLOGIC CO$_2$ STORAGE POTENTIAL

In order to estimate the CO$_2$ storage potential and to identify different sectors that should be the object of detailed assessment five geological provinces were analyzed. From north to south the geological provinces are: Burgos, Tampico-Misantla, Veracruz, Sureste and Yucatan (Figureure 7).

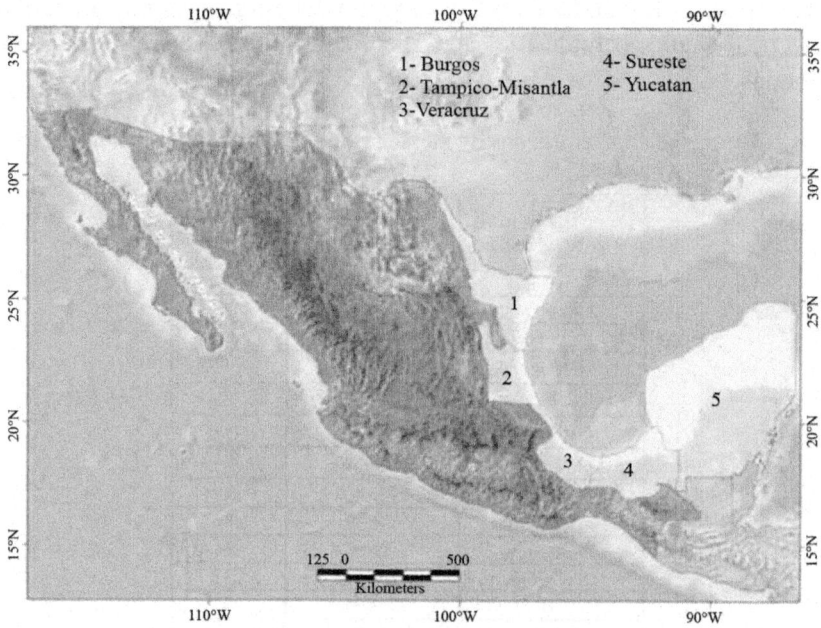

Figure. 7. Mexican geological provinces assessed for underground CO$_2$ storage.

Burgos Province

The Burgos province is located at the most northeastern portion of Mexico. This province is bordered to the north by the United States (sharing the Rio Bravo along the borderline), to the east by the Gulf of Mexico, to the south by Tampico-Misantla province, and to the west by the first exposures that form the contact between Cretaceous and Tertiary rocks [29]. The basement of the geologic province consists of metamorphic and intrusive igneous rocks [30, 31]. However, the basement geometry and its age distribution have not been well established. On top of the basement, a sedimentary evaporitic and carbonated sequence was accumulated in Mesozoic times [50, 62]. After a period of regional subsidence a thick sequence of mainly coarse to fine grained sediments was deposited starting in the Tertiary and continuing into the Quaternary. According to the geological analysis it is documented the existence of a thick terrigenous sequence composed by interbedded conglomerates, sandstones and shales of Cenozoic age [32]. These sequences have frequent lateral facies changes and abundant lenticular sand bodies which were deposited mainly in deltaic, shelf and deep marine environments. Exposures of these rock units extend from the Eocene to Quaternary (Figureure 8). Regional geological sections B1, B2, B3 and B4 were studied to estimate the CO_2 storage capacity on the continental portion on the Burgos province. All of them document similar stratigraphic units and characteristic sets of faults as a result of both extensional tectonic and sedimentological events [36]. Section B4 has no public subsurface geological information available, consequently, it was not considered during the assessment process.

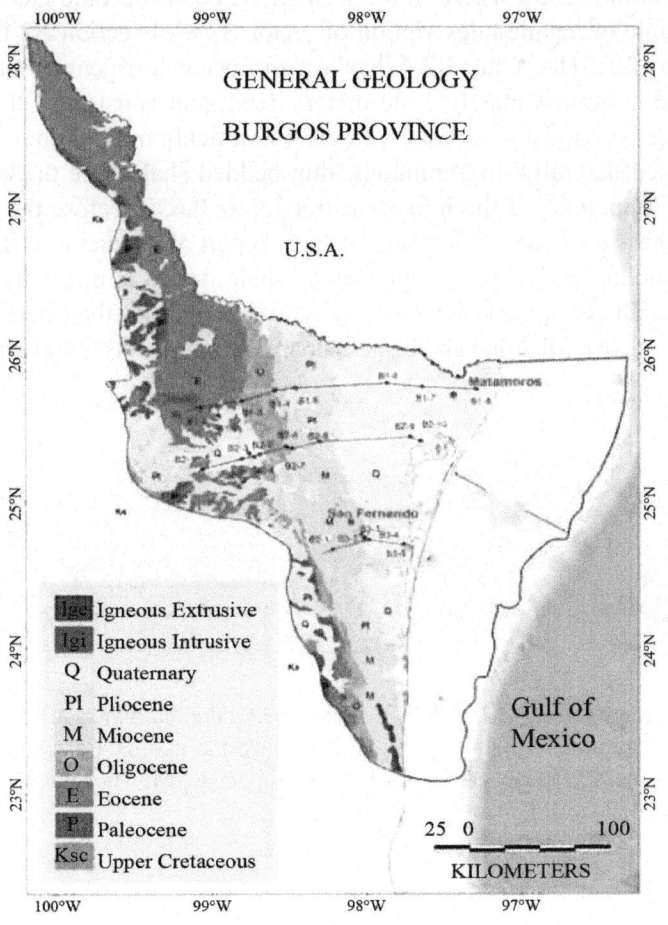

Figure. 8. Simplified geology map of Burgos province depicting geological sections and wells. After [29, 33, 34, 35, 46].

As all the sections depict similar stratigraphic and structural features, only Section B2 is presented (Figureure 9). The section B2 has approximately 150 km in length and show a basement covered by slightly deformed Jurassic and Cretaceous rocks sequences. On top of it, there is a thick tertiary sedimentary and faulted sequence of rocks. The sedimentary sequence and the fault system reveal a chronological pattern from older formations and faults on the west to younger ones on the east. Across the entire section are evident the Eocene and Oligocene rocks on the west, and Miocene formations on the east. According to the type of stratigraphical or structural trap and the lithological and petrophysical features obtained from the oil wells several extrapolations were performed along the regional geological sections in order to select the best potential sectors where saline formations could become CO_2 reservoirs. An example of detailed description of sector B2-4 of section B2 is presented (Figureure 10). The sector B2-4 displays an Eocene terrigenous sequence that is located at approximately 1500 meters depth and consists of thick bedded homogeneous sandstone layers with crossstratification and minor amounts of intercalated, laterally discontinuous, thin bedded shale. The thickness of the unit is 880 meters but the important fraction is 0.6, therefore the considered net thickness is about 528 meters. The unit is part of a structural trap in a "roll over" anticline with a seal composed of shale from the upper limit of same sequence. The Oligocene sedimentary sequence overlies the Eocene sequence and consists of a siltstone and shale that are interpreted as a seal cap-rock.

BURGOS PROVINCR
SECTION B2

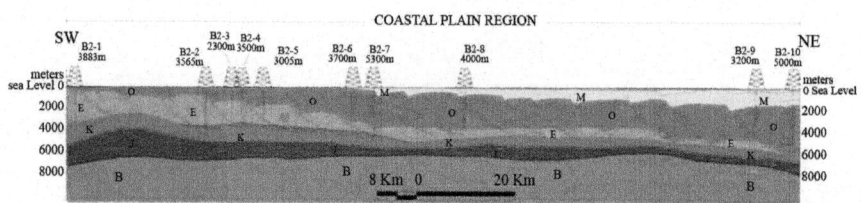

Figure. 9. Regional cross section B2. Across the section both the age of the rock units and the structural deformation are evident from west to east. B: Basement, J: Jurassic, K: Cretaceous, P: Paleocene, E: Eocene, O: Oligocene, M: Miocene, Q: Quaternary. After [31, 33, 34 y 35].

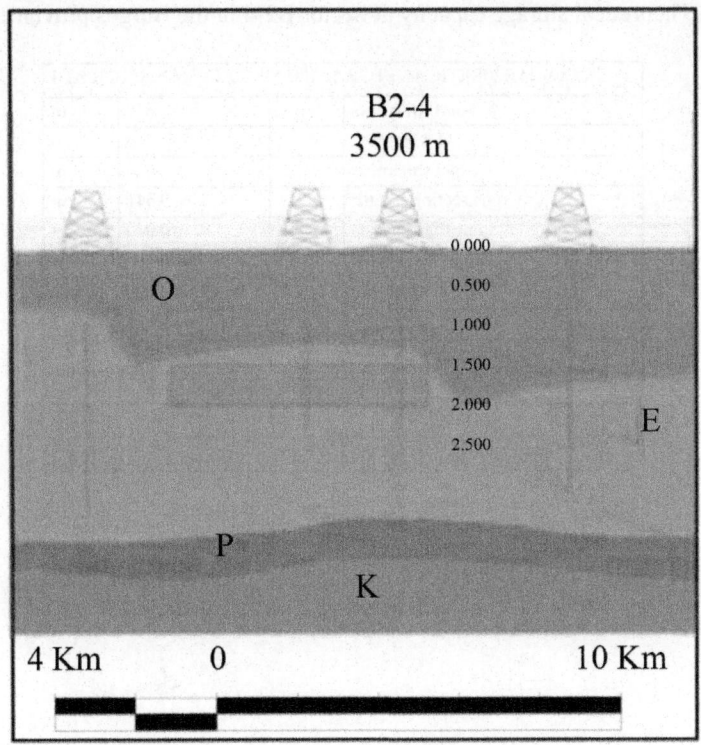

Figure. 10. Sector B2-4 from cross regional section B2. Vertical scale is in meters. K: Cretaceous, P: Paleocene, E: Eocene, O: Oligocene.

The computed petrophysical parameters are porosity 0.1, irreducible water 0.6, permeability less than 10 milidarcies (mD), density of CO_2 about 675 kg/m3. The respected volume of influence is assumed based on the lithological and petrophysical homogeneities of the rock unit supported by the extrapolation of features between oil wells, and the distances imposed by stratigraphical and structural elements. The use of these parameters in the theoretical calculation of the capacity results in 1.36 giga metric tons (Gt) of CO_2 for sector B2-4 (Table 3 and 4). The same approach was used in all sections of Burgos province giving 31 potential sectors on terrigenous sequences. Sometimes several sectors are located at the same well area of influence but at different depths. The marine zone was not computerized although several projects at the shallow marine platform in the United States point out the great potential of that zone (Figureure 11).

Table 3. Theoretical storage capacity at Sector B2-4 in the Burgos province.

CO₂ THEORETICAL STORAGE CAPACITY IN SECTOR B2-4			
Total thickness		880	m
Net fraction		0.6	m
Net thickness		528	m
Cross section length		9 541	m
Length influence		10 000	m
Area	A	95 410 000	m²
Volume	V	50 376 480 000	m³
Porosity	Φ	0.1	
Irreducible water saturation	S_{wirr}	0.6	
CO₂ Density	pCO_2	675	kg/m³
Storage capacity in volume unit	V_{CO2t}	2 015 059 200.00	m³CO₂
Storage capacity in terms of mass	MCO_2t	1.36	Gt CO₂

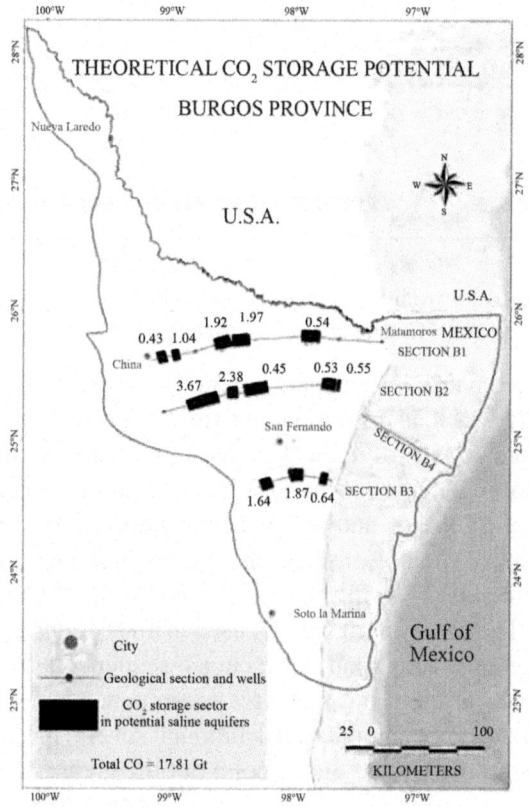

Figure. 11. Burgos province displaying the sectors (in black) of saline aquifers capable of storing CO_2. The marine zone was not quantified.

In summary, according to the geological sections, geological traps, sedimentary sequences and petrophysical parameters obtained from the Burgos province the theoretical capacity corresponds to 17.81 Gt in 31 assessed sectors (Table 4).

Table 4. Theoretical storage capacity of the Burgos province

BURGOS PROVINCE										
CROSS SECT-ION	SECTOR	TRAP (*)	TARGET SEQUENCE	SIZE		GENERAL PETROPHYSICAL PARAMETERS				Partial capacity in terms of mass (Gt)
			Terrigenous	Area (10⁶m²)	Thick-ness (m)	Ef-fective por-osity (Φ_e)	Irreducible water saturation (S_{wirr})	CO₂ Densi-ty (Kg/ m³)	Perme-ability (mili-darcies)	
B1	B1-1	Struct	E1	76.5	402	0.05	0.6	700	<10	0.43
	B1-2	Struct	P	60.5	350	0.1	0.3	700	<30	1.04
	B1-4	Struct	E7	108.64	369.2	0.1	0.5	700	<10	1.40
	B1-4	Both	O1	60.5	93.84	0.1	0.5	650	<30	0.35
	B1-4	Both	O2	115.22	59	0.1	0.5	500	<30	0.17
	B1-5	Both	O1	117.81	376.5	0.1	0.4	700	<30	1.86
	B1-5	Both	O3	140.92	13.75	0.15	0.4	650	<60	0.11
	B1-6	Both	O3	150.57	26.5	0.08	0.3	700	<60	0.16
	B1-6	Struct	O4	82.96	110	0.1	0.4	700	<10	0.38
B2	B2-2	Both	E1	95.88	30	0.05	0.6	700	<10	0.04
	B2-2	Struct	E7	77.63	97.5	0.1	0.5	600	<10	0.23
	B2-4	Struct	E1	95.41	528	0.1	0.6	675	<10	1.36
	B2-4	Both	O1	69.7	276	0.1	0.5	600	<30	0.58
	B2-5	Both	E1	85.06	94.5	0.15	0.6	700	<10	0.34
	B2-5	Both	E7	67.38	16.25	0.1	0.5	700	<10	0.04
	B2-5	Both	O1	82.52	458	0.1	0.5	675	<30	1.28
	B2-6	Both	O1	40.68	688	0.1	0.4	700	<10	1.18
	B2-7	Both	O1	46.32	741.2	0.1	0.5	700	<30	1.20
	B2-8	Both	O2	108.2	71.5	0.1	0.5	675	<10	0.26
	B2-8	Both	O3	86.33	57.75	0.08	0.3	600	<60	0.17
	B2-8	Struct	O4	67.45	10	0.1	0.4	550	<10	0.02
	B2-9	Both	O2	111.12	77	0.1	0.5	700	<30	0.30
	B2-9	Struct	O4	57.83	97.5	0.1	0.4	690	<30	0.23
	B2-10	Struct	O4	28.42	460	0.1	0.4	700	<10	0.55
B3	B3-1	Both	O1	78.1	312	0.1	0.4	700	<30	1.02
	B3-1	Both	O2	80.91	64.4	0.1	0.5	675	<10	0.18
	B3-1	Struct	O4	44.7	250	0.1	0.4	650	<10	0.44
	B3-2	Struct	O4	36	637.5	0.1	0.4	650	<10	0.90
	B3-4	Both	O2	64.85	47	0.1	0.5	700	<10	0.11
	B3-4	Struct	O4	34.56	612.5	0.1	0.4	675	<10	0.86
	B3-5	Struct	O4	59.17	257.5	0.1	0.4	700	<10	0.64
(*) Struct = Structural									TOTAL	17.81

Tampico-Misantla province

The Tampico-Misantla province lies in the central-east portion of Mexico. It is bordered to the north by the Burgos province and the Sierra de Tamaulipas mountain range, to the south by the mountainous fronts of the Sierra Madre Oriental folded-thrust belt and the TransMexican volcanic belt, and to the east by the Gulf of Mexico [29, 37]. The deep basement of the Tampico Misantla province consists of Precambrian and Paleozoic metamorphic and granitic rocks, and faults zones caused by extensional tectonic events some of which dating back to the origin of the Gulf of Mexico [38, 39]. Also, the basement pattern shows tectonic uplifts and through structures of different shapes and sizes. Overlying the basement a thick succession of sedimentary materials have been deposited ranging from Jurassic red beds and evaporites to Cretaceous carbonate sequences originated in shelf, platform and abyssal marine facies. On top of this succession a number of terrigenous sedimentary sequences were deposited concurrently with contractional tectonic events of the Laramide orogeny, since the beginning of the Cenozoic [40]. During Cenozoic times a thick terrigenous package with minor carbonates were accumulated to fulfill the coastal plain and marine regions of the west Gulf of Mexico. The surficial geology of the province exposes sedimentary rocks in parallel strips that run from the foothills of the Sierra Madre Oriental folded-thrust belt on the west to the existing coastal plain and marine platform regions of the Gulf of Mexico to the east. The older sedimentary rocks can be found on the west while the younger rocks are in the east. Some extrusive igneous rocks crop out on the northern and southern areas of the province (Figureure 12).

Five regional geologic cross sections were analyzed to understand the Tampico-Misantla province. Due to the similar geologic patterns showed along all regional sections, only section TM4 is presented. Section TM4 represents approximately 130 km in length of the subsurface regional geological profile, where basement faults and, horst and graben structures of different sizes are clearly revealed (Figureure 13). On the western portion of section TM4 are evident the folded and thrust faulted carbonate sequences of Cretaceous age, and on the eastern side is clear the minor tectonic deformation of the Cretaceous platform carbonates as well as the Cenozoic terrigenous sequences.

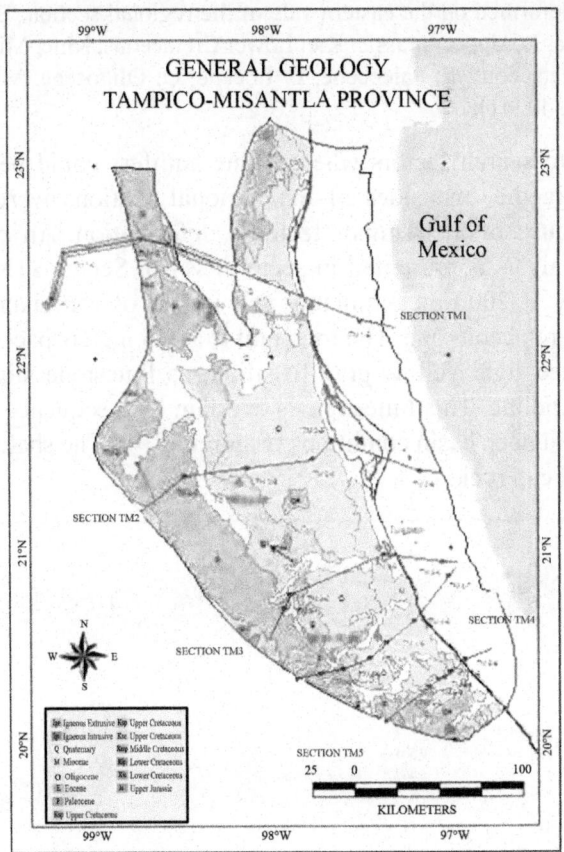

Figure. 12. Simplified geologic map of Tampico-Misantla province displaying regional cross sections and wells. After [33, 34, 35, 37, 46].

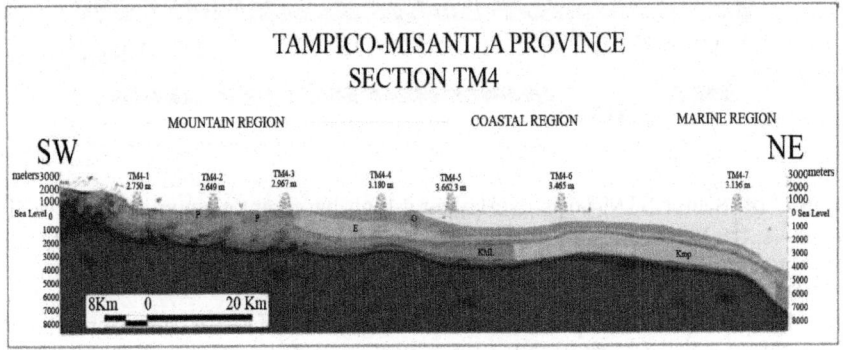

Figure. 13. Regional geologic section TM4. Mesozoic carbonate sequences are strongly deformed on the west side while Mesozoic and Cenozoic sedimentary successions

are almost undeformed on the eastern side of the regional section. B: Basement, Jm: Middle Jurassic, Js. Upper Jurassic, Kic: Lower Cretaceous, Kmc: Middle Cretaceous, Ksc: Upper Cretaceous, P: Paleocene, E: Eocene, O: Oligocene, M: Miocene. After [29, 33, 34, 35, 40, 41].

In order to search sectors where saline aquifers could become potential CO_2 reservoirs the east sides of the regional sections were preferentially assessed because of their minor tectonic deformation. An example of the performed analysis is presented in sector TM4-6. Sector STM4-6 is located approximately at 2000 meters depth, and is part of carbonate reef platform sequence of Cretaceous age. The rock unit is a 635 meters package of medium to thick bedded light yellow gray fossiliferous limestone slightly deformed as an open anticline. This limestone is overlain by a sequence of thin bedded shale formed in deep basin conditions (Figureure 14). The shales is interpreted as a good seal cap rock.

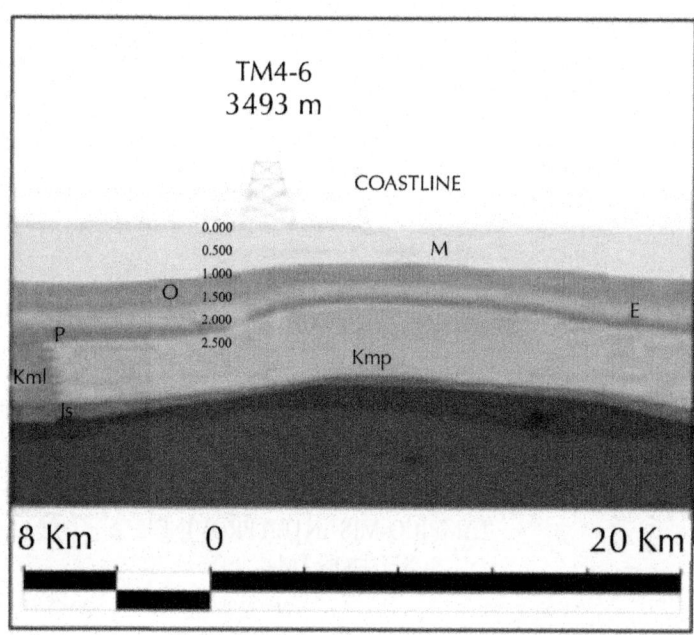

Figure. 14. Sector STM4-6 is overlaying a high basement element. Vertical scale is in meters. Pz: Paleozoic, Jm: Middle Jurassic, Js: Upper Jurassic, Kmt: Middle Cretaceous, Kmp: Middle Cretaceous, P: Paleocene, E: Eocene, O: Oligocene, M: Miocene.

The petrophysical parameters from sector STM4-6 are porosity 9%, irreducible water less than 30%, net thickness 508 meters, and CO_2 density around 693.6 kg/m3. The use of these parameters in the theoretical calculation has resulted in 1.08 Gt (Table 5).

Table 5. Theoretical storage capacity at Sector STM4-6

CO₂ THEORETICAL STORAGE CAPACITY IN SECTOR TM4-6.			
Total thickness		635	m
Net fraction		0.8	m
Net thickness		508.00	m
Cross section length		4 861.70	m
Length influence		10 000	m
Area	A	2 469 743.60	m²
Volume	V	24 697 436 000	m³
Porosity	Φ	0.09	
Irreducible water saturation	S$_{wirr}$	0.3	
CO₂ Density	ρCO₂	693.6	kg/m³
Storage capacity in volume unit	V$_{CO_2}$t	1 555 938 468.00	m³CO₂
Storage capacity in terms of mass	MCO₂t	1.08	Gt CO₂

After the analysis of the entire number of regional geological sections the Tampico-Misantla province yield 12 sectors. Four of them correspond to carbonate sequences and eight to terrigenous sequences. The total CO$_2$ capacity estimation corresponds to 9.75 Gt (Figureure 15 and Table 6).

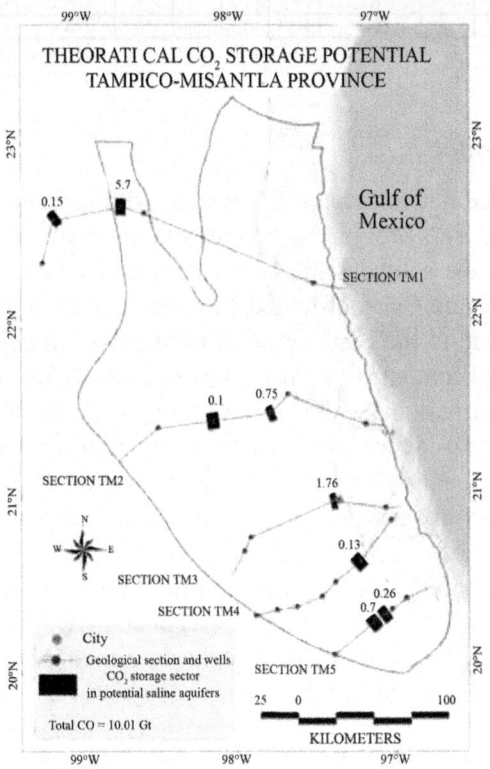

Figure. 15. Tampico-Misantla province showing sectors of potential saline aquifers capable of storing CO$_2$ (in black).

Table 6. Theoretical storage capacity of the Tampico-Misantla province.

			TARGET SEQUENCE		SIZE		GENERAL PETROPHYSICAL PARAMETERS			CO₂ Density (Kg/m³)	Partial capacity in terms of mass (Gt)
CROSS SECTION	SECTOR	TRAP	Terrigenous	Carbonate	Area (10⁶m²)	Thickness (m)	Effective porosity (Φ_e)	Permeability (milidarcies)	Irreducible water saturation (S_{wirr})		
TM1	TM1-3	Struct	Jm		59.2	784	0.20	300	0.20	696	5.70
	TM1-2	Struct	Ji		45.3	118.6	0.10	50	0.60	700	0.15
TM2	TM2-3	Struct		Jm2	77	26.1	0.10	60	0.30	702	0.1
	TM2-4	Strat		Jm2	0.3		0.10	60	0.30	702	0.15
TM3	TM3-3	Struct		Kmp	33.45	835.2	0.10	150	0.12	676	1.69
	TM3-3	Struct		P2 & P3	10.85	42.5	0.10	20	0.50	578	0.01
	TM3-3	Struct		E1, E2 & E3	29.2	42.7	0.12	300	0.30	426	0.06
TM4	TM4-6	Struct		Kmp	48.6	508	0.09	150	0.30	693	1.08
	TM4-6	Struct		P2 & P3	26.5	41.8	0.20	300	0.30	682	0.11
TM5	TM5-2	Strat		E1, E2 & E3	72.415	154.2	0.15	40	0.40	694	0.7
	TM5-3	Strat		E2 & E3	19.24	96.3	0.10	30	0.50	701	0.06
	TM5-3	Strat		O	41.94	95.4	0.10	30	0.30	701	0.2
(*) Strat = Stratigraphic, Struct = Structural										TOTAL	10.01

Veracruz Province

Veracruz province lies to the east of Mexico, sitting in the central part of the state of Veracruz. This province is bounded to the north by the Trans-Mexican volcanic belt, to the southeast by Los Tuxtlas volcanic field complex, to the west by Sierra Madre Oriental folded-thrust belt (known in this area as Sierra de Zongolica), and to the east-northeast by the Gulf of Mexico [42, 43]. The current geological context suggests a quick subsidence process along with several tectonic deformational events since Mesozoic times. The surficial geology suggests a faster subsidence process at the north of the province (Figureure 16). Six geologic sections were analyzed in order to estimate theoretical CO2 potential capacity for this province. From the subsurface point of view, the Veracruz province can be clearly divided into two geologic subprovinces. The first subprovince is the Sierra Madre Oriental folded-thrust belt and its continuation at depth known as the "Frente Tectonico Sepultado" (Buried Tectonic Front). It is characterized by folded calcareous rocks deformed by reverse faulting. The second subprovince is known as "Cuenca Terciaria de Veracruz" (Veracruz Tertiary Basin) composed by a thick succession of interbedded shale, siltstone, sandstone and conglomerate [40, 42, 47]. This terrigenous sequence has been, in turn, affected tectonically in distinctive styles and at different depths.

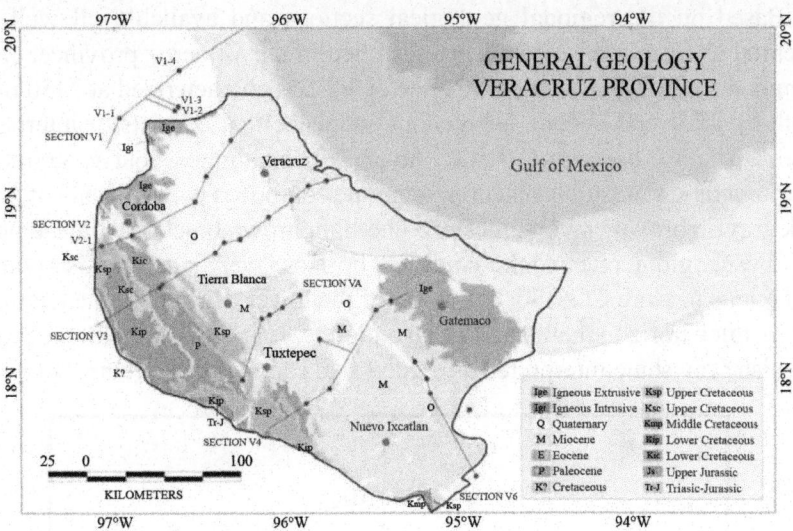

Figure. 16. Simplified geologic map of the Veracruz province, and location of regional geologic sections and wells. After [43, 33, 34, 35, 46].

For reference, Figureure 17 shows one of the regional sections that display structural features customarily found in the area. Section V3, about 180 km in length, lies in the middle of Veracruz province. The western half of the section displays calcareous sequences highly deformed by reverse faulting [42]. These sequences reveal Cretaceous facies from platform to basin environments. The eastern half of the section reflects terrigenous sequences wherein Paleocene and Eocene units expose reverse faulting folds.

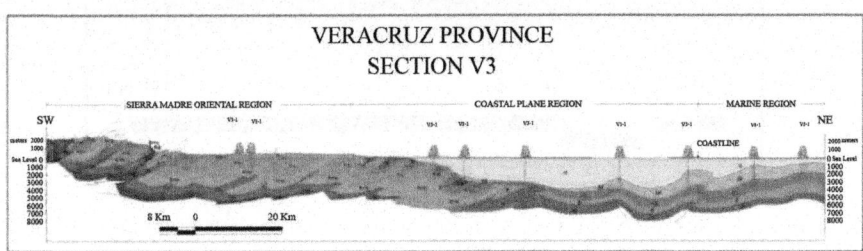

Figure. 17. Regional geological section V_3. The left hand side of the regional section shows Zongolica range's Cretaceous carbonate reverse faults as well as the buried tectonic front. The opposite side reveals early Cenozoic deformed terrigenous sequences and late Cenozoic undeformed sedimentary materials. Js: Upper Jurassic, Kip: Lower Cretaceous, Kmp: Middle Cretaceous, Ksp: Upper Cretaceous, Ksc: Upper Cretaceous, P: Paleocene, E: Eocene, O: Oligocene, Mi: Lower Miocene, M: Miocene, Q: Quaternary. After [43, 33, 34, 35, 47].

Based on the regional geological sections and available oil well data, potential CO_2 storage sectors were searched in the Veracruz province. One of them is sector V2-5 in section V2. Sector V2-5 is characterized at 2450 meters depth by a lower Miocene terrigenous sequence that consists of interbedded green to gray bentonitic shale, layers of bentonite, coarse grained to conglomeratic sandstone, and conglomerate composed by fragments of gray to dark grayish brown clayey limestone and light brown bioclastic limestone [40, 43]. The conglomerate and the sandstone horizons were interpreted as potential formations to store CO_2. So, at the top of the lower Miocene sequence is a 50 meters thick horizon that is part of an anticline. It is overlain by homogeneous greenish gray shale interpreted as a good seal cap rock (Figureure 18).

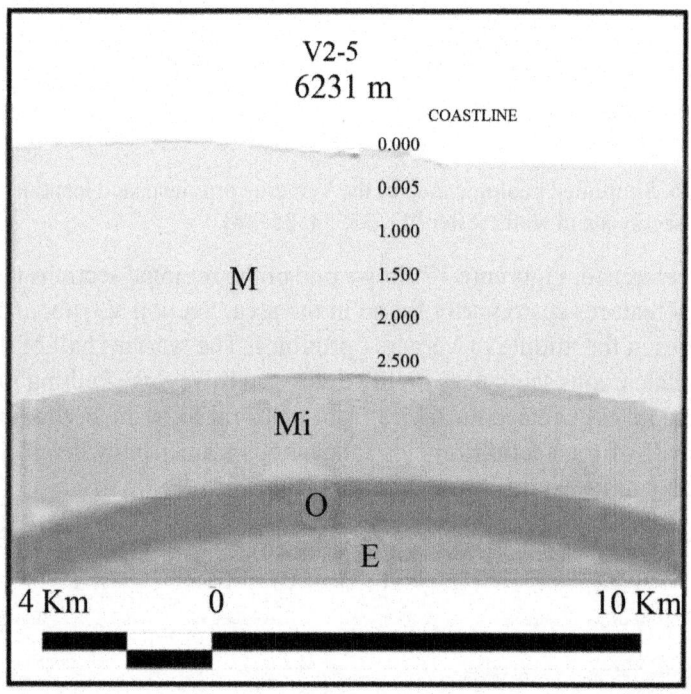

Figure. 18. Sector V2-5 showing a stratigraphic trap at the top of an anticline structure. Vertical scale is in meters. E: Eocene, O: Oligocene, Mi: Lower Miocene, M: Miocene, Q: Quaternary.

The horizon presents the following petrophysical properties, net thickness 15 meters, porosity 0.15, irreducible water 0.15, and permeability 200mD. The assumed CO_2 density for that depth of storage was 700 Kg/m3. The use of these parameters in the theoretical calculation of the capacity resulted in 0.03 Gt (Table 7).

Table 7. Theoretical storage capacity at Sector V2-5 in the Veracruz province.

CO₂ THEORETICAL STORAGE CAPACITY IN SECTOR V2-5			
Total thickness		50	m
Net fraction		0.3	m
Net thickness		15	m
Cross section length		2 500	m
Length influence		10 000	m
Area	A	25 000 000	m²
Volume	V	375 000 000	m³
Porosity	Φ	0.15	
Irreducible water saturation	S_{wirr}	0.15	
CO₂ Density	pCO_2	700	kg/m³
Storage capacity in volume unit	V_{CO2t}	47 812 500.00	m³CO₂
Storage capacity in terms of mass	MCO_2t	0.03	Gt CO₂

According to the theoretical calculations carried out in the Veracruz province resulted 21 sectors with CO_2 capacity potential (Figureure 19). Five of the sectors correspond to carbonate sequences, and the remaining 16 are terrigenous sequences. The estimated capacity targets reach 15.23 Gt (Table 8).

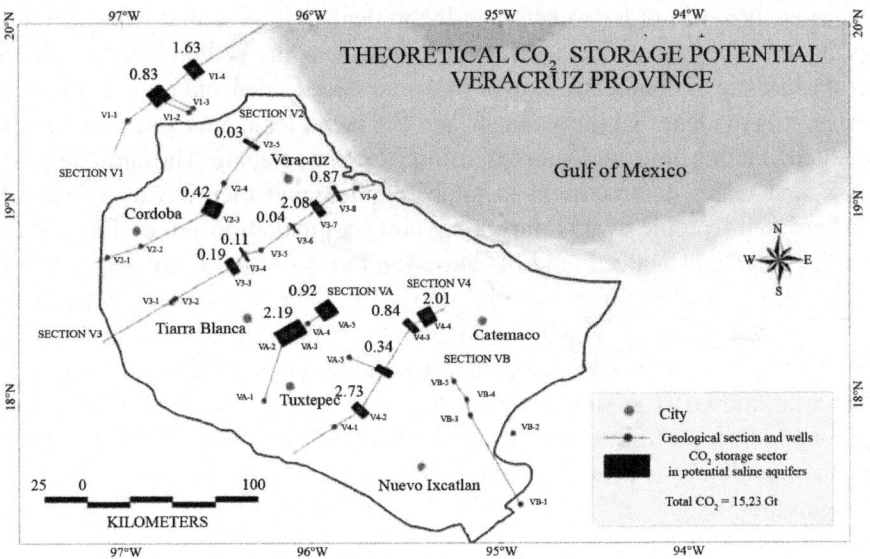

Figure. 19. Sectors with CO_2 storage potential in saline aquifers at the Veracruz province.

Sureste Province

The Sureste province is situated in the southeastern region of Mexico on the southern edge of the Gulf of Mexico. This province is bordered to the south by the Sierra de Chiapas mountainous range, to the east by the Yucatan Peninsula, to the west by the Veracruz province, and to the north and northeast by the Gulf of Mexico. The Sureste province comprises both mainland and offshore areas. In mainland the extensive geological exposures show evidence of the last episode of sedimentary infilling, therefore, most of the area is covered mainly by late Cenozoic sedimentary deposits (Figureure 20). The internal subsurface conFigureuration of the province is characterized by very deep and fragmented basement affected by different tectonic deformational events. At depth the Sureste province is divided into four subprovinces: Salina del Istmo, Comalcalco, Reforma-Akal and Macuspana [40, 44, 45]. The basement of the province consists of crystalline rocks of Precambrian and Paleozoic age [30, 49] most of which are covered by Mesozoic rock units composed of red beds, marine evaporites and carbonates of basin and platform marine facies [53]. Overlying the Mesozoic rocks are Paleogene terrigenous deposits of deep and shallow marine, deltaic, lagoonal and even alluvial facies [51, 52]. In addition, there are terrigenous sequences belonging to deltaic, lagoonal and shallow marine sedimentary facies that cover all the earlier deposits [40, 52, 54]. Six regional geologic cross sections (SE1, SE2, SE3, SE4, SE5 and SE6) were analyzed in order to estimate theoretical CO_2 potential capacity in the province. The regional cross sections show that the sedimentary sequences from Jurassic to Oligocene-Lower Miocene were folded and reversely faulted. Also, it is evident that the younger late Cenozoic terrigenous sequences were faulted, but this time, under an extensional tectonic regime. The entire province was first under contractional tectonic regimes, and then it was affected by extensional tectonic events during erosion-sedimentation stages.The position of the Sureste province could be viewed in terms of the jointly evolution of a passive continental margin associated to a strike-slip and a subduction margins both related to the plate tectonic interaction at the pacific region of Mexico. However, the complete and detailed tectonic history of the province is not yet well known. The subsurface stratigraphical and structural complexity is shown in Section SE2 which is approximately 135 kilometers long, is located in the middle of the province, and is running along a northwest-southeast line (Figureure 21).

Table 8. Theoretical storage capacity of the Veracruz province

CROSS SECTI-ON	SEC-TOR	TRAP (*)	TARGET SEQUENCE		SIZE		GENERAL PETROPHYSICAL PARAMETERS				Partial capacity in terms of mass (Gt)
			Terrige-nous	Carbon-ate	Area $(10^6 m^2)$	Thick-ness (m)	Effe-ctive porosi-ty (Φ_e)	Irredu-cible water sat. (S_{wirr})	CO_2 Dens-ity (Kg/m^3)	Perme-ability (mili-darcies)	
V1	V1-3	Strat		Kmp	52.7	202.5	0.1	0.04	700	<700	0.72
	V1-3	Strat	P		94.5	17.4	0.14	0.3	700	<60	0.11
	V1-4	Strat	E		78.15	285	0.15	0.25	650	<70	1.63
V2	V2-3	Struct		Kmp	17	27	0.07	0.04	700	<600	0.02
	V2-3	Strat		Ksp	56	387	0.03	0.7	700	<200	0.14
	V2-3	Strat	P		56	86.46	0.15	0.35	550	<40	0.26
	V2-5	Struct	Mi		25	15	0.15	0.15	700	<200	0.03
V3	V3-3	Struct		Kmp	26.3	10	0.07	0.2	700	<300	0.01
	V3-3	Strat		Ksp	43.6	147.2	0.08	0.4	600	<200	0.18
	V3-4	Strat	Mi		16	104	0.12	0.18	700	<300	0.11
	V3-6	Struct	Mi		10.4	54.9	0.12	0.18	700	<300	0.04
	V3-7	Struct	Mi		46	723.75	0.12	0.2	650	<300	2.08
	V3-8	Struct	Mi		21.65	698	0.12	0.2	600	<300	0.87
V4	V4-2	Strat	Mi		76.9	312	0.25	0.3	650	<80	2.73
	V4-3	Struct	Mi		43.75	115	0.12	0.2	700	<200	0.34
	V4-4	Struct	Mi		43.6	280	0.12	0.18	700	<300	0.84
	V4-5	Struct	Mi		83.7	348	0.12	0.18	700	<300	2.01
VA	VA-2	-	P		50	12	0.25	0.3	700	<20	0.07
	VA-3	-	Mi		100	75	0.12	0.1	700	<300	0.57
	VA-3	-	E		100	138	0.2	0.2	700	<50	1.55
	VA-5	-	Mi		100	133.5	0.12	0.18	700	<300	0.92

(*) Strat = Stratigraphic, Struct = Structural

TOTAL 15.23

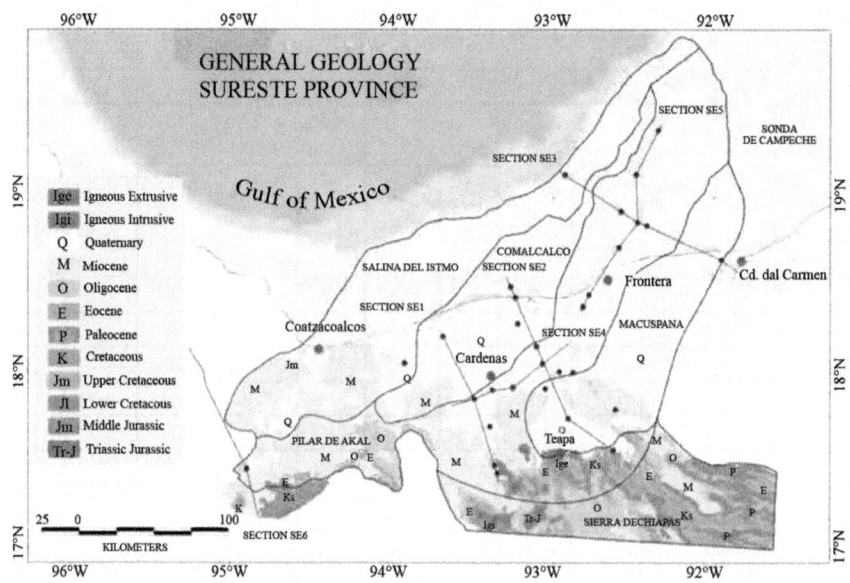

Figure. 20. Simplified geologic map of the Sureste province. It shows the location of regional geologic sections, wells, and limits of subprovinces: Salina del Istmo, Comalcalco, Macuspana and Pilar de Akal. After [33, 34, 35, 41, 44, 45, 46].

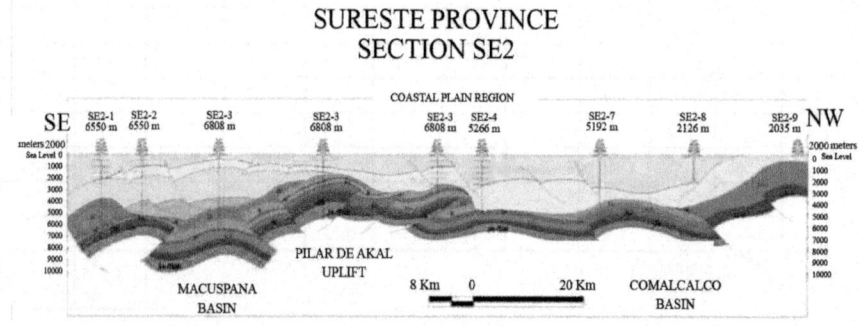

Figure. 21. Regional cross section SE₃ depicting complex tectonic deformation in the Sureste province. Js: Upper Jurassic, Ki: Lower Cretaceous, Km: Middle Cretaceous, Ks: Upper Cretaceous, P: Paleocene, E: Eocene, O: Oligocene, Mi: Lower Miocene, Ms: Upper Miocene, Pl: Pliocene, Pt: Pleistocene, Q: Quaternary. After [34, 35, 40, 51].

Section SE2 traverses the Comalcalco, Macuspana and Reforma-Akal uplift subprovinces. The Comalcalco and Macuspana are sedimentary basins separated in turn by the ReformaAkal uplift. In the three subprovinces there are from Jurassic through Oligocene folded and reverse faulted sedimentary sequences. At the Macuspana basin there are Miocene terrigenous sequences affected by both steep and gently dipping normal faults. In contrast, these terrigenous sediments are non-existent at the Comalcalco basin, therefore indicating synchronous erosion and sedimentation processes. At the Comalcalco basin the Pliocene and Plesitocene sediments can reach up to five kilometers in thickness, and the regularly spaced faults do not meet at the surface. All along the cross section is evident that the development of the basins is linked to the widespread fault systems and to subsidence mechanisms.

During the screening and selection of the sectors to estimate the CO_2 capacity, several stratigraphic and anticline traps structures were found. One of them is presented in Figureure 22 to illustrate the procedure. The sector SE2-4 consists of an anticline structure verging in northeast direction with an average axis orientation of N 300°. The anticline includes rock units from Jurassic to Oligocene times that are marked first by reverse faulting episode, and then by a regional unconformity. The unconformity is overlain by Miocene and Pliocene rock units.

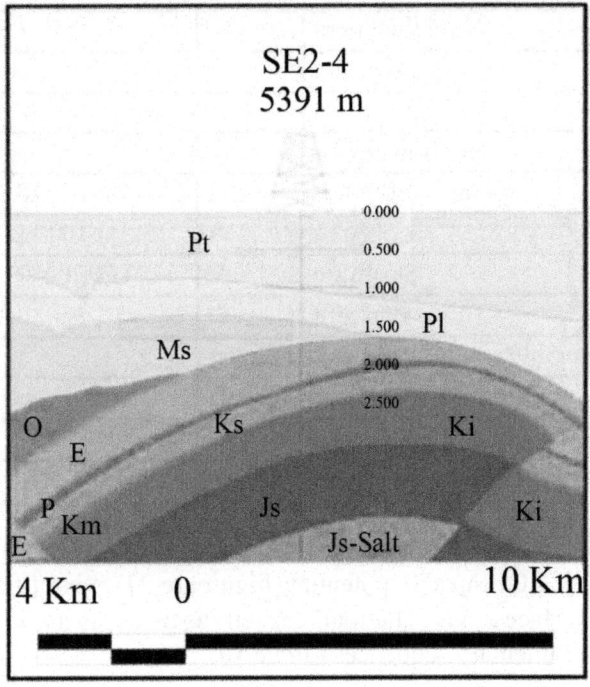

Figure. 22. Sector SE_2-4 showing the location of the CO_2 storage target in cross section SE_2 of the Sureste province. Vertical scale in meters. Js: Upper Jurassic, Ki: Lower Cretaceous, Km: Middle Cretaceous, Ks: Upper Cretaceous, P: Paleocene, E: Eocene, O: Oligocene, Ms: Upper Miocene, Pl: Pliocene, Pt: Pleistocene.

The CO_2 storage target is in a wedge of late Miocene well-bedded sequence about 280 meters thick and located 1550 meters deep. The storage sequence consists of a light gray, medium to coarse-grained, medium-bedded sandstone interbedded with occasional graygreenish shale containing mollusks and lignite fragments. The sandstone is overlain by a wide package of greenish gray shale of Pliocene age and interpreted as the seal layer. The petrophysical parameters of the sandstone target sequence are net thickness about 240 meters, clay content less than 4 %, porosity (Φe) about 30%, irreducible water saturation (Swirr) less than 20% and permeability about 60 miliDarcys (mD) (Table 9). According to the 1550 meters sandstone depth where the CO_2 density is approximately 681 Kg/m3, the theoretical storage capacity is close to 1.84 Gt (million tons of CO_2).

Table 9. Theoretical storage capacity at Sector SE2-4, in the Sureste province, is near 1.84 million tons of CO_2.

CO_2 THEORETICAL STORAGE CAPACITY IN SECTOR SE2-4.			
Total thickness		283	m
Net fraction		0.85	m
Net thickness		240.55	m
Cross section length		4 573.47	m
Length influence		10 000	m
Area	A	1 100 148.21	m²
Volume	V	11 001 482 085	m³
Porosity	Φ	0.3	
Irreducible water saturation	Swirr	0.18	
CO_2 Density	ρCO2	681	kg/m³
Storage capacity in volume unit	VCO2t	2 706 364 592.91	m³CO₂
Storage capacity in terms of mass	MCO2t	1.84	Gt CO₂

On the basis of the estimations conducted in the Sureste province resulted 17 sectors with CO_2 capacity potential (Figureure 23). Six of them are within offshore subsurface lands. The total capacity estimate is around 24.10 Gt on terrigenous sedimentary sequences (Table 10).

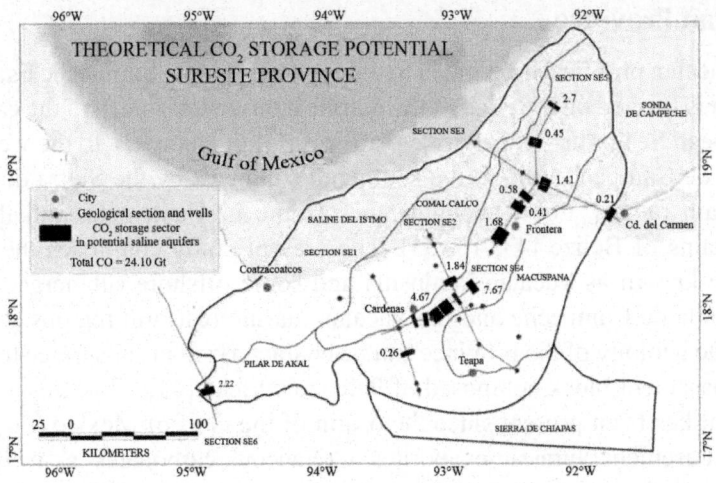

Figure. 23. Sectors shown in black with CO2 storage potential in saline aquifers at the Sureste province.

Table 10. Theoretical storage capacity of the Sureste province

SURESTE PROVINCE										
SEC-TION	SEC-TOR	TRAP (*)	TARGET SEQUENCE	SIZE		GENERAL PETROPHYSICAL PARAMETERS			CO₂ Dens-ity Kg/ m³	Partial capa-city in terms of mass (Gt)
			Terrigenous	Area (10⁶m²)	Thick-ness (m)	Effective porosity (Φₒ)	Permeability (milidarcies)	Irredu-cible water saturation (S_wirr)		
SE2	SE2-4	Struct	M	1.1	240.55	0.30	60	0.18	681	1.84
SE3	SE3-4	Struct	M	0.98		0.30	60	0.18	580	1.41
	SE3-6	Struct	O	0.3	308.70	0.05	45	0.45	591.5	0.21
SE4	SE4-1	Struct	M	0.22		0.30	60	0.18	472	0.26
	SE4-3	Struct	M	0.17		0.20	35	0.34	692.5	0.16
	SE4-3_4	Struct	M	0.25		0.20	35	0.34	682	0.23
	SE4-4	Struct	M	1.45		0.20	35	0.34	685	1.31
	SE4-4_5	Struct	M	1.72		0.30	60	0.18	688.5	2.92
	SE4-5	Struct	M	0.12		0.30	60	0.18	658.5	7.67
	SE4-6	Struct	M	4.73	811.32	0.30	60	0.18	426	0.05
SE5	SE5-2	Struct	M	0.67		0.30	60	0.18	670	1.11
	SE5-2_3	Struct	M	0.37		0.30	60	0.18	615	0.57
	SE5-3	Struct	M	0.30		0.30	60	0.18	544	0.41
	SE5-3_4	Struct	M	0.38		0.30	60	0.18	615	0.58
	SE5-5	Struct	M	0.29		0.30	60	0.18	620	0.45
	SE5-6	Struct	M	1.60	522.40	0.30	60	0.20	702	2.70
SE6	SE6-5	Struct	M	5.47	998.51	0.10	25	0.40	676.5	2.22
(*) Structural								TOTAL		24.10

Yucatan Province

The Yucatan province is bounded to the northeast by the Campeche Escarpment (which is formed on the edge of the marine continental shelf), to the east by the Caribbean Sea (where the marine platform is quite narrow), to the west by the Sonda de Campeche and to the south and southeast by the Sierra de Chiapas mountain ranges, Los Chuchumatanes Dome in Guatemala, and the Maya Mountains of Belize [43, 16, 55]. The area of study comprises the onshore portion known as Yucatan Peninsula and some offshore submerged areas in the Sonda de Campeche and the Yucatan marine platform regions (Figureure 24). The geology of the province can be characterized in subsurface terms by a huge basement block composed of Paleozoic rocks [43]. This crustal tectonic element has been present since the origin of the Gulf of Mexico [56]. On top of the basement, Jurassic evaporites, Cretaceous carbonates, as well as both Tertiary carbonates and terrigenous sedimentary sequences were deposited [57, 38, 58]. The sedimentary sequences were not under intense tectonic stress since they show a nearly horizontal depositional pattern and some minor faults. However, at the surface level, the central part of the huge province presents normal faults of considerable length that could bear testimony of extensional tectonic events which affected Mesozoic and lower Tertiary rocks. Under this geological context, four long regional geologic cross sections were analyzed to estimate the CO_2 storing capacity in the Yucatan Province.

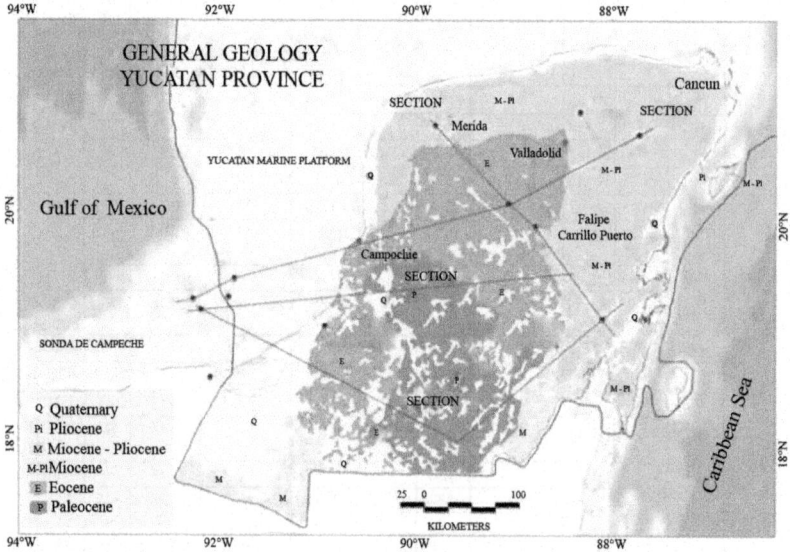

Figure. 24. Simplified geology map of Yucatan province showing regional geologic sections and wells. After [40, 43, 34, 35, 33, 55, 63].

The Yucatan province exposes a very wide and nearly horizontal sedimentary Mesozoic and Cenozoic rock sequences, where the topographic elevations rarely exceeds 200 meters above sea level. Because of this quite regular geologic homogeneity it is believed that the Yucatan peninsula remained stable throughout its geologic history. In contrast, at the edge of the basement block in the Sonda de Campeche, the offshore submerged area display Miocene contractional and extensional tectonic deformations linked to the geologic evolution of the Sureste province [59, 60]. The regional cross section Y2, approximately 400 km in length, depicts geological features frequently found in the entire province. At the offshore area within the Sonda de Campeche region gently folds structures in Mesozoic and early Cenozoic strata indicate a tectonic regime not so intense. Later, Cenozoic sequences of rocks denote normal faults systems that affected almost the complete stratigraphic column (Figureure 25). Sector PY2-1 illustrates one of the selected potential sectors where saline aquifers could eventually become CO_2 reservoirs. The Miocene terrigenous sequence is characterized by a thick succession of light colored sandstone interbedded with calcareous breccias and some layers of shale that alternate with calcareous arkoses lenses (Figureure 26). Within the Miocene sequence, only the sandstone horizons were considered for the calculations of CO_2 storage. The Miocene sequence is overlain by a thick package of Pliocene sediments composed of massive carbonaceous clay interbedded with peat layers and blue color clays. This package of sediments is interpreted as the seal rock unit.

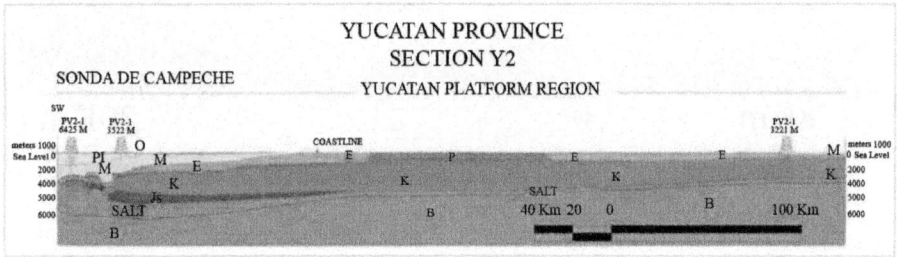

Figure. 25. Regional geological cross section Y2 showing Mesozoic sedimentary units gently deformed while the late Cenozoic sedimentary accumulations affected by extensional events within the offshore submerged region in the Sonda de Campeche. B: Basement, Js: Upper Jurassic, K: Cretaceous, P: Paleocene, E: Eocene, O: Oligocene, M: Miocene, Pl: Pliocene. After [34, 35, 33, 55].

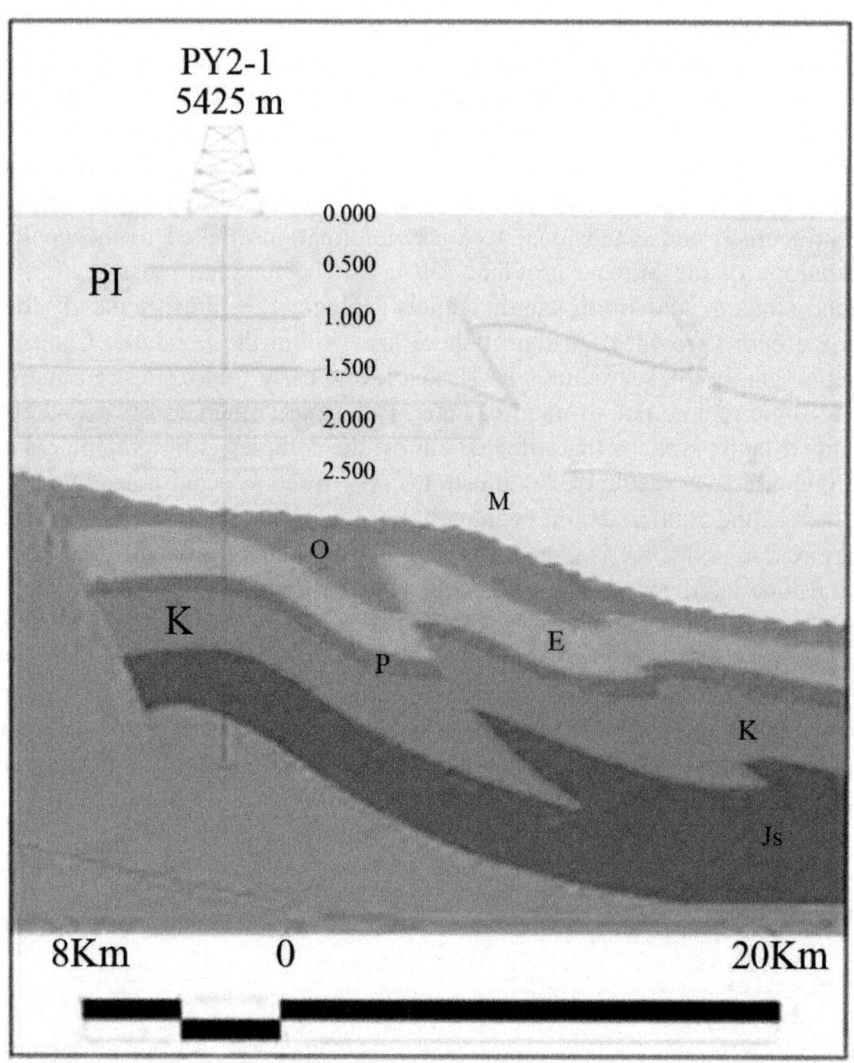

Figure. 26. Sector PY2-1 showing the location of the CO2 storage target. Vertical scale is in meters. Js: Upper Jurassic, K: Cretaceous, P: Paleocene, E: Eocene, O: Oligocene, M: Miocene, Pl: Pliocene.

The net thickness of the target sequence is about 353 meters with porosity (Φ_e) about 10% and irreducible water saturation (S_{wirr}) 30%. Based on these parameters the theoretical capacity is 3.25 Gt of CO_2 in sector PY2-1(Table 11)

Table 11. Theoretical storage capacity at Sector PY2-1 is near 3.25 million tons of CO_2.

CO_2 THEORETICAL STORAGE CAPACITY IN SECTOR PY2-1			
Total thickness		884	m
Net fraction		0.40	m
Net thickness		353.60	m
Cross section length		18 793.18	m
Length influence		10 000	m
Area	A	6 645 268.45	m²
Volume	V	66 452 684 480	m³
Porosity	Φ	0.10	
Irreducible water saturation	Swirr	0.30	
CO_2 Density	pCO2	699.2	kg/m³
Storage capacity in volume unit	VCO2t	4 651 687 913.60	m³CO₂
Storage capacity in terms of mass	MCO2t	3.25	Gt CO₂

The analyses of the Yucatan province yield seven sectors capable of storing CO_2 with a total theoretical capacity estimate of 14.44 Gt. Most of them are located in the offshore submerged lands of the Sonda de Campeche (Figureure 27). The sectors are divided in terrigenous rock sequences with 10.46 Gt and carbonate sequences with 3.98 Gt (Table 12).

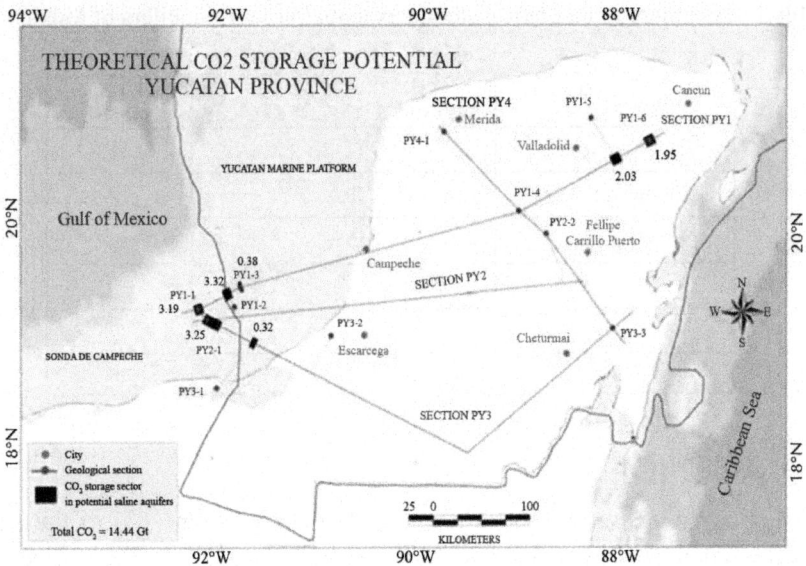

Figure. 27. Sectors (shown in black) with CO2 storage potential in saline aquifers, Yucatan province.

Table 12. Theoretical storage capacity in Yucatan province is 14.44 million tons of CO_2.

YUCATAN PROVINCE											
CROSS SEC- TION	SEC- TOR	TRAP (*)	TARGET SEQUENCE		SIZE		GENERAL PETROPHYSICAL PARAMETERS			CO_2 Den- sity (Kg/ m^3)	Partial capa- city in terms of mass (Gt)
			Terri- genous	Carbo- nate	Area (10^6 m^2)	Thick- ness (m)	Effective porosity (Φ_e)	Perme- ability (mili- darcies)	Irreducible water saturation (S_{wirr})		
PY1	PY1-1	Strat	M		6.6	760	0.10	30	0.30	692	3.19
	PY1-2	Strat	M		7.2	837	0.10	30	0.30	653	3.32
	PY1-3	Strat	M		9.5	283.12	0.10	30	0.30	575	0.38
	PY1-5	Strat		K	3.3	320	0.10	200	0.13	702	2.03
	PY1-6	Strat		K	3.2	308	0.10	200	0.13	701.5	1.95
PY3	PY2-1	Strat	M		6.6	353.60	0.10	30	0.30	699.2	3.25
	PY3-1	Strat	M		0.65		0.10	30	0.30	691.5	0.32
(*) Stratigraphic										TOTAL	14.44

In summary, the theoretical CO_2 capacity estimates in Mexico stands currently at 81.59 Gt on terrigenous and calcareous sequences located within the outlined inclusion zones. The total assessed sectors are 88 with possibilities of CO_2 storage in potential saline aquifers (Table 13). The assessed sectors in terrigenous sedimentary sequences are 77 while in carbonate sequences are 11.

Table 13. Summary of theoretical storage potential in saline aquifers of Mexico

PROVINCE	THEORETICAL CO_2 STORAGE POTENCIAL (Gt)	SECTORS ASSESSED
Burgos	17.81	31
Tampico-Misantla	10.01	12
Veracruz	15.23	21
Sureste	24.10	17
Yucatan	14.44	7
TOTAL	81.59	88

CONCLUSIONS

In Mexico the energy sector is responsible of more than 70% of the carbon dioxide emissions. In order to address the possibility of storing such anthropogenic CO_2 in deep underground geologic formations three lines of analysis were performed. First, the type, location and magnitude of CO_2 sources indicate approximately 216 Gt of CO_2 emissions coming from 1860 point sources. Second, five out of twelve geological provinces were analyzed. The assessed provinces are Burgos, Tampico-Misantla, Veracruz, Sureste and Yucatan which have the best favorable conditions for underground CO_2 storage in sedimentary rock successions of Mesozoic and Tertiary age. They are geologically well defined and located within the coastal plain region around the western portion of Gulf of Mexico. Third, theoretical storage capacities in potential saline aquifers sectors were estimated for each geological province. The theoretical CO_2 storage estimates and the number of assessed sectors are: Burgos province 17.81 Gt in 31 sectors, Tampico-Misantla province 10.01 Gt in 12 sectors, Veracruz province 15.23 Gt 21 sector, Sureste 24.10 Gt in 17 sectors and Yucatan province 14.44 Gt in 7 sectors. The total theoretical CO_2 storage potential currently stands at 81.59 Gt within 88 assessed sectors for the entire nation. During the CO_2 storage capacity estimations, it became clear that some areas yielded more and better quality data than others. Therefore, it is acknowledged that these data sets are not complete. However, it is anticipated that CO_2 storage capacity estimates, geological formation maps as well as regional geological cross sections will be updated as new information, particularly oil wells data, are acquired and methodologies for CO_2 storage capacity estimates are improved in Mexico.

REFERENCES

1.   M. Dávila, O. Jiménez, V. Arévalo, R. Castro and J. Stanley. "A preliminary selection of regions in Mexico with potential for geological carbon storage". International Journal of Physical Science, vol.5, num.5, pp.408-414, 2010.

2.   DOE (U.S. Departament of Energy). "2010 Carbon sequestration Atlas of the United States and Canada". Third edition, NETL (National Energy Technology Laboratory), 160p., 2011.

3.   RETC (Pollutant Release and Transfer Inventory) database. Secretaría del Medio Ambiente y Recursos Naturales, Mexico, Internal Report, 2008.

4.   SEMARNAT (Ministry of the Environment and Natural Resources). "Fourth National Communication to the United Nations Framework Convention on Climate Change (2006)". Instituto Nacional de Ecología, Mexico, 274p. Primera edición 2009. Available: http://www.ine.gob.mx.

5.   C.A. Hendricks and K. Blok. "Underground storage of carbon dioxide". Energy Convers Manage, vol.34, pp.949-957, 1993.

6. S. Bachu."Sequestration of CO2 in geological media: criteria and approach for site selection in response to climate change". Energy Convers Manage, vol. 41, pp. 953-970, 2000.

7. J. Bradshaw J. and A. Rigg. "The GEODETIC Program: research into geological sequestration of CO2 in Australia". Environmental Geosciences, vol. 8, pp. 166-176, 2001.

8. S. Bachu and S. Stewart. "Geological sequestration of anthropogenic carbon dioxide in the Western Canada sedimentary basin: suitability analysis". Canadian Journal of Petroleum Technology, vol. 41, num.2, pp.32-40, 2002.

9. S. Bachu. "Screening and ranking of sedimentary basins for sequestration of CO2 in geological media in response to climate change". Environmental Geology, vol. 44,pp.277-289, 2003.

10. S. Bachu, D. Bonijoly, J. Bradshaw, R. Burruss, S. Holloway, N.P. Christensen, and M. Mathiassen. "CO2 storage capacity estimation: methodology and gaps". International Journal Greenhouse Gas Control, vol.1, pp. 430-443, 2007.

11. S. Bachu. "CO2 storage in geological media: role, means, status and barriers to deployment". Progress Energy Combustion Science, vol. 34, pp. 254-273, 2008. www.intechopen.com 56 Earth and Environmental Sciences

12. A.W. Bally."Musings over sedimentary basin evolution". Philosophical Transactions of the Royal Society of London, vol. 305, pp. 325-338, 1982.

13. R. Ingersoll "Tectonics of sedimentary basins". Geological Society of America Bulletin, vol. 100, pp. 1704-1719, 1988.

14. Y.L. Leonov and Y.A. Voloz. Sedimentary basins: study methods, structure and evolution. Nauchnyi Mir, 525p., 2004.

15. Ch. French and Ch. Schenk (compilers). "Map showing geology, oil and gas fields, and geologic provinces of the Gulf of Mexico Region". USGS Open-File Report 97-470- L, 1997.

16. PEMEX (Petróleos Mexicanos). "Provincias petroleras de México". Pemex Exploración y Producción, México, Versión 1.0, 11p., 2010.

17. F. Campa and P. Coney."Tectonostratigraphic terranes and mineral resource distribution in Mexico". Canadian Journal of Earth Sciences, vol.20,pp.1040-1051, 1983.

18. F. Ortega, L. Mitre, J. Roldán, J. Aranda, D. Morán, S. Alaníz, A. Nieto. "Texto Explicativo de la Quinta Edición de la Carta Geológica

de la República Mexicana, Escala 1:2´000,000". Instituto de Geología, UNAM-Consejo de Recursos Minerales, SEMIP, México, 1992.

19. R. Sedlock, F. Ortega and R. Speed. "Tectonostratigraphic terranes and tectonic evolution of Mexico". Geological Society of America Special Paper 278, 153 p., 1994.

20. J. Gale, N.P. Christensen, A. Cutler and T. Torpe. "Demonstrating the potential for geological storage of CO2: The Sleipner and GESTCO project". Environmental Geosciences, vol.8, num.3, pp.160-165, 2001.

21. R. Tarkowski, B. Uliasz, and A. Wojcicki. "CO2 storage capacity of deep aquifers and hydrocarbon fields in Poland". Energy Procedia, vol.1, pp.2671-2677, 2009.

22. J. Bradshaw, C. Boreham, and F. La Pedalina (2005). "Storage retention time of CO2 in sedimentary basins: examples from petroleum systems". Available: http:/uregina.ca/ghgt7/PDF/papers/peer/427.pdf

23. L.G.H. van der Meer and P.J. Egberts. "A general method for calculating subsurface CO2 storage capacity". Presented at the 2008 Offshore Technology Conference. OTC 19309, May 2008.

24. S. Brennan and R. Burruss. "Specific Sequestration volumes: a useful tool for CO2 storage capacity assessment". USGS Open-File Report 03-452, 2009.

25. S. Brennan, R. Burruss, M.D. Merrill, P.A. Freeman and L.F. Ruppert. "A probabilistic assessment methodology for the evaluation of geologic carbon dioxide storage". USGS Open-File Report 2010-1127, 31p., 2010.

26. GCCSI (Global CCS Institute). "The status of CCS projects". Interim Report 2010, 26p., 2010. www.cslforum.org

27. DOE (U.S. Department of Energy). "Best practices for: Geologic Storage Formation Classification: Understanding Its Importance and Impacts on CCS Opportunities in the United States". NETL (National Energy Technology Laboratory), 54p., 2010.

28. DOE (U.S. Department of Energy). "2008 Carbon sequestration Atlas of the United States and Canada". 2nd edition. NETL (National Energy Technology Laboratory), 140p., 2008.

29. E. López Ramos. Geologia de Mexico. Tomo II. Edicion Escolar: Mexico, 454 p., 1979.

30. R.T. Buffler and D.S. Sawyer. "Distribution of crust and early history, Gulf of Mexico Basin". Gulf Coast Association Geological Societies Transactions,vol. 35, p.333-444, 1985.

31. PEMEX (Petróleos Mexicanos). "La Provincia Petrolera Burgos". Pemex Exploración y Producción, México, Versión 1.0, 27p., 2010. www.intechopen.com Geological Carbon Dioxide Storage in Mexico: A First Approximation 57

32. B. Ortiz."Interpretación estructural de una sección sísmica en la región Arcabuz– Culebra de la Cuenca de Burgos, NE de México". Revista Mexicana de Ciencias Geológicas, vol. 21, num. 2, pp. 226-235, 2007.

33. SGM (Servicio Geológico Mexicano) Cartas geológico mineras. Escala 1:250 000. Avaible: http:// mapasims.sgm.gob.mx:8399/mapasEnLinea/

34. CFE (Comisión Federal de Electricidad). "Integración de un Atlas de las principales cuencas sedimentarias de México". Technical Report. Convenio CFE-IPN-001/2009, enero 2010a.

35. CFE (Comisión Federal de Electricidad). "Geología del subsuelo de las principales zonas de las cuencas sedimentarias marinas y continentales alrededor del Golfo de México". Technical Report. Convenio CFE-IPN-001/2010, diciembre 2010b.

36. E. Lopez-Ramos and J.C. Guerrero. "Paleogeografia y tectonica del Mesozoico de Mexico". Revista del Instituto de Geologia, vol. 5, pp. 158-177, 1981.

37. PEMEX (Petróleos Mexicanos). "Provincia Petrolera Tampico Misantla". Pemex Exploración y Producción, Versión 1.0. 48 p. 2010.

38. A Salvador. "Late Triassic-Jurassic Paleogeography and Origin of the Gulf of Mexico Basin". American Association of Petroleum Geologists Bulletin, vol.71, p.419-451, 1987.

39. J.L. Pindell and J. F. Dewey. "Permo-Triassic reconstruction of western Pangea and the evolution of the Gulf of Mexico/Caribbean region". Tectonics, vol.1, p.179-211, 1982.

40. J. Santiago, J. Carrillo and B. Martell. "Geología Petrolera de México". In: Evaluación de Formaciones en México, D. Marmissolle-Daguerre, Ed. Schlumberger, 1984, p. 1-36.

41. INEGI (Instituto Nacional de Estadística, Geografía e Informática). "Atlas de Mapas Geológicos de Mexico". Ministry of Budget and Programming, Mexico, 1981.

42. PEMEX (Petróleos Mexicanos). "Provincia petrolera Veracruz". Pemex Exploración y Producción, México, Versión 1.0, 38 p., 2010.

43. E. López Ramos. Geologia de Mexico. Tomo III. Edicion Escolar: Mexico, 453 p., 1979.

44. W.A. Ambrose, T.F. Wawrzyniec, K. Fouad, S.C. Talukdar, R.H. Jones, D.C. Jennette, M.H. Holtz, S. Sakurai, S.P. Dutton, D.B. Dunlap, E.H. Guevara, J. Meneses, J. Lugo, L. Aguilera, J. Berlanga, L. Miranda, J. Ruiz, R. Rojas and H. Solís. "Geologic framework of upper Miocene and Pliocene gas plays of the Macuspana Basin, Southeastern Mexico". American Association of Petroleum Geologists Bulletin, vol.87, num.9, pp.1411-1435, 2003.

45. PEMEX (Petróleos Mexicanos). "Provincia Petrolera Golfo de México Profundo". Pemex Exploración y Producción, México, Versión 1.0, 26p., 2010.

46. PEMEX (Petróleos Mexicanos). "Provincias Geológicas de México". Pemex Exploración y Producción, México, Versión 1.0, 18p., 2010.

47. PEMEX (Petróleos Mexicanos). Provincia Petrolera Cinturón Plegado de la Sierra Madre Oriental. Pemex Exploración y Producción, Versión 1.0, 14 p., 2010.

48. SPP (Secretaría de Programación y Presupuesto). "Atlas Nacional del Medio Físico". Secretaría de Programación y Presupuesto. Gobierno de México, 224 p., 1981.

49. J.L. Pindell. "Alleghanian reconstruction and subsequent evolution of the Gulf of Mexico, Bahamas, and Proto-Caribbean". Tectonics, vol. 4, pp.1-39, 1985.

50. J. L. Pindell and L. Kennan, "Rift models and the salt-cored marginal wedge in the northern Gulf of Mexico: implications for deep water Paleogene Wilcox deposition and basinwide maturation". In: Transactions of the 27th GCSSEPM Annual Bob F. Perkins Research Conference: The Paleogene of the Gulf of Mexico and Caribbean Basins: Processes, Events and Petroleum Systems. L. Kennan, J. L. Pindell and N. C. Rosen (eds), pp. 146-186, 2007. www.intechopen.com 58 Earth and Environmental Sciences

51. F.J. Ángeles, N. Reyes, J.M. Quezada and J.R. Meneses. "Tectonic evolution, structural styles and oil habitat in the Campeche Sound, Mexico". Transactions of the Gulf Coast Associations of Geological Societies, vol. XLIV, pp.53-62, 1994.

52. R. Padilla."Evolución geológica del sureste mexicano desde el Mesozoico al presente en el contexto regional del Golfo de México". Boletín de la Sociedad Geológica Mexicana, T. LIX, num.1, p.19-42, 2007.

53. F.J. Angeles and A. Cantú, "Subsurface Upper Jurassic Stratigraphy in the Campeche Shelf, Gulf of Mexico". In: The Western Gulf of Mexico Basin: Tectonics, Sedimentary Basins, and Petroleum Systems. C.

Bartolini, R.T. Buffler and A. Cantú (eds), American Association of Petroleum Geologists Memoir 75, 2001.

54. J.Y. Narváez, J. Belenes, J. Moral, J.M. Martínez, C. Macías, O. Castillejos and M.A. Sánchez. "Bioestratigrafía de secuencias del Mioceno-Plioceno de la cuenca Macuspana, sureste del Golfo de México". Revista Mexicana de Ciencias Geológicas, vol.25, num.2, pp.217-224, 2008.

55. PEMEX (Petróleos Mexicanos). "Provincia Petrolera Plataforma de Yucatán". Pemex Exploración y Producción, México, Versión 1.0, México, 17p., 2010.

56. J. Pindell and L. Kennan. "Tectonic evolution of the Gulf of Mexico, Caribbean and northern South America in the mantle reference: an update". In The geology and evolution of the region between North and South America, K. James, M.A. Lorente and J. Pindell (eds), Geological Society of London Special Publication, 2009.

57. M. Olivas. "Aspectos paleogeográficos de la región sureste de México en los estados de Veracruz, Tabasco, Chiapas, Campeche, Yucatán y el territorio de Quintana Roo". Boletín de la Asociación Mexicana de Geólogos Petroleros, vol. XXVI, num.10- 2, pp.323- 336, 1974.

58. S. Medina. "Tertiary zonation based on planktonic foraminifera from the marine region of Campeche, Mexico". American Association of Petroleum Geologists, Memoir 75, pp.397-420, 2001.

59. R. Sánchez, "Geología petrolera de la Sierra de Chiapas". In: IX Excursión Geológica de Petróleos Mexicanos, Superintendencia General de Distritos de Exploración, Zona Sur, Libreto-Guía, 57 p. 1979.

60. M. Guzmán and J. J. Meneses. "The North America–Caribbean plate boundary west of the Motagua–Polochic fault system: a fault jog in Southeastern Mexico". Journal of South American Earth Sciences, vol.13, num.4-5, pp., 2000.

61. B. A. Méndez, "Geoquímica e isotopía de aguas de formación (salmueras petroleras) de campos mesozoicos de la Cuenca del Sureste de México: implicación en su origen, evolución e interacción agua-roca en yacimientos petroleros", Tesis Doctoral, Centro de Geociencias, UNAM, 200 p., 2007.

62. R. K. Goldhammer and C. A. Johnson, "Middle Jurassic-Uper Cretaceous Paleogeographic evolution and sequence stratigraphic framework of the northwest Gulf of Mexico rim". In: The western Gulf of Mexico Basin: Tectonics, sedimentary basins and petroleum systems. C. Bartolini, T. Buffler, and A. Cantú (eds), American Association of Petroleum Geologists Memoir 75, p. 45-81, 2001.

63. J.H. Rosenfeld. "Economic potential of the Yucatan block of Mexico, Guatemala, and Belize". In: The Circum-Gulf of Mexico and the Caribbean-Hydrocarbon habitats, basin formation, and plate tectonics: American Association of Petroleum Geologists Memoir 79, pp. 340–348, 2003.

64. S. Angus, B. Armstrong and K.M. de Reuck. International Thermodynamic Tables of the Fluid State. Volume 3. Carbon Dioxide. Pergamon Press: IUPAC Division of Physical Chemistry, 1973, pp. 266–359.

CITATION

CHAPTER 1

N. B. Raut, Dinesh Kumar Saini, and G. B. Shinde, "Environmental Informatics and Soft Computing Paradigm: Processing of Cocos Nucifera Shell Derived Activated Carbon for Treatment of Distillery Spent Wash—A Solution to Environmental Issue," Advances in Environmental Chemistry, vol. 2014, Article ID 737963, 11 pages, 2014. doi:10.1155/2014/737963

CHAPTER 2

Akindele O. Oyinloye; Geology and Geotectonic Setting of the Basement Complex Rocks in South Western Nigeria: Implications on Provenance and Evolution; http://cdn.intechopen.com/pdfs-wm/24552.pdf

CHAPTER 3

A. K. Somarin (2011). Petrography, Geochemistry and Petrogenesis of Late-Stage Granites: An Example from the Glen Eden Area, New South Wales, Australia, Earth and Environmental Sciences, Dr. Imran Ahmad Dar (Ed.), ISBN: 978-953-307-468-9, InTech, DOI: 10.5772/26024.

CHAPTER 4

Andreas Laake (2011). Integration of Satellite Imagery, Geology and Geophysical Data, Earth and Environmental Sciences, Dr. Imran Ahmad Dar (Ed.), ISBN: 978-953-307-468-9, InTech, DOI: 10.5772/27613.

CHAPTER 5

Belén Rubio, Paula Álvarez-Iglesias, Ana M. Bernabeu, Iván León, Kais J. Mohamed, Daniel Rey and Federico Vilas (2011). Factors Controlling the Incorporation of Trace Metals to Coastal Marine Sediments: Cases of Study in the Galician Rías Baixas (NW Spain), Relevant Perspectives in Global Environmental Change, Dr. Julius Agboola (Ed.), ISBN: 978-953-307-709-3, InTech, DOI: 10.5772/27772.

CHAPTER 6

George R. Ivanov, Georgi Georgiev and Zdravko Lalchev (2011). Fluorescently Labeled Phospholipids – New Class of Materials for Chemical Sensors for Environmental Monitoring, Relevant Perspectives in Global Environmental Change, Dr. Julius Agboola (Ed.), ISBN: 978-953-307-709-3, InTech, DOI: 10.5772/28978.

CHAPTER 7

Julius I. Agboola (2011). Relevant Issues and Current Dimensions in Global Environmental Change, Relevant Perspectives in Global Environmental Change, Dr. Julius Agboola (Ed.), ISBN: 978-953-307-709-3, InTech, DOI: 10.5772/39068.

CHAPTER 8

Yoshihisa Yamashita and Yukie Shibata (2011). Acid Stress Survival Mechanisms of the Cariogenic Bacterium Streptococcus mutans, Relevant Perspectives in Global Environmental Change, Dr. Julius Agboola (Ed.), ISBN: 978-953-307-709-3, InTech, DOI: 10.5772/28369.

CHAPTER 9

Ali Hammood and Zainab Radeef (2012). Characterizations of Environmental Composites, Composites and Their Properties, Prof. Ning Hu (Ed.), ISBN: 978-953-51-0711-8, InTech, DOI: 10.5772/50494.

CHAPTER 10

Oscar Jiménez, Moisés Dávila, Vicente Arévalo, Erik Medina and Reyna Castro (2011). Geological Carbon Dioxide Storage in Mexico: A First Approximation, Earth and Environmental Sciences, Dr. Imran Ahmad Dar (Ed.), ISBN: 978-953-307-468-9, InTech, DOI: 10.5772/26044.

INDEX